T0185951

Thermal System Design and Optimization

C. Balaji

Thermal System Design and Optimization

Second Edition

Ane Books
Pvt. Ltd.

C. Balaji
Department of Mechanical Engineering
Indian Institute of Technology Madras
Chennai, Tamil Nadu, India

ISBN 978-3-030-59048-2   ISBN 978-3-030-59046-8   (eBook)
https://doi.org/10.1007/978-3-030-59046-8

Jointly published with ANE Books India
In addition to this printed edition, there is a local printed edition of this work available via Ane Books in
South Asia (India, Pakistan, Sri Lanka, Bangladesh, Nepal and Bhutan) and Africa (all countries in the
African subcontinent).
ISBN of the Co-Publisher's edition: 978-93-88264-78-5

This Springer imprint is published by the registered company Springer Nature Switzerland AG
The registered company address is: Gewerbestrasse 11, 6330 Cham, Switzerland

A journey of a thousand miles begins with a single step

Chinese proverb.

To
My Parents
who did not live to see this book

Preface to the Second Edition

Nearly 8 years have passed since the first edition of the book was published. While newer algorithms to solve optimization problems continue to be developed, traditional ones are still holding fort. In this new edition, new sections are added on integer programming and multi-objective optimization. The linear programming chapter has been fortified by a detailed presentation of the simplex method. A major highlight of the revised edition is the inclusion of workable **MATLAB** codes for examples of key algorithms discussed in the book. Readers are encouraged to write their own codes for solving exercise problems, as appropriate. Needless to say, several new fully worked examples and exercise problems have been added to this edition.

I would like to thank my Ph.D. scholars Sandeep, Sangamesh, Girish, Suraj, Rajesh and Jyotheesh, and my M.S. Scholar Karthik for typing out the class notes, developing the MATLAB codes, and painstakingly proofreading the book. Thanks are due to my former Research Scholar Dr. Srikanth Rangarajan for working out the example in respect of the TOPSIS algorithm.

I would be glad to respond to queries and suggestions at *balaji@iitm.ac.in*.

Chennai, India   C. Balaji
June 2019

Preface to the First Edition

This book is an outgrowth of my lectures for the course "**Design and Optimization of Energy Systems**" that I have been offering almost continuously from 2001 to the students of IIT Madras, a great temple of learning. There are a few excellent texts on this subject and the natural question that arises in any one's mind is, Why another book on "Optimization of Thermal Systems"? The answer to this question lies in the fact that the field is rapidly advancing, newer algorithms supplement and even supplant traditional ones and the kind of problems that are amenable to optimization are ever increasing in the area of thermal sciences, all of which make it imperative to chronicle the progress at regularly decreasing intervals of time. At the same time, I wanted to write a book that reflects my teaching style, wherein "clarity" takes precedence over "extent of coverage". I am a true believer of what is said in the Upanishads—**Tejasvina Vadhi Tamastu** meaning "May what we study be well studied".

The major goals of the book are (i) to present the basic ideas of optimization in a way that would appeal to the engineering student or a practicing engineer, who has an interest or flair for trying to figure out the "best" among several solutions, in the field of thermal sciences (ii) to present only that much material that can be covered in a one semester course. By design, I have left out some optimization algorithms that are presented elsewhere and on the flip side, I have included some, which, in my opinion, are contemporary and have tremendous scope in thermal sciences.

The significant departure I have made from traditional text books is the interactive or conversational style. This, in my opinion, helps one to get across the material in an interesting way. The book is laced with several fully worked out examples with insights into the solutions. The chapters are backed up by a good number of exercise problems which, together with the example problems, should equip serious students well enough to take on optimization problems of substantial difficulty.

This book is not a one-stop-shop for learning all optimization techniques. Neither is it an exhaustive treatise on the theory of optimization nor is it a simple guide to solve optimization problems. It is an honest attempt to blend the mathematics behind the optimization with its usefulness in thermal sciences and present the content in a way that the book becomes "unputdownable". Whether it will eventually achieve this depends on the readers.

I would like to thank Prof. S. P.Venkateshan, IIT Madras, my beloved teacher, who initiated me into research and has been a trusted mentor and colleague over the last two decades.

Thanks are also due to Professors Roddam Narasimha and J.Srinivasan of IISc Bangalore, Prof. Heinz Herwig, Hamburg Institute of Technology, Germany and Prof. T. Sundararajan, IIT Madras, for supporting me at various stages of my professional life.

I cherish long discussions with Prof. Shankar Narasimhan, IIT Madras, on several topics presented in this book. I also thank Dr. Sridharakumar Narasimhan, a recent addition to IIT Madras, with whom I had several telephonic discussions on slippery aspects of optimization.

Financial assistance from the Centre for Continuing Education, IIT Madras, is gratefully acknowledged.

I would like to thank NPTEL (National Program for Technology Enhanced Learning) for giving me an opportunity to bring out a video course on Design and Optimisation of Energy Systems which served as the starting point for this book.

I also acknowledge the support and commitment of Ane Books Pvt. Ltd., for bringing this book out in a record time.

I would like to thank my wife Bharathi for painstakingly converting my video lectures to a working document that served as the basis for the first draft of the book, in an amazingly short time and for proofreading the manuscript more than once. She has been one of the greatest strengths in my life! My graduate students Ramanujam, Gnanasekaran, Chandrasekar and Konda Reddy have spent several sleepless nights, rather weeks, typing out the equations, proof reading the text and cross checking the example and exercise problems. Thanks are due to the numerous students who have taken this course over the last ten years and who, by their incisive and "stunningly surprising" questions, not only continue to humble me but also keep me on my toes.

Finally, I would like to thank my daughter Jwalika for allowing me to get committed on yet another assignment, namely, this book writing venture. I must admit that it has positively led to hijacking of quite a few weekends and holidays.

I would be glad to respond to queries and suggestions at *balaji@iitm.ac.in*.

Chennai, India   C. Balaji

Contents

About the Author

Dr. C. Balaji is currently the T.T.Narendran Chair Professor in the Department of Mechanical Engineering at the Indian Institute of Technology (IIT) Madras, India. He graduated in Mechanical Engineering from the Guindy Engineering College, Chennai, India in 1990 securing the University Gold Medal and 5 other medals for academic excellence. He completed his graduate studies at IIT Madras and obtained his M.Tech (1992) and Ph.D. (1995) in the area of heat transfer. His areas of interest include heat transfer, computational radiation, optimization, inverse problems, satellite meteorology, atmospheric sciences and climate change. He has more than 200 international journal publications to his credit and has guided 30 students so far. Prof. Balaji has several awards to his credit and notable among them include the Young Faculty Recognition Award of IIT Madras (2007) for excellence in teaching and research, K.N. Seetharamu Award and Medal for excellence in Heat Transfer Research (2008), Swarnajayanthi Fellowship Award of the Government of India (2008-2013), Tamil Nadu Scientist Award (2010) of the Government of Tamilnadu, Marti Gurunath Award for excellence in teaching (2013) and Mid-Career Research Award (2015) both awarded by IIT Madras. He is a Humboldt Fellow and an elected Fellow of the Indian National Academy of Engineering. Prof. Balaji has authored 9 books thus far. He is currently the Editor-in-Chief of the International Journal of Thermal Sciences.

Chapter 1
Introduction to Design and System Design

1.1 Introduction

The Concise Oxford Dictionary (COD) defines engineering as the *"application of science to the design, building and use of machines, constructions, etc.".* So, in a sense, one can argue that the application of science has also the potential to make us more and more lazy! We did not want to go to the movie halls, but wanted to see movies at home and hence the invention of the television. We did not want to get up from our seats while watching the television, so we invented the remote. We wanted to have the comfort of standard temperature and humidity throughout the year, so we figured out and invented the air conditioner. These days, we do not even want to get up and change the temperature and so invented the remote for the air conditioner and so on. So, we can say that engineering sometimes exploits the human weakness, the clamor for more comforts, but the positive connotation is that the ideas and concepts we conceive are also used in the design and building of machines, airplanes, cars, trucks—the list is endless. We cannot deny that engineering has positively reduced human toil, improved our lives, and allowed us to spend more time and energy on creative pursuits.

From where does the word "engineering" originate? A mechanical engineer would have very much loved to declare that engineering came from the engine. Unfortunately, it does not come from the engine, but from "injeniare" which means "contrive" or "come up with something".

If we look at the various activities that constitute engineering, important among them are analysis, design, fabrication, sales, marketing, research and development, and so on. *A "system", by definition, is a collection of components whose performances are inter-related.* Invariably, the product of an engineering enterprise is a "system". There should be at least 2 components to constitute a system, just as a thermodynamic cycle should consist of at least two processes, so that the system that goes from state 1 to 2 comes back from 2 to 1. But even this definition needs to be straightened out, because a large system like a thermal power plant or the Internet is

© The Author(s) 2021
C. Balaji, *Thermal System Design and Optimization*,
https://doi.org/10.1007/978-3-030-59046-8_1

made of several sub-systems. So we can immediately recognize that the modeling, simulation, and optimization of a very large system will be exceedingly complex and are expected to be Herculean tasks.

Suppose we want to carry out the optimization of a power plant, are there some ways of doing it? As aforementioned, a thermal power plant is very complex and is made up of several sub-systems. So, if we enlist all the variables that are involved in a thermal power plant and determine all the possible mathematical relations connecting these, one's head will spin. Notwithstanding this, the challenge is to optimize! What, then, are the options?

One possibility is to *look at the variables for each sub-system and optimize the sub-system, under the belief that if we optimize the individual sub-systems, it will lead to an overall optimum*. But that again is questionable. But just because that is questionable, we cannot be an armchair critic and say "this is not the right way to optimize" and not optimize at all. At least by dividing a very complex system into sub-systems and assigning the problem of optimization of the sub system to various groups (for example, if it is the alternator/generator sub system we want to optimize, this work can be possibly assigned to the electrical sub-systems group), there is the hope of attempting to optimize the overall design. If the system under consideration is a boiler or a heat exchanger, the optimization can be assigned to the heat transfer group. Ash handling or coal handling, for example, can be handled by the machine design group. So, each of these groups can come out with the modeling, simulation, and optimization of the sub-systems under question. When we integrate all this, we hope to get an overall optimum for the thermal power plant. Hence, we must realize that it is possible to break a system down into sub-systems; each of the sub-systems has only a finite number of variables, so that the problem is tractable or "handleable".

If we look at the design and fabrication of systems, we see that these have been around for a long time. The existence of bridges, highways, automobiles, airplanes, satellites, supercomputers, and data centers, to name a few, is ample proof that people have been designing and fabricating systems for ages. So what is the big deal? Why are we talking about it? The evolution of these systems has taken a lot of time! For example, the steam engine is only 200 years old. But civilization has been there for some millions of years. Sometimes, if we look back, in retrospect, we feel very ashamed. It has taken such a long time for us to come up with a steam turbine, jet plane, and all that. But we are happy that we at least have these systems now, because there is an accumulated knowledge for so many generations; it should not take that long a time to come up with improved designs or for that matter, new developments.

Now if we look around, many of the systems around us are certainly not working at the optimum conditions. We do not know whether the water pump in our homes is optimally designed. Did we ever try to figure out whether it is the optimum? Is it the best pump? Why are we not able to optimize that? Some reason should be there, right? The electric motor and pump arrangement in our apartment or bungalow, why have we not thought about it?

So the erstwhile goal was basically to design something that would work! "Optimization is not for us" could have been a refrain! But now, because of cost-consciousness and also because there are several competing designs and models which will do the same job and so on, the original goal (of designing something that will just work!) is no longer acceptable, as this approach, more often than not, leads to designs that are conservative and costly.

In the past, improved systems were designed only after a break even was achieved. Break even usually represents the time required to recover the original investment in a project. Optimization was considered as a costly add-on. Many people did not want to invest the time and effort required for optimizing a system.

Now, from what we have discussed thus far, it is clear that the key point is there are several ways of accomplishing the same task. This is one of the cornerstones in our quest for an optimum. *Optimization is afforded by the "choice" or "space" the design variables provide us.* One has, more often than not, a choice of various designs. It is different from the evaluation of an integral in calculus, where we think that everybody should get the same answer. Design is certainly not like that!

There are many ways of accomplishing the same task, and we will have to pick and choose the one that best suits us. Again, some are better than the others. The word "better" has to be necessarily within parenthesis, because what is better or not has to be decided by the user or the analyst, based on what the "objective function" is and accordingly, he/she goes ahead and optimizes the design.

Suppose we want to join sheets of paper, there are several ways of doing this. We can use a stapler or we can use a bell clip or simply bind the sheets. So, even a seemingly innocuous task like joining sheets of paper opens up several possibilities of accomplishing the same task.

So, first, the choice is available to us in engineering design, just as it is in life, in general. Secondly, the design of complex systems requires large calculations, often repetitively, for various combinations of the design variables. Let us take the problem of temperature control of the ubiquitous desktop. The goal here is to arrange all the components of the desktop so that we get optimal cooling. We have a choice here, no doubt, but it is not like we have infinite choices because there are certain positions that are fixed. We do not have much leeway in the location of certain components or in the way they have to be positioned close to each other. These are known more formally as "constraints" in an optimization problem. But we still have a choice of where we want to have slits, where we would like to place the fans, and so on. There are usually two fans in a desktop. One fan that sits on top of the heat sink, which in turn sits right on top of the processor. The other fan is very close to the panel on the back side to facilitate the movement of hot air out of the cabinet. The fan, which is dedicated to the processor will turn on only when the temperature reaches a certain level. When we initially start working, this fan usually will not turn on. When several programs are concurrently running and the outside is also hot, this fan will turn on. We can now consider this as a Computational Fluid Dynamics (CFD) problem to determine the maximum temperature for a particular arrangement of the components. If there are 10 components, there are a million ways of arranging these. Each time we need to create a solid model on the computer, get the converged CFD solution, and look at the maximum temperatures in the system. How many times do we have to do this? Is it possible to optimize using such a helter skelter approach? The example we are considering is not a very complex system. We are not talking about optimizing a data center or a big cluster of many computers. Even so, the task of obtaining the optimum configuration appears to be daunting!

So we need to reiterate that the design of complex systems requires large calculations, often repetitively, for various combinations of the design variables. But fortunately for the last 40 years, there have been really big improvements in computers and computation. In order to get a sneak peek, imagine that one arbitrarily chooses a combination of the variables in the desktop cooling problem and instead of doing the calculations a million times, one may do it only a hundred times. It is then possible to develop what is called an artificial neural network (ANN), which will give a correlation between the positions of the various components and the maximum temperature. Now, out of the 100 cases, one can use 80 or 90 cases to come up with the neural network and the remaining 10 cases can be used to test whether the neural network model developed gives us the right temperatures for the chosen set of variables. Now if this model works fine, we can do away with the CFD model and can instead repeatedly run the neural network model for various positions of the components with a view to determining the geometric configuration that results in the lowest maximum temperature. This is a possible smart way of solving the optimization problem under consideration.

The computers of today can perform exceedingly complex calculations and process large amounts of data very efficiently. Engineering design has largely benefited from this revolution. Better systems can now be designed by analyzing various options in a short time. We just saw an example of how better systems can be designed. However, this is just one possible approach.

1.2 Design Analysis Through a Flowchart

Now we look at analysis and design using flow charts. Let us look at Fig. 1.1.

From the figure, it is clear that first, we have to state the need. This is followed by a techno-economic feasibility study. A project needs to be both technically feasible and economically viable. If it is not feasible, we drop the project. 20–30 years back, desalination was technically feasible, but economically unviable for a developing country like ours. But now there is a desalination plant in Chennai. Once the techno-economic feasibility is established, detailed engineering is done and this is followed by manufacturing. Distribution and consumption are again critical activities that contribute largely to the success of a product.

The last few steps involved in the design are the incorporation of user feedback and modifications to the existing design. Research and Development (R&D) may supplement these efforts or sometimes fresh ideas may come directly from R&D.

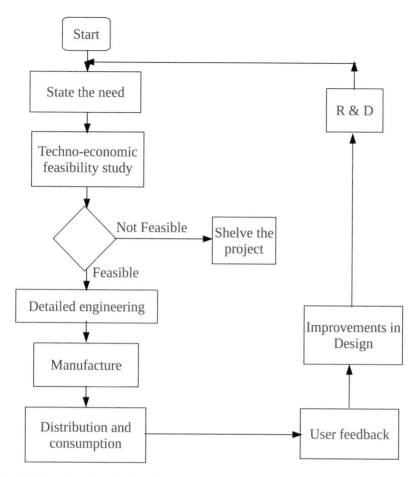

Fig. 1.1 Flow chart for analysis and design

1.2.1   Identifying the Need

From the preceding discussion, it is clear that *identifying the need is the first step in any design*. Stating the need is not always so straightforward. For accomplishing a job, the device we have in mind may be simple or extremely complex. Let us consider the previously cited example of joining sheets of paper. We can come out with various options, like using a stapler or punching a hole and tying with a tag or binding the sheets.

But if the problem is complex, like for example, if we want to solve the water problem of a city like Chennai, some may propose that there is a need to enlarge one or more existing reservoirs, desilt them, make them deeper, and let them store more water, whenever there is rain during the north-east monsoon season during October, November, and December. So the need is to enlarge the reservoirs. But unfortunately, the way the need has been stated now is incorrect because that is the solution! Sometimes, if there are various possible solutions, we incorrectly state one of the solutions as the need. One can still solve the water problem of Chennai without enlarging the reservoirs because there are other ways of doing it. We can do desalination, or charge a very high tax for water consumption which is exponentially increasing with consumption. We can have a water meter for every apartment. So for x liters a month, the cost is Rs.y; in the next slab, it will go for 1.25y, 1.5y for the next x, and so on. That is one way of doing it. Then we can have projects to fetch water from a neighboring state such as the Krishna water project. We can get water through pipes from some other reservoir, like we have done for Chennai from the Veeranam[1] lake. So, there are various ways of solving the problem. Therefore, *the need has to be stated in such a way that it is very broad. More importantly, the need should not be stated in such a way that some solutions are already not even considered.*

Sometimes, *there is a need to come out with new products* lest the customers get tired with the existing range of products of the company. Sometimes, a company is forced to come out with a new product if its competitor introduces one that also is successful.

Needs or opportunities arise in the renovation or expansion of facilities to manufacture or distribute a current product. For example, there is a heavy waiting time for certain cars now. Some companies can be happy that people are waiting for their cars. But it could be an opportunity for the competitors. If somebody is able to make something similar, which can be delivered with minimum waiting time, people can switch loyalties.

A new product may be developed intentionally or accidentally. If we look at how Velcro was invented, it is interesting. George de Mestral observed how Cockleburs, an herbaceous annual plant species get attached to jeans and this keen observation and inspiration led to the invention of the ubiquitous Velcro. The "Post-It" sticky note or prompt was the outcome of a failed product because it defied all the conventional properties of an adhesive. Dr. Spencer Silver, an American chemist, discovered an adhesive which did not stick after some time. Mr. Arthur Fry, an American inventor, figured out that if it did not stick for a long time it could be used as a bookmark, because it could be removed without leaving any trace of having used that. So, by ignoring conventional wisdom, a "failed adhesive" became a very successful office product.

Now let us take a quick look at some aspects of the flowchart, not usually considered by engineers, in detail.

1.2.2 Some Key Considerations in the Techno-Economic Feasibility Study

Success Criterion

The Cumulative Grade Point Average (CGPA) of a student is one criterion of success in an academic setting. However, it may not be the sole criterion for success. In

---

[1] Veeranam is an irrigation lake located 200 km south of Chennai in South India.

commercial organizations, the usual criterion for success shows a profit. Commercial enterprise means there should be a return on the investment. One invests Rs. 100 and the key question is, at the end of the day, how much do we get back? This, as a fraction of how much we have invested, is called the rate of return. In a big project, say, a power plant, one cannot start getting returns at the end of the first year. At the end of 5 years, maybe, we reach the break-even point. After we reach the break-even point, what is the return on investment? So this is the figure that all the commercial enterprises look at, primarily.

In public projects, such as building a new terminal in the airport or a new flyover, the criterion of success is whether the people are happy with that. So sometimes it can become very subjective and qualitative. However, the projected earnings of a proposed commercial project exert a lot of influence on whether to go ahead with the project. However, these concerns are also moderated by social considerations to a greater degree, for example, whether we want to have a dam over Narmada, whether we want to have the Tehri-Garhwal project, and whether we want to stop the Alaknanda river in the fragile Uttarakhand state of North India. There are other considerations which will temper our drive to keep the return on investment as our only criterion of success. However, emergency decisions are based on reasons outside the realm of economics. If a nation wants to declare war on another for legitimate reasons, then there is no consideration of economics!

Figure 1.2 shows a typical return on investment as a percentage at various stages of a typical plant. It can be seen that during the preliminary design stage, the uncertainty band is high. After a detailed design, we see that the uncertainty is decreasing.

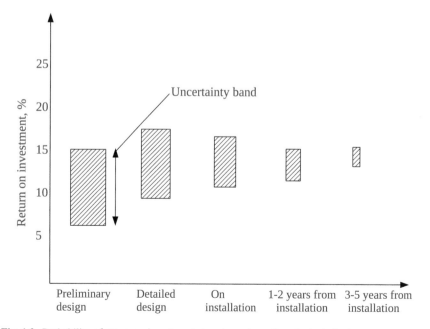

Fig. 1.2 Probability of return on investment at various stages for a typical plant

This shows that our confidence in design increases as the design evolves. When the uncertainty (σ) is low the return on the investment will be within a tight band, as according to a normal distribution, the probability of x being with x $\pm$ 3 σ is 99%. First, the curves are very diffuse. After construction, it becomes better. After 5 years, we have a good idea, assuming that the market does not change dramatically. This is basically the probability at various stages of decision-making. Please look up Stoecker Stoecker (1989) for a detailed discussion on this.

The uncertainty reduces as we move from preliminary design to complete design to construction. After 1 year, it is lower. After 3 or 5 years, the rate of return is known exactly. The probability degenerates into a small uncertainty. The probability eventually looks like a Dirac delta function. But the prediction of future behavior is not deterministic. We have to factor in various things like the future market conditions, the inflation associated with this, the fluctuation in interest rates, and so on. Therefore, the design will change from being deterministic to probabilistic.

Stochastic or probabilistic design is very popular nowadays. For example, the fluctuations in the stock prices of various companies (P) can be treated as

$$P = \bar{P} + P' \tag{1.1}$$

where $\bar{P}$ is the mean and P' is the fluctuation. We can model it along the lines of turbulent fluid flow. And then we can come out with stochastic partial differential equations, solve them, and have a predictor model for stock prices. Such activities are part of the new discipline called "*Financial engineering*".

When a flock of birds is trying to catch food, the probability of getting food increases if all the birds stay closest to the leader. The leader is the one who is closest to the food. We can write equations for this and work it out. People have modeled all this and have applied it to solve practical engineering problems. Such a technique is known as *Particle Swarm Optimization or PSO*. The foraging behavior of ants or how ants search for their food has been studied by various optimization groups. If ants go in search of food, when they return after eating, they leave a trail of a chemical called pheromone. This pheromone concentration will, of course, be stronger if more food is available at the place where they went. This concentration will also exponentially decay with time. So the ants which follow this will look at the pheromone trail and wherever the pheromone concentration is very weak, they avoid that path. Even along the path, where there is a pheromone trail, if the signal is very feeble, the ants know that the last ant that went must have done so 2–3 days back and most probably by this time, the food would have got exhausted! Birds or ants are not taking any optimization course or a 101 course on Introduction to finding food. Yet, they are doing it well all the time!

Market Research Typically, sales volume is inversely proportional to the product price (see Fig. 1.3a). So one expects that as the price increases, sales will decrease. This is not always true. Can we think of some goods which will defy this? Gold! Gold will buck this trend because when the gold price increases, people will get more scared that it will rise further and more people will go and buy it. That will

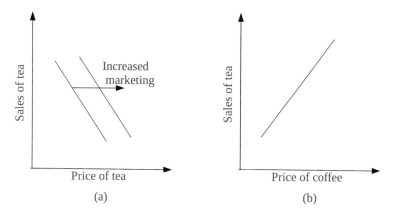

Fig. 1.3 **a** Elasticity and **b** cross-elasticity of demand

cause increased demand, which will cause a short supply and will further increase its price. So, vanity goods have the opposite trend. The demand for essential goods like milk, bread, and rice is price insensitive. We call this inelastic demand. The demand for some goods like tea and coffee displays some elasticity of demand. In fact, there is a possibility that the consumption of tea will increase when the price of coffee increases. This is known as cross-elasticity of demand. For example, if we plot the price of coffee versus the sales of tea, the curve may look like what is shown in Fig. 1.3b.

In Fig. 1.3a, we have a bunch of lines. They represent different sales and advertising efforts. So, to a limited extent, it is possible to employ aggressive marketing and make people remember a brand and increase the sales. But after some time, the return on investment, because of this advertising effort, will decrease. That is called the law of diminishing marginal returns in micro-economics.

Let P denote the price and Q denote the quantity. The elasticity of demand E_d is given by

$$E_d = \frac{\Delta Q/Q}{\Delta P/P} = \frac{P}{Q}\frac{\partial Q}{\partial P} \tag{1.2}$$

A feasibility study, of which market research is a critical aspect, is done to enable a person or a company to make a decision on whether a project is feasible or not. Some projects may be technically feasible but economically unviable. Even if a project is technically feasible, if it is not economically viable, we have to shelve it.

1.2.3 Research and Development

The results of research and development may be an important input to the decision process. Research efforts may result in an original idea or an improvement of the existing idea, whereas, development requires a pilot study. This is the difference between research and development. It is first research and then development and then we need to transform it into a product. So if we look at working models or pilot plants, it is development work. The idea need not always originate from within the organization. It may also come from a rival organization or competition.

1.2.4 Flow Loop and Iterations

From the flowchart (Fig. 1.1), it is evident that the process of design is essentially a trial and error or more formally, an iterative procedure. Each pass through the loop improves the amount and quality of information. What is flowing through the flowchart is basically information, which gets refined. After we go through sufficient iterations and we are confident, we stop the process and make a decision on whether to go ahead with the project or not. The stopping criterion is a prerogative of the design engineer and depends on several factors.

1.3 Optimization

The flowchart in Fig. 1.1 involved no optimization. However, we have user feedback, followed by improvements possible in design. These improvements can also serve as an impetus to the Research and Development design. What we have discussed is a conventional design. The original design was based on some parameters, but during the operation of the plant, many parameters will change. One becomes interested in identifying the set of parameters when the design/plant or machine will work at its "best". However, a better strategy would be to integrate the optimization with the design itself.

1.4 Analysis and Design

The difference between engineering design and analysis is that the analysis problem is concerned with determining the behavior of an existing system. Let us look at an example, a steam condenser, a schematic of which is shown in Fig. 1.4. On the shell side, we have steam entering, which condenses and comes out as hot water. This will be an isothermal process, as far as the steam in the shell side is concerned. This condensation is accomplished by cold water circulated in the tubes.

The cooling water will enter at a temperature $T_{c,in}$ and will go out at a temperature $T_{c,out}$. So its temperature will increase.

Shown in Fig. 1.5 is the temperature—length or the T-x diagram of the heat exchanger. It does not matter if we have a parallel flow or a counter flow because

Fig. 1.4 Schematic of a steam condenser

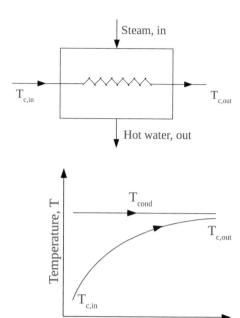

Fig. 1.5 Typical T-x diagram of a steam condenser

one of the fluids is changing phase. There are several ways of working this problem out. One possibility is that we want to condense, say, "m" kg of steam, which is at a particular temperature and pressure, into "m" kg of water. We have cooling water at 30 °C. Now we want to come up with the design of a steam condenser. So how do we go about designing this? Is energy balance alone enough? What are the equations? Let us start with the following energy balance equation:

$$\dot{m}_w C_p (\Delta T_{cold}) = \dot{m}_s h_{fg} \tag{1.3}$$

With just this alone, we cannot design the heat exchanger. That is the limitation of thermodynamics. Thermodynamics will tell us what the resulting temperatures are, and how much heat will be transferred eventually between the hot fluid and the cold fluid. But it will not tell us what the area of the heat exchanger required to accomplish this task is.

1.4.1 Difference Between Design and Analysis

What do we want to find out actually in the previous example? The design solution will be how many tubes are required in the shell, what should be the size of the shell, what should be the diameter of the shell, whether we will have a U-tube kind of situation, i.e., whether it has several passes. Accordingly, the diameter will get reduced. So there is an issue of surface area. How much surface area is required for accomplishing this? What is the equation for this? This is basically Q of the heat exchanger or the heat duty of the heat exchanger.

$$Q = U A \, \Delta T_{LMTD} \tag{1.4}$$

ΔT_{LMTD} is the logarithmic mean temperature difference. U is the overall heat transfer coefficient. A is the area available for the heat exchanger, which is directly related to its cost.

In the analysis problem, on the other hand, we already ("a priori") have some heat exchanger or condenser available with us. We want to study its thermal performance, when used as a steam condenser, for the given pressure and temperature.

What do we mean by thermal performance? It is the temperature rise of the cold water. We are not trying to come up with the design of an equipment, because here UA is known. If UA is known, there is nothing to design. We do not design the outlet temperature. We analyze and find out what the outlet temperature is. But in design, we have a choice as to whether we can use a plate heat exchanger or a shell and tube heat exchanger, how many number of tubes, what type of tubes, what will its thermal conductivity be, and what will its heat transfer coefficient be.

Let us take one of the above design parameters, say the overall heat transfer coefficient. Even this can change with time.

Why should the heat transfer coefficient vary with time? Because of scaling and fouling, there will be a buildup of biological matter, because of which the resistance to heat transfer increases. When the resistance to heat transfer increases, the heat transfer coefficient will exponentially decay with time. Therefore, we may have a situation where we may have heat transfer coefficient (h) varying with time (t) as shown in Fig. 1.6.

To cut a long story short, the design of heat exchangers and heat transfer equipment for life is not such a straightforward task. Suppose we want to design it for 15 years, we have to consider all this. There may be something like an asymptotic fouling factor which can be determined. The design will involve all this. The straightforward analysis problem, which one learns in a heat transfer course, will be one where all the dimensions are given, and invariably, we work out the temperatures or flow rates.

Fig. 1.6 Exponential decay of heat transfer coefficient with time

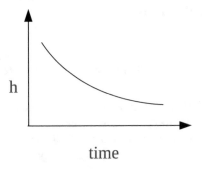

1.4.2 Constraints in Design

Sometimes, the solution is very tricky, since the LMTD, which is the logarithmic mean temperature difference, is not known and we will have to perform tedious iterations to get this temperature in many situations.

What then is a design problem? Let us say we want to design the steam condenser for the 500MW fast breeder reactor coming at Kalpakkam.[2] India will build it up in the next few years. We know the conditions, the pressure, and the temperature of the steam. Also, we know the temperature of the cooling water, which is got from the sea. So how do we go about designing this? Can we take water at 30 °C and send it out at 80 °C? What is the problem? Thermal pollution! So there is a limit or a cap on the maximum ΔT that is allowed. What is the value? It is about 5 °C. Otherwise, it severely disturbs the nearby aquatic life. A constraint is already coming up in our design. We are allowed a ΔT of only 5 °C.

Of course, the sky is the limit for the flow rate, and we can have an infinite flow rate. What will be the size of the pump? We can have a beautiful and wonderfully green design. ΔT is only 5 °C. But what will be the size of the equipment? Heat transfer equipment works best when there is a temperature difference. So we want more temperature difference. But we are forced to have a maximum ΔT of 5 °C. So there is a constraint. There is also a constraint which is coming in the form of losses. The power required for the pump should not be more than the output of the 500MW plant! It will and should not happen but if we come up with some simple design based on the most fantastic fluid mechanics solution, we go home with negative power output from the plant.

Are there more constraints? Of course, yes. The saltwater will corrode normal steel. So we have to go for stainless steel or titanium. So what innocuously started as a harmless simple equation seems to get increasingly tricky. So the design problem is

$\dot{m}_{steam} \rightarrow 221$ fixed

T, p $\rightarrow 221$ fixed

Q $\rightarrow 221$ known

A $\rightarrow 221$ unknown.

For the given Q, what the size of the equipment will be, that is the design problem! For the given U and A, what the Q is, that is the analysis problem. Q is not known a priori in the analysis problem. Q is known, to begin with, in the design problem. So for a simple problem like this, we can branch out into design and analysis. But originally when people tried to design, they would usually follow a nomogram or a data book, which would list out the thickness and diameter of the shaft for a given load, and they would have a factor of safety and decide. So, analysis was divorced from design. But now, we have tools like finite element analysis and therefore, analysis is an integral part of design. The idea behind this is that alternative designs can be analyzed and we can find out whether stresses are within limits, without having

---

[2] Kalpakkam is a coastal town 70km south of Chennai, India, where a 500 MW Fast Breeder Nuclear reactor plant is coming up.

to build new prototypes, on the computer itself, using simulations. Hence, we can study the performance of several choices and choose the best. This is essentially the difference between analysis and design.

The sizes and configurations of the parts are known a priori for the analysis problem. For the design problem, we need to calculate the sizes and shapes of the various parts of the system to meet performance requirements. In this system, suppose we change the U or A or we just change the $\dot{m}$, what happens? From the design flow rate of steam, let us say we deviate, $\pm 10\%$ or $\pm 15\%$, what happens? So we will have to do what is called a *sensitivity study* too!

So the design of a system is essentially a trial and error procedure.

- We will select a bunch of components, assemble them together, and that becomes a system.
- Then we will use all the equations known to us and see if the system is working.
- If the system is working, it is a feasible or an acceptable design.
- If it is not working, we will turn around and try to change some of these components or parameters, and
- We will keep on doing it till we have an acceptable or a feasible design.

This is how the design procedure is done. In both these cases, we must have the capacity or ability to analyze designs to make further decisions. Therefore, analysis has to be embedded and integrated into the design process. Design and analysis are two sides of the same coin. So, we should not simply do some design based on some thumb rules or formulae. Of course, for a simple problem like the determination of the size of the pump required for an apartment, we do not have to conduct a complicated finite element analysis. We go to our plumber, who will give us a quick fix solution and say that "5hp pump will work". But there, we are not talking about optimum. We are trying to choose something which works. That is fine for this simple problem as we are not looking at a costly equipment, wherein optimization really matters.

1.5 Workable System and Optimum System

For a design problem, there are several possible solutions. But all solutions are not equally desirable. Some solutions are better than the others. Whether it is better or not, how some solutions are better than the others depends on the objective function, on the analyst, on the user, and what he/she wants to call desirable. Actually, when an objective function criterion is defined like cost, size or weight, heat transfer or pumping power, and so on, invariably only one solution will be an optimum. The goal of this book is to help us identify the optimum in a given situation. Sometimes, the optimum may not exist, and that is fine. Sometimes, the optimum may be infeasible. Sometimes, the constraints themselves may fix the solution and so on. It is sufficient to say at this stage, just as there exists a difference between design and analysis, there exists a difference between a workable system and an optimum system.

An optimum system is necessarily a workable system. But a workable system need not necessarily be the optimum system. We are not making fun of the workable system. We are not making fun of the plumber. After all, he gives us a pump which works, given that some of us may not even know what the ratings available in the market are. The approximate power consumption of a refrigerator is around 100W or 150W. That is why if refrigerator companies advertise saying they have come up with a refrigerator that reduces consumption by 10%, we do not really care. But if some company says that they have come up with an air conditioner with a 10% reduction in consumption, we will go and buy it, because in the air conditioner, the power consumption is generally much higher compared to a refrigerator and over a 5-year period, it will amount to a lot of money and a lot of savings.

A workable system is definitely preferable over a non-workable system. A workable system is infinitely superior to something that is beautifully designed but does not work. First, we have to make something work and then we have to go for optimization. Furthermore, the cost of engineering time, human resources, computer systems, and software which are required to optimize large systems may not warrant the use of additional efforts to progress toward an optimum system. So there may be cases where we know that the system is underperforming but we are prepared to live with it. But we cannot do this for a 500 MW power plant, for example.

Example 1.1 *Select the pump and piping for a system to convey 4 kg/s of water from point A (in the sump) to point B (in the overhead tank). B is located 10 m higher than A, and the total length of the pipe from A to B is 300 m.*

Solution

The depiction is given in Fig. 1.7. There is a sump and an overhead tank. We are taking water from the sump to the overhead tank. The height difference is 10m and the length is 300 m.

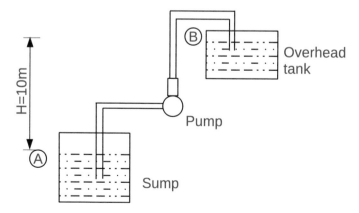

Fig. 1.7 Schematic of a simple pump and piping system

Given data: $\dot{m} = 4\,\text{kg/s}$; Length $L = 300$ m; $H = 10$ m.
Four bends exist along the pipeline.
The head loss in a bend is given by

$$h = kv^2/2g, \text{ where } k = 0.15$$

The friction factor for turbulent flow is given by

$$f = 0.182Re^{-0.2}$$

Properties of water are

$$\rho = 1000\,\text{kg/m}^3$$
$$\mu = 8 \times 10^{-4}\text{Ns/m}^2$$

We now need to determine the pump rating and the diameter of the pipe. The length of the pipe is known. So it is a design problem and is open ended.
Let us assume a pipe diameter d of $1\frac{1}{2}$" or 38 mm

$$\dot{m} = \frac{\rho\pi d^2}{4}v$$
$$4 = 1000 \times \frac{\pi}{4} \times (0.038)^2 \times v$$

Therefore, v = 3.53 m/s.
The Reynolds number is given by $Re = vd/\nu$

$$Re = \frac{\dot{m}}{\frac{\rho\pi d^2}{4}}\frac{d}{\nu} = \frac{4\dot{m}}{\pi d\mu} = 1.7 \times 10^5$$

Hence, the flow is turbulent (flow in a pipe, $Re_D > 2300$, transition to turbulence). Therefore, the friction factor is

$$f = 0.182 \times (1.7 \times 10^5)^{-0.2} = 0.016$$
$$\Delta P = h\rho g = 8.90 \times 10^5 \text{Pa}$$
$$\Delta P = (\frac{flv^2}{2gd})\rho g$$
$$h_L = (\frac{flv^2}{2gd}) = 80.4\,\text{m}$$

Loss in four bends $= 4 \times 0.15 \times (3.53)^2/2g = 0.38$ m.
We add 10 m height to take care of the elevation difference between A and B.
The total head loss $= 80.4 + 10 + 0.38 = 90.78$ m.

For calculating the power, we use the formula:

$$\text{Power} = \text{Discharge} \times \text{Pressure loss}$$
$$= Q \times \Delta P$$
$$= \frac{\dot{m} \times \Delta P}{\rho}$$

If $\eta \neq 1$, then

$$\text{Power} = \frac{\dot{m} \times \Delta P}{\eta \rho}$$

Let $\eta = 0.8$ (reasonable), hence power required = 4.4 kW $\approx$ 6B HP.

So we can go to the market and buy a 6 BHP pump and a 1.5" pipe, call the plumber, fix it, and switch it on. The system will work. However, this is not the optimum design. It is "a" design, and not "the only" design. Maybe a 1" pipe will have a lower cost but as the diameter decreases, what happens to the pressure drop?

There will be a complex interplay between these three parameters. Now if we add a constraint that the tank capacity is 4000 liters and we want the tank to be filled up in 30 minutes, then we will have to do some additional stuff and after all this, see if the design works. If it does not, we need to change our design. There may be other factors like how many individual apartments are there, what the usage of each is, how the usage differs in the morning and evening, etc. So what started out as a simple problem can now be written as a thesis!

1.5.1 Features of a Workable System

A workable system is one which satisfies our performance requirements. In the case of a pump and piping example, it should be able to pump the water. Is it enough if it satisfies the performance requirements?

What about the cost? Right now, we do not know if it is the cheapest or not. However, *the design we come out with should have a reasonable cost.* For example, in this problem that we worked out, had we taken a 6mm or $\frac{1}{4}$" diameter pipe, we could have come out with an answer, where the rating of the pump would be 10 or 25 kW. Immediately, we would have realized that although, technically, this pump would also do the job, there is something fundamentally incorrect about the value of the diameter, which was assumed in the first place. We then turn around and try to correct it. Therefore, even though several designs may work, all these may not be workable systems, not because they do not satisfy our performance requirements, but because they do not satisfy our criterion of a reasonable cost. What a reasonable cost is cannot be taught. For this case, if we have a pump and piping system that costs between Rs. 5,000 and Rs. 10,000, it is alright. But suppose it were to cost

Rs. 75,000 or a lakh,[3] we know that something is going wrong. It means some of the fundamental dimensions we have assumed are not correct. This cost consists of two components: fixed cost and running cost, which includes the cost of power, maintenance cost, etc. Both of these should be reasonable.

Is that all? If a system satisfies these two criteria, does it become a workable system? We are almost there but not yet! *The system should satisfy all our constraints.* The constraints can be in the form of pressure, temperature, pollution guidelines, material properties, and so on. For example, asbestos is now banned. Let us say we are trying to come out with an insulation system made of asbestos. It will satisfy our performance requirements and the cost factor. But it will fail as it will violate some regulation which is in place.

So the features of a workable system are that it

1. Satisfies the performance criteria;
2. Satisfies the cost requirements—fixed and running costs included;
3. Satisfies the constraints—pressure, temperature, pollution guidelines, weight, size, material properties, etc.

Now that we have designed the workable system, and also know its characteristics and attributes, when we carried out this design of the workable system, and how we went about doing it? What were the steps involved? The requirement was given. What did we first choose? *First we selected the concept.* It may sound silly, but every day, we can hire somebody, (if that somebody is available and affordable) pay the person some money, give 2 or 3 buckets, and ask that person to fill up the tank. Technically, that is also a solution. The person will take buckets full of water, climb the stairs, and pour it into the overhead tank. We first decided that there will be a centrifugal pump or a jet pump and that there will be a piping and pump arrangement that will do this job more efficiently. *Then we went about fixing the pertinent parameters.* For example, in this case, we fixed the diameter and performed the calculations. The final step is to *choose the nearest "available solution".* So if we get the pump rating as 3.486 kW, if 3.5 kW is available, we use it else we go for the nearest rating which may be 4 or 5 kW. Normally in the markets, they still follow the horsepower system. So a pump will normally be available in multiples of 0.5 horsepower—1 hp, 1.5 hp, and 2 hp. We round off our calculated pump rating to the nearest horsepower motor and decide that this is the solution. This basically gives us an overview of how we go about designing a simple system. If we want to design a complex system, we may have to do a lot more calculations. We may write a computer program or take recourse to commercially available software.

---

[3]One lakh of rupees = Rs. 1×10^5; 1 USD $\approx$ Rs. 70 (as of April 2019).

1.6 Optimum Design

But it is now that the actual story starts. Each of us can come out with a particular design. But all of these are not equally desirable. When we say all of these are not equally desirable, we are actually getting more specific. We are introducing what is called the *objective function*. The objective function is mathematically denoted by "y" and has to be declared upfront. For example, in a problem like what we discussed a little while ago, the "y" is very straightforward; it is basically the cost.

However, in social projects, "y" is very nebulous and cloudy. For the case of a flyover, what is the objective function? It is whether the flyover has eased the traffic problem. It could be at several levels. The users can subjectively say yes or no. One can do a survey or referendum and find out whether they are happy.

So we do not have to assume that the agreement on the objective function "y" is trivial and straight forward. Oftentimes, the definition of this objective function requires a lot of time and effort. For example, if we want to design a heat exchanger, it is not always that we want to design it at a minimum cost. Because, more often than not, we may end up with a minimum cost heat exchanger which is highly desirable, no doubt, but one that results in a lot of pumping power which is certainly not desirable. So, shall we say that we want to have maximum thermal performance divided by the pumping power? The ratio of the heat transfer rate to the pumping power or the pressure drop $Q/\Delta P$, is it a performance criterion?

So there need not be a consensus on what the objective function is. This is invariably left to the analyst. He/She decides what the objective function is and then goes about minimizing or maximizing the "y".

The question now is: What could be the objective function in the case of the pump and piping problem? How do we optimize the system under question? First and foremost, we need to agree on what is to be optimized. In this pump and piping system, what is to be optimized? We have the following costs:

1. Cost of the pump,
2. Cost of the piping,
3. Cost of installation,
4. Maintenance and running costs.

So the objective now is to minimize the lifetime total cost of the system. That is a fair objective. Now we will have to enumerate the various components that constitute the lifetime total cost and try to see if we can write down all these costs in terms of some numbers or equations. We then have to assign values to the various constants available in these equations and then use an optimization technique of our choice to determine the optimum operating conditions for the system under question.

The costs involved are

1. Cost of pump,
2. Cost of pipe (and fittings),
3. Lifetime running cost.

Let us leave out the maintenance cost for the present. Now, is it possible for us to write all these costs in terms of the fundamental quantities which are involved in this problem? The fundamental variable in this problem is ΔP or the pressure change. Let us start with the cost of the pump. The cost is proportional to ΔP and the volumetric flow rate Q. But the volumetric flow rate is fixed in this problem, and hence does not enter the problem.

$$\text{Pump cost} = a + b\,\Delta P \qquad (1.5)$$

How do we get a and b? It cannot be obtained from textbooks as a and b depend on local conditions and may have some values in Ahmedabad, may be different in Chennai, may be different in New York or whatever, and are easy to obtain. We just go to a nearby hardware shop and get the cost of a 1 hp to 10 hp pump, and using curve fitting, a and b can be determined.

Lifetime running cost is proportional to the energy consumed. Energy consumed is proportional to the power × time. Assuming that the performance of the machine does not deteriorate with time, to the extent that time required for operating the pump does not change, it is a fair assumption that the time need not be considered as a variable in the problem. Therefore, the cost is proportional to the power.

$$\text{Power} = \Delta P\, Q/\eta$$

We assume that efficiency is constant over this period and if we assume that the discharge is constant, the power is proportional to ΔP. Therefore,

$$\text{Lifetime running cost} = c + d\,\Delta P. \qquad (1.6)$$

Getting c and d again is not that straightforward as the cost of electricity changes with time. If the pump and piping system needs to work for 15 or 20 years, we have to factor in the inflation. The cost of electricity is never constant; it changes. So, a fair approach would be to look at the trend in the last 5 years, assume the average inflation rate, and then factor in the inflation. Things get complicated if we are trying to design and optimize a power plant; just as the costs are going up, the revenue will also keep changing with inflation. On the other hand, for the loan taken, interest needs to be paid up, which will basically be based on a diminishing balance. At the end of 5 years, say, a part of the principal would have been paid. A large program is required to solve this problem. For the example chosen, we will stick to this story. Nevertheless, we can appreciate that eventually the problem could get really messy.

The cost of the pipe is directly proportional to the mass or assuming that it is operating under one g, we can say that the cost is proportional to the weight. The length is pretty much fixed as we know the length and how many bends there are.

$$\text{Mass of pipe} = \rho \pi dtL$$

where t—thickness, d—diameter, and L—length.

Assuming that we are not going to have designs varying from 1 hp to 200 hp, if thickness "t" is considered more or less constant, which is a questionable assumption and L is maintained a constant, the cost is directly proportional to the diameter. This does not help us much. Piping cost is proportional to diameter, but we have already seen the other two costs in terms of ΔP. If we can write this cost also in terms of ΔP, we have formulated the optimization problem where the whole thing is dependent only on ΔP. We solve this problem and obtain the optimum ΔP, so that the cost is minimized.

Now we have to use all our knowledge of fluid dynamics to get this relationship. There is a friction factor "f" and this "f" is related to diameter. For turbulent flow, f is given by

$$f = 0.182 \, Re^{-0.2} \tag{1.7}$$

Head loss $= flv^2/2 \, gd$, where v is the velocity.

$\dot{m} = \rho av$, where a is the cross-sectional area given by

$$a = \pi d^2/4 \tag{1.8}$$

Cost of pipe:

$$\text{cost} \propto \text{weight}$$
$$\propto \text{volume} \times \text{density}$$
$$\propto \pi dtl \, \rho$$
$$\propto d$$

There is an assumption involved in the $\pi \, dtl$. What is this d now? There is a d_i and a d_o, the inner and outer diameters, so this is actually the average diameter. It is normally called the nominal diameter. But the nominal diameter will usually not be the arithmetic average of d_i and d_o.

$$\Delta P = \Delta P' + \text{static head loss}$$
$$\Delta P' = \Delta P - \text{static head loss}$$
$$\Delta P' = (fLv^2/2gd)\rho g$$

We do not know v directly because we have specified only the mass flow rate.

$$\dot{m} = \rho A v = \rho \frac{\pi d^2}{4} v$$

$$v = \frac{4\dot{m}}{\pi d^2 \rho}$$

$$\Delta P' = \frac{8fL\left(\frac{4\dot{m}}{\pi d^2 \rho}\right)^2}{2gd} \rho g$$

We are not done yet because f is also a function of the Reynolds number. Of course as a practicing engineer, we may neglect it and get on with it but we will get to the bottom of it. Assuming that the flow is turbulent,

$$f = 0.182\, Re^{-0.2}$$

$$f = 0.182 \left(\frac{vd}{\nu}\right)^{-0.2}$$

$$f = 0.182 \left(\frac{4\dot{m}d}{\pi d^2 \rho \nu}\right)^{-0.2}$$

$$f = 0.182 \left(\frac{4\dot{m}d}{\pi d^2 \mu}\right)^{-0.2}$$

$$\Delta P' \propto \frac{C}{d^5}\left(\frac{1}{d}\right)^{-0.2}$$

$$\Delta P' \propto d^{-4.8}$$

But since cost $\propto$ d, we have

$$\text{cost} \propto (\Delta P)^{-\frac{1}{4.8}}$$

The total cost is given by

$$\text{Total cost} = a + b\Delta P + c + d\Delta P + e(\Delta P - \text{static head loss})^{-0.21} \qquad (1.9)$$

$$\text{Total cost} = f + g\Delta P + e(\Delta P - \text{static head loss})^{-0.21} \qquad (1.10)$$

Now we have written the total cost in terms of ΔP. ΔP is the pressure rise taking place in the system and which is under our control. So for different values of ΔP, the total cost goes up with ΔP, if we look at term no. 2 on the right-hand side of Eq. 1.10. Term no. 3 goes down with ΔP. Because there is one term that increases with ΔP and there is another term which decreases with ΔP, there is hope for us to optimize and there should be a particular value of ΔP at which the total cost will be extremized, that is, the first derivative of the total cost will be 0. However, we do not know if this will be the maximum or minimum cost. Even so, after doing all this, intuitively we can guess that it will be the minimum cost. However, as a purist or a mathematician, if we are not able to appreciate this, we can go ahead and do the second derivative test. As engineers, if we try to optimize the total cost of the power plant, put all the factors together, get all these numbers, have a complicated optimization program running, and get our final answer, why would the final solution be the maximum cost of the power plant? There is little chance of this happening!

Reference

Stoecker, W. F. (1989). *Design of thermal systems*. Singapore: Mc Graw Hill.

Chapter 2
System Simulation

2.1 Introduction

System simulation basically mimics an actual system. It can be defined as the calculation of operating variables such as pressures, temperatures, concentrations, and mass flow rates of fluids in a thermal or energy system, operating in a steady state or under transient conditions. For example, in a power plant, most of the time we are interested in its steady-state operations but there are also issues like starting up and shutting down of the power plant, which is very critical especially for a nuclear power plant. When an accident occurs in a nuclear power plant, there is an emergency shutdown of the reactor. However, nuclear fission will continue to proceed and if all systems break down, we will not have pumps to take the heat from the fission reaction and therefore we will not be able to dissipate all the heat to the ambient. In this case, natural convection will take over and so the system should be designed such that even in the case of such an accident, even under natural convection, we do not have catastrophic consequences. Nuclear safety is thus an important part of nuclear plants and is critically linked to the transient behavior of the reactor. So, as thermal engineers or mechanical engineers, though most of the time we are interested in the steady state, we are also interested in the transient one. A chemical engineer on the other hand is more interested in transients, and process control is a big deal in chemical engineering.

Much in the same way as we calculated the diameter of the pipe and the rating of the pump in the example considered in Chap. 1, we should be able to calculate the other operating variables too in a problem involving several variables. It could be possible that we actually have a heat exchanger and we want to work out the outlet temperatures of the fluids. Now we are talking about an analysis problem. The design problem on the other hand answers questions like given the flow rate and other constraints, what will the heat exchanger size be, how many tubes, etc.? We have already seen the difference between the analysis and design problems. *So simulation is more concerned with the analysis rather than the design.* This definition is basically for a thermal system and for us to do system simulation, we need to

© The Author(s) 2021
C. Balaji, *Thermal System Design and Optimization*,
https://doi.org/10.1007/978-3-030-59046-8_2

know the (i) performance characteristics of all components and (ii) equations for all thermodynamic properties of the working substances.

Now, in order to do system simulation, it is not practical to use property tables. Properties must preferably be in the form of equations. Therefore, regression is required for this. So one needs to be good at statistics. For example, if the enthalpy (y) is given as a function of pressure (x_1) and temperature (x_2), we should be able to construct equations such that y is a function of x_1 and x_2.

Similarly, if one knows the operating variables of a compressor or a blower, we should be able to calculate the efficiency in terms of these operating variables. Therefore, knowledge of regression and the knowledge of how to represent thermophysical properties, system performance, and so on in the form of equations is imperative. This information has got to be embedded, otherwise, we cannot do simulation.

The equations for performance characteristics of the components and thermodynamic properties along with the mass and energy balance will form a set of simultaneous equations that relate the operating variables. If we solve the set of simultaneous equations, we will be able to solve for all the parameters under question. This is basically system simulation. There may be situations involving a large number of variables, so we will have to do matrix operations for such large systems.

System simulation is thus a way of mimicking the performance of a real system. So instead of performing a real experiment, we try to find out how the output will change when each of the inputs is changed. We have a computer model or a mathematical model of the system and carry out numerical experiments upfront, even before we design. By way of this, we get an idea of how the performance of the system changes when the operating variables are changed. This is the key goal of system simulation. Needless to say, the end product of system simulation is a workable design or workable designs. In view of this, system simulation becomes a precursor to optimization.

2.2   Some Uses of Simulation

1. At the design stage, simulation is extremely useful in evaluating alternate designs and deciding which of these is better.
2. System simulation may be applied to an existing system to explore possible modifications. If we want to work on an existing system and improve it, then system simulation comes in handy.
3. Simulation is usually needed for studying performance at off-design conditions. This is very critical because most of the time, systems work at off-design conditions. There is a change in some variable or the other, all the time, in a real system. For example, for a power plant, during evening hours, there will be heavy demand while during early mornings there is not much demand. There will be a design load, but most of the time we are off the design load. So how the system performs under off-design conditions is an important question that can be answered if we can develop a model and simulate the system.

Why are we interested in all these? Because in a power plant or an air conditioning system for an auditorium or in a heat exchanger, where a lot of costs are involved, we are not only trying to simulate but also optimize. The costs are going up, and so simulation and optimization are being taken more seriously lately. Accompanying this is the fact that we have powerful computers and software programs that can do all these analyses very quickly. By combining these analyses with an optimization technique, we can attempt to arrive at an optimum.

2.3   Different Classes of Simulation

(a) Dynamic or transient and steady state

In transient simulations, we study the changes in the operating variables with reference to time. The goal is to avoid unstable operating conditions.

Figure 2.1 shows a typical temperature–time response of a steady-state system. So in a power plant or when one starts a motor car or bike or an airplane, there will be a startup phase involved.

In a nuclear power plant, there is the stage where we have a start-up and then we reach a steady state and then there is a shutdown. Shutdown could be because of maintenance or breakdown or even a very serious accident like a Tsunami. The recent failure at the Fukushima nuclear power plant near Sendai in Japan is a classical case of inadequate cooling strategies for removing decay heat in a nuclear reactor, under station blackout conditions.

(b) Deterministic or stochastic

There are several cases possible where the input conditions are not known precisely and the probability distribution may be given instead, with a dominant frequency, an average, and an amplitude of variation. Then we cannot use a deterministic simulation but have to do a stochastic simulation.

Let us say a corporate hospital is hiring a consultant to do simulations to determine the optimum deployment of an operation theater. For example, the bypass surgery and the kidney transplant may take a long time whereas, the gall bladder removal

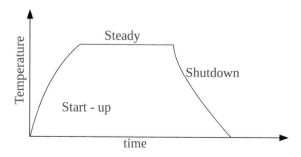

Fig. 2.1 Typical temperature–time history of a thermal system

may take 2 or 3 h only. The hospital administration wants to study this and get to its bottom. So what are the various steps involved? For example, in the bypass surgery, the patient is first wheeled in. Then the anesthetist moves in. Then they make measurements (of the chest, where the sternum has to be incised). Surgery is largely engineering nowadays! X-rays and echocardiogram reports are used to see its size and then mark the place where the cut is to be made. Then the main surgeon comes and anesthesia is administered; the main surgery begins wherein they hook the patient onto a heart-lung machine or nowadays they do a beating heart surgery (the heart is not technically stopped here). Then the incision is closed and the patient is taken to the ICU where he is monitored.

Each of these events can have a Gaussian distribution. The normal time taken by the anesthetist is, say, 30 minutes with a σ of 5 or 8 minutes. The normal time taken for a bypass to be done on a patient with 3 blocks will be, say, 3 h with a σ of 15 minutes. Suppose one wants to do a Monte Carlo simulation, he/she will generate a random number. If it is between 0 and 50, then the anesthetist will finish his/her job exactly at (mean + σ). If it is between 50 and 70, he/she will take (mean $-\sigma$). If it between 70 and 90, something else. Like this, we pre-assign numbers. We come out with a model and using sequential random numbers, and we add up all this and find out the total time taken. Let us say the total time is 218 minutes for one case. We can run this Monte Carlo simulation several times and can calculate the average time and take the variance and that will give us an idea of the average time a bypass surgery takes in that hospital. That is one information we can have. Like that, if we do for all surgeries, we can determine the total number of operations that can be performed in a day and also how they can all be scheduled.

Another example that comes to our mind is from the hospitality industry where corporate hotels are interested in the average time taken by a guest between his/her entering the hotel and his/her entering the room. Can we optimize this? How much time is taken for checkout? These are all related to queueing theory in operations research. In all these cases, including the hospital case, we cannot exactly say that all surgeons will complete the surgery in exactly 2.5 h. They may have some unexpected thing, or they may find something new. This is not deterministic but stochastic because the variables can change with time.

In a deterministic simulation, all the variables are known a priori with certainty. We can also have a continuous and discrete simulation. Oftentimes in thermal systems, we are interested in the continuous operation of power plants, air conditioning systems, IC engines, and so on. The flow of fluid is assumed to be continuous. We do not encounter discrete kinds of systems in thermal sciences often. Simulation of discrete systems is of particular relevance to manufacturing engineering where we need to look at individual items as they go through various metallurgical and/or manufacturing processes.

2.4 Information Flow Diagrams

The information flow diagram is a pictorial way of representing all the information which is required for simulating the overall system by looking at the information pertinent to a particular component.

The information flow diagram tells us what the inputs to this block are, what the outputs from the block are, and more importantly, the equations relating these quantities. Figure 2.2 shows the information flow diagram for a typical heat exchanger. The inputs here are the inlet temperatures of the hot and cold fluids, $T_{h,i}$ and $T_{c,i}$, respectively, and the flow rates of the hot and cold fluids, $\dot{m}_h$ and $\dot{m}_c$, respectively. In fact, if three of the four quantities are known, the fourth one gets fixed automatically by energy balance.

$$(\dot{m}C_p)_{hot}\Delta T_{hot} = UA(\Delta T)_{LMTD} = Q \qquad (2.1)$$

$$(\dot{m}C_p)_{cold}\Delta T_{cold} = UA(\Delta T)_{LMTD} = Q \qquad (2.2)$$

$$Q - UA(\Delta T)_{LMTD} = 0 \qquad (2.3)$$

where

$$Q = (\dot{m}C_p dT)_{hot/cold} \qquad (2.4)$$

Shown in Fig. 2.3 is the information flow diagram of an air compressor. The inputs are the mass flow rate, $\dot{m}$, the inlet pressure, p_1, and the output is the outlet pressure, p_2. The transfer function f is $f(\dot{m}, p_1, p_2)$ and gives the performance of the compressor.

$$\text{Power} - \frac{\dot{m}\Delta P}{\eta} = 0 \qquad (2.5)$$

Now let us look at the information flow diagram for a vapor compression refrigeration system. The thermodynamic cycle for the above system on T-s coordinates is given in Fig. 2.4. The various thermodynamic processes associated with this reversed Rankine cycle are

Fig. 2.2 Information flow diagram for a heat exchanger

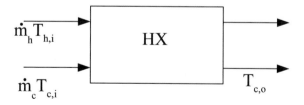

Fig. 2.3 Information flow
diagram for a compressor

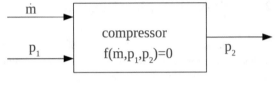

Fig. 2.4 T-s diagram of a
vapor compression
refrigeration system

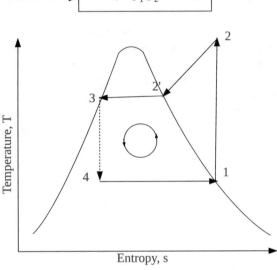

1-2 → compression

2-2'-3 → condensation

3-4 → throttling/expansion

4-1 → Evaporation

The key quantity of interest for this system is the *coefficient of performance* (COP)
given by

$$COP = \frac{h_1 - h_4}{h_2 - h_1} \tag{2.6}$$

In Eq. 2.6, h_1 corresponds to the enthalpy at state 1, h_2 is the enthalpy at 2, h_3 is
the enthalpy at 3, and h_4 is the enthalpy at 4. Process 3-4 involves throttling and is
an isenthalpic process. Hence, $h_3 = h_4$. An information flow diagram for the vapor
compression system will be as shown in Fig. 2.5.

This is called a **sequential arrangement**. The output of one component becomes
the input to the other component. Now, if we know the equations for each of the
components that come from the manufacturer or if we have data to generate the
performance curves, then it is possible to perform a system simulation and predict
the performance of the system for a set of operating variables. Please note that the
isenthalpic nature of process 3-4 is tacitly embedded in the information flow diagram.

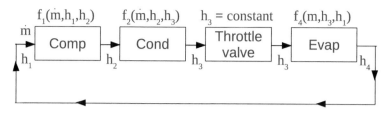

Fig. 2.5 Information flow diagram for a vapor compression refrigeration system

Example 2.1 A closed plastic container, used to serve coffee in a seminar hall, is made of two layers with an air gap between them. List all the heat transfer processes associated with the cooling of the coffee in the inner plastic vessel. Draw an information flow diagram for the system. What steps will you take for a better container design, so as to reduce the heat loss and keep the coffee as hot as possible for a long period of time?

Solution

We do not worry about the free surface. We first draw a schematic, as shown in Fig. 2.6. Let us assume that the container is fully closed.

The problem under consideration is an unsteady one. Assuming that the coffee is well mixed and that there are no temperature gradients within the coffee, the temperature of the coffee varies with time exponentially as shown in Fig. 2.7.

Let us start with drawing the blocks involved. We can now list all the heat transfer processes associated with this problem.

$$\text{Inner vessel—layer 1} \;\rightarrow\; \text{Convection}$$
$$\text{Layer 1—air gap} \;\rightarrow\; \text{Conduction}$$
$$\text{Air gap—layer 2} \;\rightarrow\; \text{Natural convection}$$
$$\text{Layer 2—ambient} \;\rightarrow\; \text{Convection + Radiation}$$

Fig. 2.6 Heat transfer process in a coffee flask

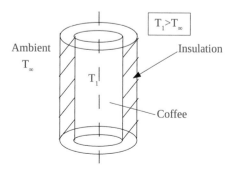

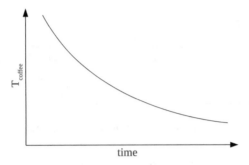

Fig. 2.7 Temperature of coffee with time (Example 2.1)

Even this apparently simple problem has so many heat transfer processes associated with it. The corresponding information flow diagram is given in Fig. 2.8. We can write the pertinent differential equations, and solve them to obtain the instantaneous temperature of the coffee. However, what we want to do now is to model it even before we start the simulations. So modeling precedes simulation and simulation precedes optimization. This is the story we are going to look at throughout this book—modeling, simulation, followed by optimization. The temperature T of the coffee is generally not under our control. We cannot overheat the coffee as the water boils at 100 °C at 1 atmosphere pressure. So probably the temperature of the coffee may be 60 or 70 °C. The temperature at which we normally drink coffee may be around 45 or 50 °C. The ambient temperature is 30 or 35°C. Both of these are not under our control. Now that we have drawn the information flow diagram, the possible steps to reduce heat loss are

1. Low emissivity coating on the outside.
2. For the same volume, we can make a tall container so that the air layer is slender and tall so that we reduce the natural convection. If we look at natural convection

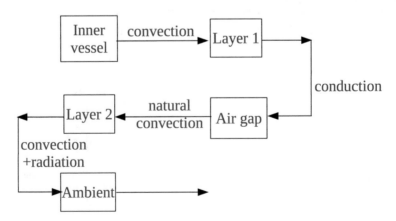

Fig. 2.8 Representation of various heat transfer processes taking place in the coffee flask

from a cavity like the air gap in this problem, we see that there is a heated wall on the inside, a cooled wall on the outside, and the top and bottom walls are adiabatic (an assumption). If we go in for a tall flask, it results in a multicellular pattern, which resists the flow of heat and the convective heat transfer coefficient reduces. Hence, another option is to redesign the shape.

3. The third option is to use better insulation material.
4. We can try to maintain a vacuum between the layers.

Between the air gap and layer 2, we can also have radiation. As we start probing deeper and deeper, this problem itself starts getting messier and messier. We have to just leave it at some stage as is the case with any problem. The perfect mathematical model may be elusive!

For this problem, the initial temperature of the coffee is known and is an input. The temperature of the coffee at various instants is what we desire using the modeling. A question that arises is that the temperature of the coffee varies with time and depends on all the boxes. So is one right in saying that the information is flowing in one direction?

The answer to this question is that there is no feedback loop because the inside of the flask is hotter than the outside. All the heat is flowing from the inside to the outside. There is no question of something coming back. This is quite different from the vapor compression system where we have a closed loop. Here, there is no refrigerant that flows continuously. That is the difference between a sequential arrangement and an arrangement like this.

2.5 Techniques for System Simulation

By system simulation, we mean solving a set of equations which controls the phenomenon under consideration. If the system consists of more than one component, then we have to assemble the equations which govern the performance of each of these components. Now this can be at various levels. We can have just algebraic equations which give us the performance of the system or these can be governed by ordinary differential equations or linear partial differential equations or even nonlinear partial differential equations. The latter arise in problems involving fluid flow and heat transfer (continuity equation, the Navier–Stokes equations, and the equation of energy). What we have listed above represents escalated levels of difficulty.

The central theme of this topic, in so far as thermal system design is concerned, is looking at situations where more than one component is involved. Hence, we restrict our attention to algebraic equations and make the analysis realistic, more meaningful, and also to add a little spice, we will look at nonlinear algebraic equations. In many of the components like compressors, turbines, and pumps, the dependent variable can be related to the independent variable in terms of simple linear/nonlinear algebraic equations.

We now look at two techniques, namely the successive substitution and Newton–Raphson method.

2.5.1 Successive Substitution Method

Consider an equation like this:

$$x = 2 \sin x \qquad (2.7)$$

where x is in radians and is positive. Can it be analytically solved? The answer is NO! It is a transcendental equation and has to be solved iteratively. So we can rewrite the equation as

$$f(x) = x - 2 \sin x \qquad (2.8)$$

The goal is to make $f(x) = 0$ and find out where this happens. These are then the roots of the equation.

In successive substitution, we write an algorithm for this procedure and can either do the calculations using a pen and paper or write a program and solve it on the computer. What will the algorithm for this be?

$$x_{i+1} = 2 \sin x_i \qquad (2.9)$$

We make a Table, as shown in Table 2.1, where the first column is the serial number or iteration number, then comes x_i, the third column is x_{i+1}, and the fourth column is $(x_{i+1} - x_i)^2$, which is some kind of a norm. Whenever the fourth column goes on to a reasonably small value, we can stop the iterations and be happy that we have got a solution to the problem. We take an initial guess of x_i and solve this. Taking an initial value of x_i as 1, the first 3 iterations are as shown in the table. As we continue doing this, let us consider the 8th iteration.

As we can see, after a few iterations the method of successive substitution works. The problem is that if there is a function that changes very rapidly, then there is a possibility that the method of successive substitution may miserably fail. If the gradients are very sharp and/or we start with a wrong guess, we may end up with a solution that diverges and we will be unable to proceed further. So the method of successive substitution is not a universal cure or panacea that can be used to solve all kinds of problems. Even so, simple problems can be solved with the help of this method. We can also demonstrate this with the help of a spreadsheet (left as an exercise to the student). A graphical depiction of this example is given in Fig. 2.9.

Table 2.1 Successive substitution method for solving x=sin(x)

S. No.	x_i	x_{i+1}	$(x_{i+1} - x_i)^2$
1	1	1.6829	0.47
2	1.6829	1.987	0.09
3	1.987	1.83	0.022
⋮	⋮	⋮	⋮
8	1.902	1.891	1×10^{-4}

Fig. 2.9 Depiction of the successive substitution technique for solution of Eq. 2.7

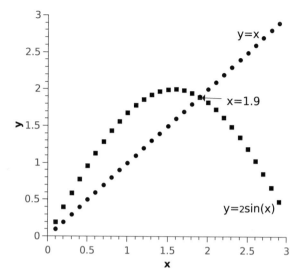

Example 2.2 Consider a conductor of a circular cross section of emissivity 0.6 that is electrically heated. Under steady-state conditions, the electrical input to the conductor is 900 W/m² of the surface area of the conductor. The heat transfer coefficient to the ambient h is 9 W/m²K. The ambient temperature is 300 K (see Fig. 2.10). What is the energy equation governing the steady-state temperature of the conductor? Using the method of successive substitution, determine the steady-state temperature of the conductor. Decide on an appropriate stopping point.

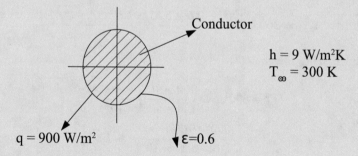

$q = 900$ W/m² $\varepsilon = 0.6$

Fig. 2.10 Figure for the electrically heated wire problem

Solution

The wire is losing heat by both convection and radiation. The surface emissivity and the local heat transfer coefficient are given. The surface area is not required. Power density is given. There is no need to draw an information flow diagram for this problem. It is a single component system, and enlisting of the various heat transfer processes in the system is relatively straight forward. First, we will draw a sketch of the system indicating all the heat transfer processes taking place (see Fig. 2.10). We then write down the governing equation and start solving. The governing equation is

$$q = h(T_w - T_\infty) + \epsilon\sigma(T^4 - T_\infty^4) \tag{2.10}$$

When we write an equation like this, first we state the inherent assumptions as follows:

1. Steady state,
2. No temperature gradients in the conductor: this is not really correct because it is generating heat at the rate of 900 W/m². Where will the conductor be the hottest? At the center. But we are not considering this because we are not looking at this level of detail here!
3. Emissivity of the surroundings is 1. The ambient is considered to be a black body at T_∞. The most important thing is that the ambient temperature for free convection is the same as the ambient temperature for surface radiation. Oftentimes, without realizing, we just take this for granted. It is possible that in a room, we may have a T_∞ for convection and because of reflectors and other stuff, we may have a different ambient temperature, as far as radiation is concerned. But unless otherwise stated, the T_∞ for radiation is the same as that for convection. Even so, it is good to remember that it need not always be true.

Two types of information flow diagrams are possible. Using one information flow diagram causes the solution to diverge. Then we have to turn around and use the other information flow diagram.

Information flow diagram (a)

$$9(T_w - 300) = 900 - 0.6 \times 5.67 \times 10^{-8}(T_w^4 - 300^4)$$
$$T_{i+1} = 300 + \frac{[900 - 0.6 \times 5.67 \times 10^{-8}(T_i^4 - 300^4)]}{9} \tag{2.11}$$

Information flow diagram (b)

$$(T_w^4 - 300^4) = \frac{[900 - 9(T_w - 300)]}{0.6 \times 5.67 \times 10^{-8}}$$
$$T_{i+1} = \left\{ \frac{[900 - 9(T_i - 300)]}{0.6 \times 5.67 \times 10^{-8}} + 300^4 \right\}^{\frac{1}{4}} \tag{2.12}$$

The above is a highly nonlinear problem, due to the presence of T^4 term on the right-hand side. We will not investigate why one algorithm is not working. Even so, suffice it to say that because of T^4, one of the two algorithms will not work. The errors will propagate for one of the two algorithms. We know that it is a physical problem: a conductor is cooling, it has got emissivity, it has got convection and radiation, and it has to attain a physical temperature. 900 W/m² is a reasonable flux; emissivity, h, and ambient temperature are all reasonable values and hence we "should" get a reasonable value for the temperature. If we, however, get an unrealistic solution, the problem is surely not with the physics but with the algorithm!

When we try both, we see that the second one fails. So we find that even a one-component system is not as simple as we thought. Suppose we have a two-component system and it is nonlinear. The solution will be a lot more formidable.

In the second algorithm, the moment T_i on the right side exceeds 400 K, and is gone. If we take the slope and investigate it mathematically, we can see this, but it is not central to our discussion here. So, if we solve a problem using successive substitution, encounter difficulty and get an answer that is negative or diverging and is not able to proceed further, then we should not declare that this problem has no solution. Because from the data given to us, there must be a solution to the given problem. We should immediately think of the other information flow diagram.

There are two information flow diagrams. So in the first algorithm, we are saying that

$$Q_{conv} = Q_{total} - Q_{radn}$$

and in the second algorithm, we are saying that

$$Q_{radn} = Q_{total} - Q_{conv}.$$

It so happens that one of the two is not working.

An initial value of 360 K for T_i seems reasonable. Assuming T_i to be 360 K, let us use the method of successive substitution and determine T_{i+1}.

The solution is $T_{wire} = 91.2\,°C$ or 364.2 K. The steps involved in obtaining the final solution are presented in Table 2.2. If we had started the second algorithm with

Table 2.2 Successive substitution method for Example 2.2

S. No.	T_{i+1}	T_i	$(T_{i+1} - T_i)^2$
1	360.00	367.13	50.82
2	367.13	361.95	26.83
3	361.95	365.74	14.40
4	365.74	362.98	7.64
5	362.98	365.00	4.09
6	365.00	363.53	2.17
7	363.53	364.60	1.16
8	364.60	363.82	0.62
9	363.82	364.39	0.33
10	364.39	363.97	0.18
11	363.97	364.28	0.09
12	364.28	364.06	0.05
13	364.06	364.22	0.03
14	364.22	364.10	0.01

364 or 365 K as the initial guess, it may have worked. But, had we started with an initial guess of 300K or 400K, it may not have worked.

Now we will go to a real thermal system, namely the fan and duct system. Let us consider a two-component thermal system, like a ducting system for the air conditioning of an auditorium. We want to decide on the fan capacity. We are not talking about the chiller capacity and how many tons of refrigeration is required. We want to send in chilled air and have to decide on the fan and ducting system. It is possible for us to calculate the total length of the duct, how many bends are there, and so on. If we can calculate the total pressure drop as we did before, using formulae, this multiplied by the discharge and divided by the fan efficiency will give us an idea of the power required.

For overcoming a particular head, we will also know what the discharge is from the manufacturer's operating characteristics. These are called fan curves. The fan curve is obtained from the manufacturer's catalog. But how would this have been obtained? Basically through experiments. Someone must have first done an experiment with almost zero discharge and then will have slowly increased the discharge and determined the head and he/she may have got some points. But if we want to do simulation and we just have these points, it is no good for us. We want a curve. Hence, we will have to convert all these points into an approximate and best-fitting curve using principles of regression.

Now if we have to design the fan and ducting system, the fan characteristics of the manufacturer have to match the load characteristics of the particular design. When these two intersect at a particular point, this is called the **operating point**. This is graphically depicted in Fig. 2.11. The head is usually in meters and the discharge in m^3/s. Using system simulation, we want to calculate this operating point.

The duct curve can easily be plotted if we know the friction factor and other associated quantities.

Subsequently, we have to determine the sensitivity of the operating point. It is possible that with time, our fan curve goes down (it cannot possibly go up!). The duct curve may also change depending on the factors that affect it like age, friction

Fig. 2.11 Typical fan and duct curves

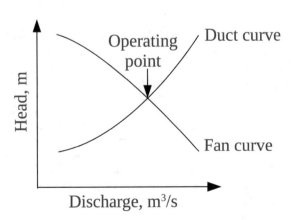

factor, etc. So we will have to study the sensitivity of the performance of the system or equipment with the change in design operating conditions.

Example 2.3 It is required to determine the operating point of a fan and duct system. The equations for the components are

$$\text{Duct: } P = 100 + 9.5Q^{1.8} \qquad (2.13)$$
$$\text{Fan: } Q = 20 - (4.8 \times 10^{-5})P^2 \qquad (2.14)$$

where P is the static pressure in Pascals and Q is the discharge in m^3/s.

(a) Draw the information flow diagram for the above system.
(b) Using successive substitution, determine the operating point.

Solution

An important point here is the initial guess. We can either take P to be 250 Pa or Q to be 10 m^3/s. So we first draw the information flow diagram for the two components, the fan and the duct. We draw two rectangular boxes, where information goes in and comes out. This is a sequential arrangement because the output of the fan becomes the input to the duct and the output of the duct is the input to the fan. Two types of information flow diagrams are possible. We can use the first equation to determine P if we know Q. But there is no hard and fast rule about this.
   So we can say

$$Q^{1.8} = \frac{(P - 100)}{9.5} \qquad (2.15)$$

$$Q = \left[\frac{(P - 100)}{9.5} \right]^{\frac{1}{1.8}} \qquad (2.16)$$

Equation 2.13 can be used to calculate P, given Q or vice versa. By the same token, the second equation (Eq. 2.14) also need not necessarily be used only to calculate Q; it can also be used to calculate P.
   So if we want to calculate P and Q as shown in the equations, it is one information flow diagram. If we want to swap the calculations of P and Q in the equations, it is another information flow diagram. And since the equations are nonlinear (we have P^2 and $Q^{1.8}$), one of the two information flow diagrams will not work. At least, this is our belief reinforced by our bitter experience with the previous example.
Total head = Static head + dynamic head

Fig. 2.12 First information flow diagram for the fan and duct system

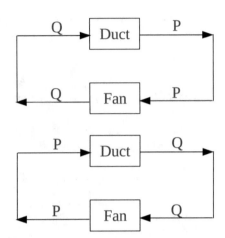

Fig. 2.13 Second information flow diagram for the fan and duct system

$$\text{Dynamic head} = 0.182 Re^{-0.2} \frac{flv^2}{2gd}$$

$$= 0.182 \left(\frac{vd}{\nu}\right)^{-0.2} \frac{flv^2}{2gd} \qquad (2.17)$$

The dynamic head becomes v^2 multiplied by $v^{-0.2}$ which gives us $v^{1.8}$. Then what does the value 100 indicate in the first equation (Eq. 2.13)? It is the static pressure. Using regression, we can get that. Equation 2.14 cannot be got from physics, but the manufacturer does experiments and gives it in the manufacturer's catalog. It is called a fan characteristic. This is the fundamental experiment the manufacturer has to do and give the user.

How did we get this $Q^{1.8}$? Normally the dimensions are fixed. The diameter of the pipe or the hydraulic diameter of the rectangular duct is fixed. P is the head.

We now use the fan characteristic to determine Q, knowing P and the duct characteristic to get P from Q. So at the end of the exercise, whatever P we get, we compare it with the P we assumed and started with. If they are equal, we have the solution. In general, they will not be equal, so we keep working on it. What goes through this is the information on P and Q. As we proceed with the iterations, it will stabilize and whatever is coming out will be the same as what was sent in. That will result in a situation where $(P_{i+1} - P_i)^2$ will be very small. Is this enough for this problem? What should be the stopping criteria for this problem? No, this is not sufficient. We need to incorporate the Q factor too. So we write the stopping criterion ϵ as

$$\epsilon = (\frac{Q_{i+1} - Q_i}{Q_i})^2 + (\frac{P_{i+1} - P_i}{P_i})^2 \,;\, \epsilon \leq 10^{-7} \qquad (2.18)$$

Using the first information flow diagram:

We use Eqs. 2.13 and 2.14 and start iterations with $P_i = 250 Pa$. We see that the solution diverges very swiftly (see Table 2.3).

Table 2.3 Successive substitution method for the fan and duct problem (using the first information flow diagram) Fig. 2.12

S. No.	P_i, Pa	Q_i, m³/s	P_{i+1}, Pa
1	250	17	1657
2	1657	−111.91	Diverges

Table 2.4 Successive substitution method for the fan and duct system (using the second information flow diagram) Fig. 2.13

S. No.	P_i, Pa	Q_i, m³/s	P_{i+1}, Pa	$(\dfrac{Q_{i+1} - Q_i}{Q_i})^2 + (\dfrac{P_{i+1} - P_i}{P_i})^2$
1	250	4.62	566.1	2.367
2	566.1	8.67	485.8	0.0299
3	485.8	7.81	503.9	2.04×10^{-3}
4	503.9	8.01	499.8	9.11×10^{-5}
5	499.8	7.97	500.6	4.1×10^{-6}
6	500.6	7.98	500.4	2×10^{-7}
7	500.4	7.98	500.4	0

Hence, we have to use the other information flow diagram given in Fig. 2.13.

$$Q = \left[\frac{P - 100}{9.5} \right]^{0.555} \tag{2.19}$$

$$P = \left[\frac{20 - Q}{4.8 \times 10^{-5}} \right]^{0.5} \tag{2.20}$$

We see that the solution is rapidly converging (see Table 2.4). The method of successive substitution converges rapidly, but when it starts diverging, that too will be rapid. Success or failure is guaranteed and is immediate.

The stopping criterion cannot be 10^{-8} when we are rounding it off to two decimal places. If we are retaining up to the fourth decimal, then it can be of the order of 10^{-4}. So the solution to the problem is $P = 500.4$ Pa and $Q = 7.98$ m³/s.

Let us now examine the stability of the system at its operating point. For this, we start with an initial guess which is close to the correct answer, but is to the left or right of the correct answer, and see whether it is proceeding to the same operating point. We have to go back to the equations and check if we had overshot. If we start P as 490 or 510, after 3 iterations, does it come to 500.4 and 7.98? That is one way of checking it. The other way is to start with Q as 7.5 or 8.5 and see if it approaches the same value.

2.5.1.1 What Is the Big Deal About the Stopping Criterion?

We may think that this is not the most important part of the problem, while getting the algorithm and iteratively solving it is the most important. However, the stopping criterion is also equally important because in real life, no one will tell us what the stopping criterion is. It is for us to figure out and for us to get convinced that we have reached the level at which the solution will not change. This is the level of accuracy we are comfortable with. If we know what the true value is, we can compare our result with the true value. If we get 1% close to the true value, it is fine. But, when we are doing CFD computations or simulations, the true value is elusive. We do not know what the true value is. We are looking at the difference between successive iterations. A word of caution may be necessary here. *Convergence can be highly misleading* as we can finally converge into a beautiful but totally incorrect solution.

So if we have reached convergence, it does not necessarily mean that we have got an accurate solution. Convergence means, with the present numerical scheme, if we do 100 more iterations, we cannot expect a significant change in the answer. But whether that answer itself is correct or not, convergence will not tell us! So as an additional measure, *we will have to validate our solution.* We need to apply our numerical technique to some other problem for which either somebody else has done experiments or for which analytical solutions are possible. Then for that case, after we run through many iterations, after we incorporate our stopping criterion, if we get a solution exactly the same as other people have reported or the analytical solution, then we can say, for this particular problem, when we apply this stopping criterion, our solution works. Therefore, when we apply it to an actual problem, for which we do not know what the true solution is, there is no reason why it should misbehave.

Why do some information flow diagrams not work?

Let us try to figure this out from a numerical point of view. If we are trying to get the roots of the equation $g(x) = 0$ around $x = \alpha$, the goal is

$$g(\alpha) \approx 0 \tag{2.21}$$

Then α is a root of the equation. It won't be 0.0000, but within limits, it is alright. The requirement now is

$$|g'(\alpha)| < 1 \tag{2.22}$$

We can do a Taylor series expansion and take an example and show when $|g'(\alpha)| > 1$, and the whole thing will diverge.

Let us try to answer the question of why the information flow diagram does not work. Consider the same Example 2.3. We substitute the initial condition values and determine dP/dQ and dQ/dP. In one information flow diagram, one of the derivatives is very small compared to 1 and will not grow. The other one is trying to create "some mischief" and after a few iterations, will reach a balance. But if we look at the other information flow diagram, we get very very large values for both the derivatives. This justification does give us an approximate understanding of why

Fig. 2.14 Load and engine characteristics for a truck at a particular gear setting (Example 2.4)

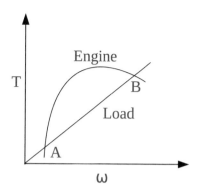

certain information flow diagrams do not work. As we know, since P and Q are continuously changing, with every iteration, these will fluctuate. But one of the derivatives is showing promise. If we work these derivatives out for the third or fourth iteration, the value will be less than 1 and that is how it eventually converges. In the previous information flow diagram, if we look at the derivative, it is large and that is why it is diverging. Here, though one of the two derivatives is more than 1, the subsequent iterates reduce and die down.

Already, we are not allowing the errors to propagate as we are raising the system to a power of 0.55. When we raise something to the power of 1.8 or 2, mischief is guaranteed. We do not know if the other information flow diagram works when one starts very close to the final solution. This is a good exercise, in the sense that it tells us how to rewrite the governing equations in such a way that even when we are way off from the final answer, because we are raising something to the power of 0.55, it can constrain or bound the errors. We can work it out (readers are encouraged to do this exercise).

We will see one last example, which concerns the movement of a truck/lorry on a ghat[1] road. There are two characteristics involved in this—the load and engine characteristics, as shown in Fig. 2.14. These two have to match in order that we obtain the operating point. For different transmission settings, we will have different curves and different operating points. When a vehicle is climbing uphill, sometimes when the vehicle is in the fourth gear or sometimes even in third gear, it may produce some odd sounds. This means that the engine is telling us that we have to downshift. We reduce to the appropriate gear so that the load curve and engine curve match and we stay very close to the operating point. This example will highlight how we employ the successive substitution for such a two-component system.

---

[1] In India, ghat refers to a winding hill road.

Example 2.4 Consider a lorry/truck climbing up a ghat (hill) road slowly. Given below are the torque speed characteristics of the engine and the torque speed characteristics of the load. Assume no gear change during the climb. Using the successive substitution method, determine the operating condition of the truck.

$$T = 18 + 11\omega - 0.12\omega^2 \text{ Nm - Engine} \tag{2.23}$$

$$T = 0.7\omega + 185 \text{ Nm - Load} \tag{2.24}$$

Solution

The load characteristics is a straight line while the engine characteristics is a parabola as it is quadratic in ω. It has two operating points. We can use two possible information flow diagrams. We can determine T from ω using the equation

$$T = 0.7\omega + 185 \tag{2.25}$$

or we can write it as

$$\omega = \frac{T - 185}{0.7} \tag{2.26}$$

and use it to determine ω from T. The two information flow diagrams should lead to two possible solutions. Let us rewrite the equations to find T and ω:

$$\omega = \frac{-(-11) \pm \sqrt{(-11)^2 - (4)(0.12)(T - 18)}}{2 \times 0.12}$$
$$T = 0.7\omega + 185$$

Let us assume an initial value of Torque as 220 Nm. The $\pm$ appearing in the equation for ω gives rise to two values corresponding to the points A and B in the figure.

After 5 iterations, we find that the solution converges to a torque of 229.9 Nm and speed of 64.14 rps (see Table 2.5). Let us do the stability analysis around point A. We will start with T as 300Nm and if we do two quick iterations using the above algorithm, we will see that the algorithm hits the same solution for T and ω.

Table 2.5 Successive substitution method for the truck problem

S. No.	T_i, Nm	ω_i, rps	T_{i+1}, Nm	ω_{i+1}, rps	$(\frac{T_{i+1} - T_i}{T_i})^2 + (\frac{\omega_{i+1} - \omega_i}{\omega_i})^2$
1	220	66.25	231.4	63.78	0.004
2	231.4	63.78	229.65	64.18	9.65×10^{-5}
3	229.65	64.18	229.93	64.13	2.09×10^{-6}
4	229.93	64.13	229.89	64.13	3.03×10^{-8}
5	229.89	64.14	229.89	–	–

2.5.2 The Newton–Raphson Method

Consider the equation y = f(x). The goal is to determine the values of x at which y = 0.

Let us draw a tangent at the point on the curve corresponding to x_i and extend it to cut the x-axis and the point at which it intersects the x-axis is taken as x_{i+1}. The procedure is graphically elucidated in Fig. 2.15. The algorithm for this method can be written as

$$f'(x_i) \approx \frac{f(x_i) - 0}{(x_i - x_{i+1})} \tag{2.27}$$

$$x_i - x_{i+1} = \frac{f(x_i)}{f'(x_i)} \tag{2.28}$$

$$x_{i+1} = x_i - \frac{f(x_i)}{f'(x_i)} \tag{2.29}$$

This is one of the most powerful and potent techniques in solving single-variable problems and can also be used for multivariable problems. The next step is to go to x_{i+1}, extend it to intersect the curve, draw the tangent at that point, and get x again. Hopefully, it will converge to the correct solution within a few iterations. *The major difference between this method and the successive substitution method is that we are using the information on the derivatives.* Since we are using the information of the derivatives, the convergence should be very fast. But we cannot say that the convergence is guaranteed. The advantages are obvious that it converges fast but let us look at the limitations that are not so obvious.

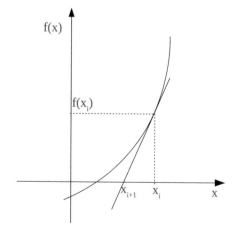

Fig. 2.15 Graphical depiction of the Newton–Raphson method

1. If $f'(x) = 0$, it is a real disaster for the method.
2. If $f'(x) = 0$, it causes the solution to diverge. Even if it is not 0, if it is very small, the solution oscillates too much. So $f(x)/f'(x)$ should be decent such that we get a new x_{i+1}.
3. Sometimes it may get trapped in local minima or maxima.

For each of these, we can take a particular example and illustrate. How do we get the same algorithm using the Taylor series?

Derivation using Taylor series

$$f(x_{i+1}) = f(x_i) + f'(x_i)(x_{i+1} - x_i) + \frac{f''(\zeta)}{2}(x_{i+1} - x_i)^2 + \cdots \qquad (2.30)$$

where ζ is between x_i and x_{i+1}. Neglecting terms beyond the linear term, we get

$$f(x_{i+1}) \approx f(x_i) + f'(x_i)(x_{i+1} - x_i) \qquad (2.31)$$

When x_i and x_{i+1} are both close to the true solution, higher order terms do not matter. Now it becomes a mathematical question of what this ζ is, where we want to evaluate the second derivative, etc. Let us not worry about that. So we get the same solution as before. What are we trying to make 0 here? We force $f(x_{i+1}) = 0$.

$$0 = f(x_i) + f'(x_i)(x_{i+1} - x_i) \qquad (2.32)$$

$$x_i - x_{i+1} = \frac{f(x_i)}{f'(x_i)} \qquad (2.33)$$

$$x_{i+1} = x_i - \frac{f(x_i)}{f'(x_i)} \qquad (2.34)$$

So we see that both the Taylor series expansion and the graphical method give the same result. Let us start with an example where the Newton–Raphson method is not powerful at all.

Example 2.5 Determine the roots of the equation $f(x) = x^8 - 1$ using the Newton– Raphson method, with an initial guess of $x = 0.5$.

Solution

Before drawing the tabular columns, we write down the algorithm.

$$x_{i+1} = x_i - \frac{(x_i^8 - 1)}{8x_i^7} \qquad (2.35)$$

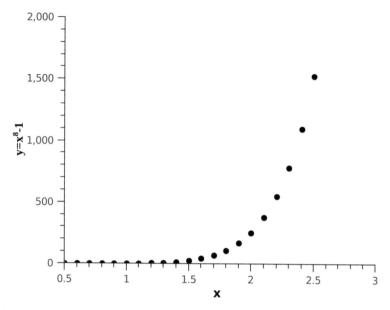

Fig. 2.16 Plot of f(x) versus x for Example 2.5

Table 2.6 Newton–Raphson method for Example 2.5

S. No.	x_i	x_{i+1}	$(x_{i+1} - x_i)^2$
1	0.5	16.44	254.1
2	16.44	14.39	4.22
3	14.39	12.59	3.23
4	12.59	11.01	2.45
5	11.01	9.63	1.89
⋮	⋮	⋮	⋮
24	1.000	1.000	–

Let us plot the function and see how it looks. A plot of the function is given in Fig. 2.16. Now we try to figure out why is it misbehaving. We started at 0.5, which seems a reasonable guess. The problem is that when the slope is very gentle, it takes forever to converge.

We see from Table 2.6 that the convergence is very slow. So the conventional wisdom that this method is very fast because it uses the derivative does not hold true at all times! Now let us work out a problem in which the Newton–Raphson method really works.

Example 2.6 Use the Newton–Raphson method to determine the first positive root of the equation $f(x) = x - 2sinx$, where x is in radians, with a starting guess, $x_0 = 2$ radians.

Solution

Let us first write the algorithm.

$$x_{i+1} = x_i - \frac{x_i - 2sin(x_i)}{1 - 2cos(x_i)} \qquad (2.36)$$

Figure 2.17 shows the graphical representations of the 2 functions f(x) = x and f(x) = 2 $sinx$. The solution is x = 1.895. The curve with diamond symbols shows the difference between the two functions, which is actually f(x). One can see where f(x) becomes zero. We are looking at the first positive root. It will also have negative roots because f(x) will keep going on the other side too.

The method is so fast and there is no comparison with successive substitution. If our initial guess is in the right region and the function is not changing very slowly, there is a good chance that we will get convergence within 3–4 iterations (see Table 2.7 for the solution).

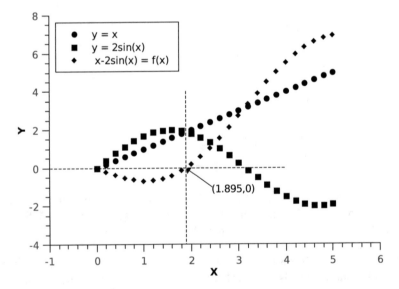

Fig. 2.17 Graphical representation of Example 2.6

Table 2.7 Newton–Raphson method for Example 2.6

S. No.	x_i	x_{i+1}	$(x_{i+1} - x_i)^2$
1	2.0	1.9	0.01
2	1.9	1.895	2.5×10^{-5}
3	1.895	1.895	–

Convergence characteristics of the Newton–Raphson method

The starting point of the Newton–Raphson method is the Taylor series expansion given by

$$f(x_{i+1}) \approx f(x_i) + f'(x_i)(x_{i+1} - x_i) \tag{2.37}$$

Instead of x_{i+1}, we can use x_t, where x_t is the true solution.

$$f(x_t) = f(x_i) + f'(x_i)(x_t - x_i) + \frac{f''(\zeta)}{2!}(x_t - x_i)^2 \tag{2.38}$$

Equations 2.38–2.37 gives

$$(x_t - x_{i+1})f'(x_i) + \frac{(x_t - x_i)^2}{2!}f''(\zeta) = 0 \tag{2.39}$$

Here, $x_t - x_i = E_{t,i}$ is the error in the true value in ith iteration. By the same token, $x_t - x_{i+1} = E_{t,i+1}$ is the error in the true value in the i+1th iteration. Let us rewrite the above equation in terms of $E_{t,i}$ and $E_{t,i+1}$.

$$E_{t,i+1}f'(x_i) + \frac{E_{t,i}^2}{2!}f''(\zeta) = 0 \tag{2.40}$$

$$E_{t,i+1} \approx -\frac{f''(\zeta)}{2f'(x_i)}E_{t,i}^2 \tag{2.41}$$

Anyway we are expanding f(x) very close to the solution. Therefore, we say that

$$f'(x_i) \approx f'(x_t) \tag{2.42}$$

$$f''(\zeta) \approx f''(x_t) \tag{2.43}$$

$$E_{t,i+1} \approx -\frac{f''(x_t)}{2f'(x_t)}E_{t,i}^2 \tag{2.44}$$

Around the true value, $f'(x_t)$ and $f''(x_t)$ are both fixed. Therefore

$$E_{t,i+1} \propto E_{t,i}^2 \tag{2.45}$$

So the error in the $i + 1$th step is proportional to the square of the error in the ith step because the terms $f'(x_t)$ and $f''(x_t)$ are a constant as we are looking at values close to the true value. We are saying that x_i, x_t, x_{i+1} are all very close. Hence if $E_{t,i}$ is 0.1, $E_{t,i+1}$ will be 0.01, then 0.0001 in the next step, and so on. Therefore, the Newton–Raphson method exhibits quadratic convergence. People who are interested can perform a similar exercise for successive substitution and can see that it has only a linear convergence.

Let us revisit Example 2.2 and try solving it using the Newton–Raphson method.

Example 2.7 Consider an electrically heated conductor of emissivity 0.6 operating under steady-state conditions. The energy input to the wire is 900 W/m^2 of surface area and the heat transfer coefficient $h = 9$ W/m^2. The ambient temperature is 300 K (see Fig. 2.18).

1. What is the governing equation for the steady-state temperature of the conductor?
2. Using the Newton–Raphson method, solve the governing equation and determine the steady-state temperature of the conductor.
3. Employ an appropriate stopping criterion.

Solution

We are looking at a steady state, energy is input to the conductor, and the conductor is radiating heat by both natural convection and surface radiation. The ambient for radiation is the same as the ambient for convection. We had a small discussion about it earlier where we said that the ambient for convection need not be the same as that for radiation, though normally both are the same.

Governing equations for the steady-state condition of the conductor, as already worked out in Example 2.2.

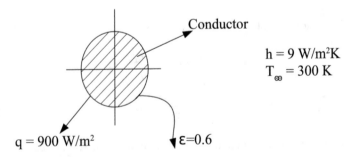

Fig. 2.18 Schematic for Example 2.7 (electrically heated conductor)

Table 2.8 Newton–Raphson method for the conductor problem

S. No.	T_i	$f(T_i)$	$f'(T_i)$	T_{i+1}	$(T_{i+1} - T_i)^2$
1	400	595.4	17.7	366.4	1130.8
2	366.4	34.73	15.69	364.15	4.9
3	364.15	0.12	15.57	364.15	0.00006
4	364.15	0	15.57	364.15	–

$$q = h(T_w - T_\infty) + \epsilon\sigma(T^4 - T_\infty^4) \tag{2.46}$$

$$f(T_w) = -900 + 9(T_w - 300) + (0.6)(5.67 \times 10^{-8})(T^4 - 300^4) \tag{2.47}$$

$$f'(T_w) = 9 + 1.36 \times 10^{-7} T_w^3 \tag{2.48}$$

Algorithm:

$$T_{i+1} = T_i - \frac{f(T_i)}{f'(T_i)} \tag{2.49}$$

The solution is presented in Table 2.8. We see that $f'(T_i)$ is extremely stable here. It took 16 iterations for us to get the solution using the method of successive substitution, as seen in Table 2.2. But here, it is just 4 steps. So this has to be quadratic convergence while the successive substitution had linear convergence. Hence, the Newton–Raphson method is really a powerful one. If the problem is well posed, it will not give us any trouble.

2.5.3 Newton–Raphson Method for Multiple Unknowns

How do we extend the Newton–Raphson method for multiple unknowns? The problem of multiple unknowns has great practical relevance, as invariably any thermal system will have more than two components.

Let us consider a three-variable problem, where the variables are x_1, x_2, and x_3, something like pressure P, temperature T, and density ρ. Three independent equations connecting x_1, x_2, x_3 are required to close the problem mathematically.

$$f_1(x_1, x_2, x_3) = 0 \tag{2.50}$$

$$f_2(x_1, x_2, x_3) = 0 \tag{2.51}$$

$$f_3(x_1, x_2, x_3) = 0 \tag{2.52}$$

Expanding $f_1(x_{1,i+1}, x_{2,i+1}, x_{3,i+1})$ around $x_{1,i}, x_{2,i}, x_{3,i}$, we have

$$f_1(x_{1,i+1}, x_{2,i+1}, x_{3,i+1}) = f_1(x_{1,i}, x_{2,i}, x_{3,i}) + \frac{\partial f_1}{\partial x_{1,i}}(x_{1,i+1} - x_{1,i})$$

$$+ \frac{\partial f_1}{\partial x_{2,i}}(x_{2,i+1} - x_{2,i}) + \frac{\partial f_1}{\partial x_{3,i}}(x_{3,i+1} - x_{3,i})$$

$$+ O(h^2) \qquad\qquad (2.53)$$

We can obtain similar looking equations for f_2 and f_3. The goal is to make the left-hand side equal to 0, because we are seeking the roots to the equations f_1, f_2, and f_3. Therefore, we force $f_1 = 0$, $f_2 = 0$, and $f_3 = 0$. If we do that, we get

$$\begin{bmatrix} \frac{\partial f_1}{\partial x_1} & \frac{\partial f_1}{\partial x_2} & \frac{\partial f_1}{\partial x_3} \\ \frac{\partial f_2}{\partial x_1} & \frac{\partial f_2}{\partial x_2} & \frac{\partial f_2}{\partial x_3} \\ \frac{\partial f_3}{\partial x_1} & \frac{\partial f_3}{\partial x_2} & \frac{\partial f_3}{\partial x_3} \end{bmatrix} \begin{bmatrix} x_{1,i+1} - x_{1,i} \\ x_{2,i+1} - x_{2,i} \\ x_{3,i+1} - x_{3,i} \end{bmatrix} = \begin{bmatrix} -f_1 \\ -f_2 \\ -f_3 \end{bmatrix}$$

The algorithm is written as shown here. The left-hand side is equivalent to what is called a **Jacobian matrix, which is a sensitivity matrix**. The partial derivatives with respect to the various variables are referred to as the sensitivity or the Jacobian. Unfortunately for us, the Jacobian matrix is not fixed because all the elements of the matrix keep changing with iterations. There are certain derivatives which get fixed.

For example, in the engine problem

$$T = 0.7\omega + 175$$

and so

$$f_1 = T - 0.7\omega - 175$$

- Its derivative $\frac{\partial f_1}{\partial T}$ will be 1 and it will remain 1 throughout. Similarly, $\frac{\partial f_1}{\partial \omega}$ will remain -0.7. In a typical problem, the solution will proceed as follows.
- We take initial values for x_1, x_2, x_3.
- We substitute in the respective equations for f_1, f_2, f_3 and calculate the values of f_1, f_2, f_3 at the initial starting point. So the forcing vector or the column vector on the right-hand side is known.
- Once x_1, x_2, x_3 are known, the partial derivatives at this point can be evaluated. Now all the elements of the Jacobian or the sensitivity matrix are known.
- This system of equations can be solved to obtain $\Delta x_1, \Delta x_2, \Delta x_3$.
- x_1, x_2, x_3 can now be updated.

The above algorithm is easily programmable. But the moment we have more than 10 or 12 variables, there are some issues with handling the matrix. Matrix inversion will not work very efficiently for more than a certain number of variables. However, the procedure described above can be extended to any number of variables. In our fan and duct problem or the truck problem, we will have to solve only a 2 × 2 matrix. Compared to successive substitution, the Newton–Raphson method is extremely fast. The Jacobian matrix also gives us the sensitivity of the system to changes in the variables as it contains information about the partial derivatives of the functions. Let us now revisit the truck problem (Example 2.4).

Example 2.8 Consider a lorry/truck climbing up a ghat (hill) road slowly. Given below are the torque speed characteristics of the engine and the torque speed characteristics of the load. Assume no gear change during the climb. Using the Newton–Raphson method, determine the operating condition of the truck.

$$T = 18 + 11\omega - 0.12\omega^2 Nm \text{ - Engine}$$
$$T = 0.7\omega + 185 Nm \text{ - Load}$$

Solution

Let us rewrite the equations in terms of f_1 and f_2.

$$f_1 = T - 11\omega - 18 + 0.12\omega^2 \tag{2.54}$$

$$f_2 = T - 0.7\omega - 185 \tag{2.55}$$

$$\frac{\partial f_1}{\partial T} = 1 \tag{2.56}$$

$$\frac{\partial f_1}{\partial \omega} = -11 + 0.24\omega \tag{2.57}$$

$$\frac{\partial f_2}{\partial T} = 1 \tag{2.58}$$

$$\frac{\partial f_2}{\partial \omega} = -0.7 \tag{2.59}$$

Newton–Raphson for two variables

$$\begin{bmatrix} 1 & (-11 + 0.24\omega) \\ 1 & -0.7 \end{bmatrix} \begin{bmatrix} \Delta T \\ \Delta \omega \end{bmatrix} = \begin{bmatrix} -f_1 \\ -f_2 \end{bmatrix}$$

For the first iteration, let $\Delta T = 200$ Nm and $\omega = 60$rps. The error is given by

$$error = \left(\frac{T_{i+1} - T_i}{T_i}\right)^2 + \left(\frac{\omega_{i+1} - \omega_i}{\omega_i}\right)^2 \tag{2.60}$$

After the first iteration, we see that the error is 936! If we look at this, it looks "hopeless". But Newton–Raphson is quadratically convergent and will converge quickly. Table 2.9 shows the iterations involved and corresponding error after each iteration.

Table 2.9 Newton–Raphson method for two variables (the truck problem)

S. No.	T	ω	f_1	f_2	$\frac{\partial f_1}{\partial T}$	$\frac{\partial f_1}{\partial \omega}$	$\frac{\partial f_2}{\partial T}$	$\frac{\partial f_2}{\partial \omega}$
1	200	60	−46	−27	1	3.4	1	−0.7
2	230.24	64.63	2.55	−0.001	1	4.5	1	−0.7
3	229.9	64.14	0.0328	0.002	1	4.39	1	−0.7
4	229.89	64.13	3.9e-4	2e-4	1	4.39	1	−0.7
	S. No.	T	ω	ΔT	$\Delta \omega$	Error		
	1	200	60	30.24	4.63	0.0288		
	2	230.24	64.63	−0.34	−0.49	6.07×10^{-5}		
	3	229.9	64.14	−0.006	−0.006	8.37×10^{-9}		
	4	229.89	64.13	−2.3e-4	−3.8e-5	–		

The error at the end of the second iteration has reduced to 0.37! We can see the quadratic convergence if we plot the error as a curve. We will see a nice curve that falls off rapidly.

If we are solving a transient problem, the kind of error we have considered may not be sufficient. If we have an extremely small step, we are simply not allowing the variables to change, and then saying that the stopping criterion is 10^{-6} or 10^{-8} is meaningless. When we have a microsecond or nanosecond as our time step, we just do not allow the variables to change. For this kind of a time step, we should allow it to run for a million time steps. We perform additional tests like grid independence study, mass balance check, and energy balance check to confirm convergence!

MATLAB code for Example 2.8

```
1   clear;
2   clc;
3
4   T_initial = 200;    % Initial guess for T
5   w_initial = 60;     % Initial guess for w
6
7   syms T w            % defining variables T & w
8   f1 = T − 11*w − 18 + 0.12*(w^2); % first function
9   f2 = T − 0.7*w − 185; % second function
10
11  f1_T=diff(f1,T);    % f1 differentiated with T
12  f1_w=diff(f1,w);    % f1 differentiated with w
13  f2_T=diff(f2,T);    % f2 differentiated with T
14  f2_w=diff(f2,w);    % f2 differentiated with w
15  T_old=T_initial;
16  w_old=w_initial;
17
```

```
18   J=zeros(2,2);
19   count_max=4; % Maximum no.of iterations needed
20   errTol=10^-20; % Error tolerance
21   count=0;
22   label=0;
23
24   while label==0
25
26        count=count+1;
27   % calculating value of function f1 for T, w
28        f1_value = single(subs(f1,{T,w},{T_old,w_old}));
29   % calculating value of function f2 for T, w
30        f2_value = single(subs(f2,{T,w},{T_old,w_old}));
31
32        forcing_vector(1,1)=-f1_value;
33        forcing_vector(2,1)=-f2_value;
34   % calculating 1st element value of Jacobian for T, w
35        J(1,1)=single(subs(f1_T,{T,w},{T_old,w_old}));
36   % calculating 2nd element value of Jacobian for T, w
37        J(1,2)=single(subs(f1_w,{T,w},{T_old,w_old}));
38   % calculating 3rd element value of Jacobian for T, w
39        J(2,1)=single(subs(f2_T,{T,w},{T_old,w_old}));
40   % calculating 4th element value of Jacobian for T, w
41        J(2,2)=single(subs(f2_w,{T,w},{T_old,w_old}));
42
43        diff=J\forcing_vector;%finding DELTA T and DELTA w
44        T_new=T_old+diff(1,1);
45        w_new=w_old+diff(2,1);
46
47        % finding residue
48        err=((T_new-T_old)/T_old)^2+((w_new-w_old)/w_old)^2;
49
50        if err < errTol || count==count_max
51            T_final=T_new;
52            w_final=w_new;
53            label=1;
54        else
55            T_old=T_new;
56            w_old=w_new;
57        end
58        % Print
59        prt = ['Itr = ',num2str(count) ,...
60            ', T = ',num2str(T_new) ,...
61            ', w = ',num2str(w_new) ,...
62            ', err = ',num2str(err)];
63        disp(prt)
64   end
```

The output of the program is

> *Itr = 1, T = 230.2439, w = 64.6341, err = 0.028833*
> *Itr = 2, T = 229.8978, w = 64.1397, err = 6.0775e-05*
> *Itr = 3, T = 229.8938, w = 64.134, err = 8.3746e-09*
> *Itr = 4, T = 229.8938, w = 64.134, err = 0*

2.5.4 System of Linear Equations

Where do we encounter a system of linear equations in engineering?

1. In thermal engineering, we encounter a system of linear equations in fluid flow circuits. A fluid or a mixture of several fluids comes out through one pipe, then branches, flows out and branches again, and so on. We are trying to determine the mass flow rates and pressure drop characteristics.
2. In chemical reactors—for example, in a fractionating column—the total mass of petroleum products must be equal to the difference between whatever is sent in and whatever is left behind. So this must be equal to the amount of naphtha + diesel + petrol + so on. We may eventually try to solve an optimization problem as to what should be the optimal product mix to maximize the profits. It will mathematically be posed as a system of linear equations.
3. Conduction heat transfer—Consider a composite insulation made of different materials A, B, C, D, and so on up to G which are of various thicknesses and thermal conductivities. We have the free stream temperatures $T_{1,\infty}$ on the left side and $T_{2,\infty}$ on the right side (see Fig. 2.19).

Fig. 2.19 Conduction heat transfer across a composite slab

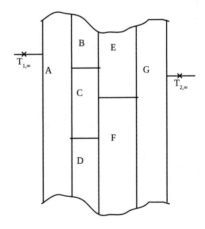

We can write these as resistances in series and parallel and can combine them. This is one possible way of approaching the above conduction problem.

From the foregoing examples, it is clear that there are several examples where a system's performance can be mathematically represented as a system of linear equations. So we need to know how to solve them. Solving a system of equations having just 2 variables is very trivial. Let us try and solve one such system.

Example 2.9 Solve the following system of equations.

$$2x + 3y = 13 \tag{2.61}$$
$$3x + 2y = 12 \tag{2.62}$$

Solution

How many ways are there of solving this? We can use the method of elimination by substitution, where in one equation we substitute for one variable from the other equation and solve for it. The second method that we use is matrices, and a third possible solution could be graphical.

Let us look at matrix inversion, as the other two are elementary.

$$[A][X] = [B] \tag{2.63}$$

$$[A] = \begin{bmatrix} 2 & 3 \\ 3 & 2 \end{bmatrix}$$

$$[A]^{-1} = \frac{1}{D} \begin{bmatrix} 2 & -3 \\ -3 & 2 \end{bmatrix}$$

in which D denotes the determinant value of the A matrix.
$D = 2 \times 2 - 3 \times 3 = -5$
Therefore,

$$[A]^{-1} = \frac{-1}{5} \begin{bmatrix} 2 & -3 \\ -3 & 2 \end{bmatrix}$$

$$\begin{bmatrix} X \\ Y \end{bmatrix} = \frac{-1}{5} \begin{bmatrix} 2 & -3 \\ -3 & 2 \end{bmatrix} \begin{bmatrix} 13 \\ 12 \end{bmatrix}$$

$$\begin{bmatrix} X \\ Y \end{bmatrix} = \frac{-1}{5} \begin{bmatrix} (26 - 36) \\ (-39 + 24) \end{bmatrix} = \begin{bmatrix} 2 \\ 3 \end{bmatrix}$$

Matrix inversion is important. In any system of up to 100 variables, one can invert the matrix without difficulty. Beyond 100 variables, it becomes a problem. Sometimes a matrix could be singular, where the determinant is 0. If the determinant is 0, it cannot be inverted. Sometimes there are an infinite number of solutions. Other techniques like Gaussian elimination, LU decomposition, Jacobian method, Gauss–Seidel method, and so on have been developed.

Before we look at that, let us see how even a two-variable problem can give us trouble. Where are the situations where one can have singular and ill-conditioned systems? Let us consider some examples.

Example 2.10 Solve the following system of equations.

(a) Consider the following system of equations:

$$2x + 3y = 12 \tag{2.64}$$
$$4x + 6y = 28 \tag{2.65}$$

They are parallel lines and never meet. There is no solution as $D = \begin{bmatrix} 2 & 3 \\ 4 & 6 \end{bmatrix} = 0$ and this is a singular system.

(b) Consider a system of equations similar to Eqs. 2.64 and 2.65

$$2x + 3y = 12 \tag{2.66}$$
$$4x + 6y = 24 \tag{2.67}$$

These have an infinite number of solutions. This is also a singular system, where $D = \begin{bmatrix} 2 & 3 \\ 4 & 6 \end{bmatrix}$

(c) Consider the following system of equations:

$$2x + 3y = 12 \tag{2.68}$$
$$1.9x + 3y = 11.8 \tag{2.69}$$

The determinant is very close to zero and the two lines are so close to each other that we cannot find out exactly where the two lines meet. This is called an ill-conditioned system. So $D \begin{bmatrix} 2 & 3 \\ 1.9 & 3 \end{bmatrix} = 0.3$ is small in this case, but nonzero. If we have several variables and the system is ill conditioned, the roundoff errors will eventually propagate and prevent us from obtaining the true solution.

From Fig. 2.20, we see that the two lines are almost indistinguishable and hence solving this problem graphically is not possible, though we can get the solution algebraically. Very small changes in the coefficient can lead to large changes in the result. The roundoff errors become very critical. For large systems, when

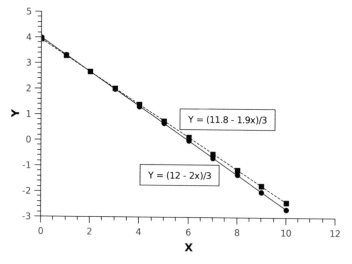

Y = (11.8 - 1.9x)/3

Y = (12 - 2x)/3

Fig. 2.20 Graphical solution to the ill-conditioned problem

we are solving a nonlinear partial differential equation, if the coefficients are
such that the system is ill conditioned, the roundoff errors become very critical.
That is why we use double precision, wherein instead of 8 decimal places, we
use 16 decimal places and so on. So this is an example of a simple two-variable
ill-conditioned system.

Now we consider several variables in the system of equations. The general representation is

$$a_{11}x_1 + a_{12}x_2 + \cdots + a_{1n}x_n = b_1 \tag{2.70}$$

$$a_{21}x_1 + a_{22}x_2 + \cdots + a_{2n}x_n = b_2 \tag{2.71}$$

$$\vdots$$

$$a_{n1}x_1 + a_{n2}x_2 + \cdots + a_{nn}x_n = b_n \tag{2.72}$$

In matrix form, we can write these equations compactly as

$$\begin{bmatrix} a_{11} & a_{12} & \cdots & a_{1n} \\ a_{21} & a_{22} & \cdots & a_{2n} \\ \vdots & \vdots & \ddots & \vdots \\ a_{n1} & a_{n2} & \cdots & a_{nn} \end{bmatrix} \begin{bmatrix} x_1 \\ x_2 \\ \vdots \\ x_n \end{bmatrix} = \begin{bmatrix} b_1 \\ b_2 \\ \vdots \\ b_n \end{bmatrix}$$

$$[A][X] = [B] \tag{2.73}$$

We can now write an expression for [X] as

$$[X] = [A]^{-1}[B] \qquad (2.74)$$

Of course, we have to reiterate that when the number of variables exceeds a certain number of components, it is very difficult to invert the matrix.

Gauss–Seidel Method

The Gauss–Seidel method is an iterative method while the matrix inversion is a one-shot procedure. By one shot, we mean we invert the matrix and we automatically get values for x_1 to x_n. In the Gauss–Seidel method, we have to start from some initial values and proceed and stop iterations only when the solution converges. One may start with 0 or 1. There should be something which is nonzero which will start driving the system. Then we should see how it converges.

The algorithm goes like this.

$$x_1 = \frac{[b_1 - a_{12}x_2 - a_{13}x_3 - \cdots - a_{1n}x_n]}{a_{11}} \qquad (2.75)$$

$$x_2 = \frac{[b_2 - a_{21}x_1 - a_{23}x_3 - \cdots - a_{2n}x_n]}{a_{22}} \qquad (2.76)$$

$$\vdots$$

$$x_n = \frac{[b_n - a_{n1}x_1 - a_{n2}x_2 - \cdots - a_{n(n-1)}x_{n-1}]}{a_{nn}} \qquad (2.77)$$

Needless to say, this system method will fail when $a_{ii} = 0$. When we are given a system of equations, it is important for us to rearrange the system of equations such that a_{ii} is the highest.

For example, $2x + 3y + 6z = 14$ can be used to solve for z as the expression on the right will be divided by 6, which ensures that the errors do not grow.

- The first step in the Gauss–Seidel method is to rearrange the equations so that diagonal dominance is satisfied and the algebraic manipulations required to do this are permitted.
- We have to start with guess values for all the variables.
- We substitute the guess values for all the variables and get the value of x_1 first.
- Now when we go to x_2, for all variables except x_1, we will use the initial guess value, while the latest determined value will be substituted into x_1.
- When we go to x_3, the guess values will be used for all variables except x_1 and x_2, for which the updated values will be used. Therefore, there is a dynamic updating of the variables in the Gauss–Seidel method.
- It is also perfectly OK if we use the guess values in all the equations of the algorithm and get the values for x_1 to x_n without using the dynamically updated values in any of these. This is called the Jacobi iteration and needless to say, it will be slow. So the key point in Gauss–Seidel is dynamic updating.

Stopping criterion:

$$R^2 = \sum_{i=1}^{N}(x_{i,k+1} - x_{i,k})^2 \tag{2.78}$$

k refers to iteration number

$$R^2 < \epsilon$$

where ϵ is a pre-defined value.

Diagonal dominance:

$$\forall i, \left\{ |a_{ii}| > \sum_{\substack{j=1 \\ j \neq i}}^{N} |a_{ij}| \right\}$$

Diagonal dominance is a sufficient condition, but it is not a necessary condition. The absence of diagonal dominance does not mean that the system will not converge. There are many systems which are not diagonally dominant, but may converge. However, *if diagonal dominance is there, convergence is assured*. So it is not a necessary and a sufficient condition. Fortunately, most engineering systems are diagonally dominant!

Are there methods to control convergence? The logical question is why to control convergence as what we really want is accelerated convergence! Not necessarily! Many times, if we try to accelerate convergence, the solution diverges, particularly in CFD systems because of the nonlinear equations. A typical radiation equation has $\epsilon \sigma T^4$ and is a highly nonlinear equation. We use the concept of relaxation to control convergence.

$$x_{new} = \omega x_{new} + (1 - \omega)x_{old} \tag{2.79}$$

Let us say at the end of the fifth iteration, we have the value of x and we also have the value of x at the end of the fourth iteration. It is possible to assign $\omega = 1$, such that we care only about the new value of x and not the old value. This may be because we have much faith in the new iterate! We can go ahead with this! Such a scheme is called a **fully relaxed numerical scheme**.

Suppose $\omega = 0.5$, then we are giving 50% weight to the new iterate and 50% weight to the old iterate. So we are essentially **under relaxing**. Why would one want to do that? Sometimes, there are a lot of oscillations of the variables and we want to dampen these. The system may be going haywire. Therefore, we deliberately slow down the system so that convergence is guaranteed.

It is also possible for us to have $\omega > 1$, such as $\omega = 2$. Yes! In this case, we are giving too much weightage to the present iterate. It is called over-relaxation or **successive over-relaxation** and works well for pure diffusion problems (as for example, the Laplace or Poisson equation in heat conduction).

Fig. 2.21 Schematic of a
stirred chemical reactor

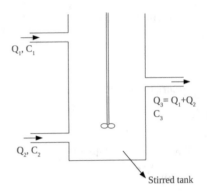

Whether we choose to under relax, over-relax, or fully relax, we have to choose some ω. For a problem, nobody will tell us what the value should be. Through the hard route, we have to find out which values of ω will work and which do not. These are very specific to the problem and have to be determined only through bad or good experience.

Where do we encounter a system of linear equations in mainstream engineering?

Chemical reactors are a good example.

Consider a stirred chemical reactor shown in Fig. 2.21. If we have two streams entering a vessel, the first flow rate is Q_1 while the concentration is C_1 and for the second, the flow rate is Q_2 and the concentration is C_2, and the flow rate of the liquid flowing out is

$$Q_3 = Q_1 + Q_2 \tag{2.80}$$

We do not have to specify C_3 because whenever Q_1 and Q_2, along with C_1 and C_2, are specified, Q_3 is automatically fixed. This is a stirred tank having some chemical inside. The key point here is that the mixture from this tank goes into another, from there again to the third, and so on. Hence, we can say

$$Q_1C_1 + Q_2C_2 = Q_3C_3 \tag{2.81}$$

and under steady state,

$$Q_1 + Q_2 = Q_3 \tag{2.82}$$

Likewise, if we have multiple reactors and the output of one or two reactors is going to the third, the third is going to the fourth, fifth, and so on, we can come up with the system of equations and solve for the concentrations in the various reactors. Apart from getting the concentrations, because we have declared that under steady-state conditions $Q_1 + Q_2 = Q_3$, this C_3 which is coming out will be the concentration of the chemicals within the reactor itself. This is additional information we are getting, consequent upon the fact that the stirred reactor is in a steady state and is well mixed.

Example 2.11 The mass balance for 3 chemicals x, y, and z in a chemical reactor is governed by the following equations.

$$-6x + 4y + z = 4 \tag{2.83}$$
$$2x - 6y + 3z = -2 \tag{2.84}$$
$$4x + y - 9z = -16 \tag{2.85}$$

x, y, and z are in appropriate units. Solve the system of equations using the Gauss–Seidel method to determine x, y, and z. Rearranging the equations is allowed, if required. Do you expect convergence for this problem? Initial guess, x = y = z = 1.

Solution

We write down the algorithm first.

$$x = \frac{(4 - 4y - z)}{-6} \tag{2.86}$$
$$y = \frac{(-2 - 2x - 3z)}{-6} \tag{2.87}$$
$$z = \frac{(-16 - 4x - y)}{-9} \tag{2.88}$$

Does it satisfy diagonal dominance? Yes, because diagonal dominance does not depend on − and +, but only the modulus. So the convergence is guaranteed. We can use $\omega = 1$, i.e., a fully relaxed scheme. The Gauss–Seidel solution is presented in Table 2.10.

The final answer is x = 0.98, y = 1.88, and z = 2.42.

Table 2.10 Gauss–Seidel method for Example 2.11

Iter	x	y	z	Error $\epsilon = \sum(x_{i+1} - x_i)^2 + \sum(y_{i+1} - y_i)^2 + \sum(z_{i+1} - z_i)^2$
0	1	1	1	
1	0.167	0.889	1.951	1.611
2	0.251	1.392	2.044	0.269
3	0.602	1.556	2.218	0.181
4	0.741	1.689	2.295	0.043
5	0.842	1.761	2.348	0.018

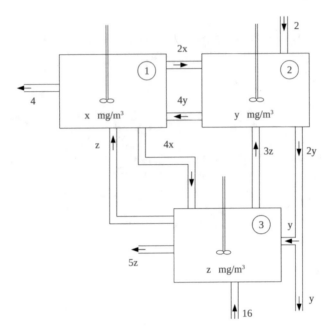

Fig. 2.22 Steady-state analysis of a series of reactors

We can employ $\omega = 1.8$ and solve the problem with a view to accelerating the convergence (left as an exercise to the reader).

We see that Gauss–Seidel method is a lot simpler than the Newton–Raphson method for 2 variables. But if the system of equations is not diagonally dominant, it can be very painful or can proceed very slowly.

What would the physical representation of the reactors in Example 2.11 look like?

We now turn around and see how one can get these equations. Let us look at the steady-state analysis of a series of reactors. Consider 3 reactors all of which are well stirred and we call these reactors 1, 2, and 3, as shown in Fig. 2.22. The concentration of the species in each of the reactors is x, y, and z, respectively, in let us say mg/m³. These reactors are interconnected by pipes. There is uniform concentration.

In the figure, if we look at the quantity 2x between reactors 1 and 2, 2 denotes the flow rate in m³/s while x is the concentration in mg/m³. So the concentration multiplied by the flow rate gives us the mass flow rate. Now if we do the steady-state mass balance for the 3 reactors, will we get the 3 equations stated in Example 2.11?

Let us start with reactor 1. 2x and 4x are going out from it which makes it −6x. 4y is coming inside, hence it becomes +4y. Since z is also coming in, it becomes +z. Since 4 is again going out, the mass balance equation for reactor 1 can be written as

$$- 6x + 4y + z - 4 = 0 \tag{2.89}$$

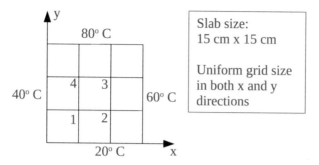

Fig. 2.23 Details of the geometry for problem 2.7

Please check the first equation we had in Example 2.11. Alternatively, we can state the problem as "The steady-state analysis of a series of reactors is shown here, whose concentrations are x, y, and z. Set up the governing equations for this problem and using the Gauss–Seidel method, and determine x, y, and z". We can also verify that the other 2 equations are obtained thus.

$$2x - 6y + 3z = -2 \tag{2.90}$$
$$4x + y - 9z = -16 \tag{2.91}$$

The beauty of this example is that we should satisfy mass balance and the system must also be diagonally dominant. Now one can understand why the Gauss–Seidel method will be more useful or in a more straightforward way, where we can expect a linear system of equations in engineering—when we are doing chemical reactor engineering, fluidized beds, fractionating columns, and petroleum engineering.

Recall an earlier example where we designed the pump and piping system for an apartment complex, with several branches. Suppose we are designing the distribution system for the city itself, there will be places where it will be pumped again and so on, and we may end up with a situation similar to Example 2.11.

Problems

2.1 An oil cooler is designed to cool 2 kg/s of hot oil from 80 °C, by using 3 kg/s of water entering at 30 °C ($T_{c,i}$) in a counter flow heat exchanger. The overall heat transfer coefficient of the exchanger is 2000 W/m²K, and the surface area available for heat transfer is 3 m².

(a) Write down the energy balance equations for the above exchanger.
(b) Using the method of successive substitution with an initial guess of the hot water outlet temperature, $T_{h,o} = 60$ °C, determine the outlet temperatures. Perform at least 8 iterations.

(Properties: C_p(oil) = 2.1 kJ/kgK, C_p(water) = 4.182 kJ/kgK)

2.2 The operating point of a centrifugal pump is to be determined. The pump performance curve and the system load characteristics are given as follows.

$$\text{Pump: } \Delta P = 240 \times 10^3 - 43.8 \times 10^2 Q^2$$
$$\text{Load: } \Delta P = 40 \times 10^3 + 156.2 \times 10^3 Q^{1.8}$$

where ΔP is the static pressure rise in Pa, and Q is the discharge in m³/s.

- Using the successive substitution method for 2 unknowns, determine the operating point of the pump.
- Decide on your own stopping criterion and start with an initial guess value of $Q = 0.5$ m³/s.
- At the operating point, if the pump efficiency is known to be 85%, what is the electrical power input required for the pump?
- For the same electrical power input, if over time, the pump efficiency drops to 82%, what will be the operating point for the pump?
- Can you quantify the sensitivity of the discharge with respect to the pump efficiency?

2.3 The demand for an engineering commodity in thousands of units follows a curve:

$$Q = 1500(0.97)^P$$

where Q is the demand (in thousands of units) and P is the price in Rupees. This is frequently referred to as the demand curve in micro-economics. The supply of the quantity in the market (again in thousands of units) varies with price

$$P = 10 + 1.1368 \times 10^{-4} Q^2$$

The above is known as the supply curve.

(a) Determine the equilibrium price and the equilibrium quantity by using the Newton–Raphson method for multiple unknowns.
(b) Decide on your own stopping criterion and choose P = Rs. 70 or Q = 1000 as the initial guess.
(c) If because of technological improvements, the company has a better way of making the product and the new supply curve becomes

$$P = 7 + 9.25 \times 10^{-5} Q^2$$

what will be the new equilibrium price and the new equilibrium quantity?

2.4 Solve the problem of determining the operating point of the centrifugal pump (problem # 2.2) by using the Newton–Raphson method for multiple unknowns with the same initial guess and stopping criterion.

2.5 Steam at the rate of 0.12 kg/s, at a temperature of T_s bled from a turbine, is used to heat feed water in a closed feed water heater. Feed water at $80\,^\circ$C (T_i) flowing at the rate of 3 kg/s enters the heater. The U value for the feed water heater is 1500 W/m^{2}K. The area of the heat exchanger (heater) is 9.2 m^2.

The latent heat of vaporization of steam is given by $h_g = (2530 - 3.79T_s)$ kJ/kg where T_s is the temperature at which steam condenses in $^\circ$C (which is also the same as the temperature at which steam is bled assuming no heat losses along the way).

(a) Set up the energy balance equations for the steam side and the water side.
(b) Write down the expression for the heat duty of the feed water heater.
(c) Identify the relationship between (a) and (b).
(d) Using information from (a), (b), and (c) and the method of successive substitution, determine the outlet temperature of the feed water (T_0) and the condensing temperature of steam(T_s).
(e) Start with an initial guess of $T_0 = 130\,^\circ$C and perform at least 4 iterations.

2.6 The mass balance of three species (x_1, x_2 and x_3 all in kg/s) in a series of interconnected chemical reactors is given by the following equations:

$$1.4x_1 + 2.2x_2 + 4.8x_3 = 4.8$$
$$6x_1 - 3x_2 - 2x_3 = 22$$
$$2x_1 + 4x_2 + x_3 = 24$$

Using the Gauss–Seidel method, with initial guess values of 1.0 for all the three variables, determine the values of x_1, x_2, and x_3. Perform at least 7 iterations and report the sum of the squares of the residues of the three variables at the end of every iteration.

2.7 Two-dimensional, steady-state conduction in a square slab with constant thermal conductivity, $k = 50$ W/m K, and a uniform internal heat generation of $q_v = 1 \times 10^6$ W/m^3 is to be numerically simulated. The details along with the boundary conditions are given in Fig. 2.23. For simplicity as well for demonstrative purposes, the number of grid points is intentionally kept small for this problem.

(a) Identify the equation that governs the temperature distribution T(x,y) for the given problem.
(b) Using the Gauss–Seidel method, estimate T_1, T_2, T_3, and T_4. Start with an initial guess of $50\,^\circ$C for all the four temperatures. Do at least 7 iterations.
(c) What is the approximate center temperature? What would have been your crude guess of the center temperature? Are these two in agreement?

Chapter 3
Curve Fitting

3.1 Introduction

Next, we will move on to an equally interesting topic, namely, curve fitting. Why study curve fitting at all? For example, if the properties of a function are available only at a few points, but we want the values of the function at some other points, we can use **interpolation**, if the "unknown points" lie within the interval, where we have information from other points. Sometimes **extrapolation** is also required. For example, we want to forecast maybe the demand of a product or the progress of a cyclone. Whether the forecasting is in economics or weather science, we need to establish the relationship between the independent variable(s) and the dependent variables to prove our proposal of extrapolation. Suppose our data is collected over 40 years, we first develop a model using the data of, say, 35 years and then test it using the data for the 36th, 37th year, and so on. Then we can confidently say we have built a model considering the data from 0–35 years, and in the time period 36–40 years, it is working reasonably well. So there is no reason why it should misbehave in the 41st year. This is an extremely short introduction to, say, a "101 course on Introduction to Forecasting".

In thermodynamics, oftentimes, thermodynamic properties are measured at only specific pressures and temperatures. But we want enthalpy and entropy at other values of pressures and temperatures for our calculations, and so we need a method by which we can obtain these properties.

This is only one part of the story. Recall the truck problem of Chap. 2. We wanted to get the operating point for the truck climbing uphill. Here, we started off with the torque speed characteristic of the engine and the load. These curves have to be generated from limited and finite data. We want functional forms of the characteristics because these are more convenient for us to work with. Therefore, the first step is to look at these points, draw a curve, and get the best function. So when we do curve fitting, some optimization is already inherent in it because we are only looking at the "best fit".

© The Author(s) 2021
C. Balaji, *Thermal System Design and Optimization*,
https://doi.org/10.1007/978-3-030-59046-8_3

What is the difference between the best and some other curve? For example, if we have some 500 data points, it will be absolutely meaningless for us to have a curve that passes through all the 500 points, because we know that error is inherent in this data. When an error is inherent in the data, we want to pass a line which gives the minimum deviation from all the points. So we get a new concept here that curve fitting does not mean that the curve has to pass through all the known points. But there can also be cases when the curve passes through all the points.

So in the light of the above, **when will one want to have an exact fit and when will one want to have an approximate fit?** The answer is when the number of parameters and also the number of measurements are small and we have absolute faith and confidence in our measurements, we can go in for an exact fit. But when the number of parameters and measurements is more and if we are using a polynomial, then the order of the polynomial keeps increasing. The basic problem with higher order polynomials is that they tend to get very oscillatory and if we want to work with higher order information like derivatives and so on, it gets very messy. Therefore, if we are talking about high accuracy and limited amount of data, it is possible for us to do an exact fit. But if we are dealing with a large number of data points that are error prone, it is fine to have the best fit.

In the light of the above, curve fitting can be of two types

(i) best fit and
(ii) exact fit.

3.1.1 Uses of Curve Fitting

- To carry out system simulation and optimization;
- To determine the values of properties at intermittent points;
- To do trend analysis;
- To make sense of the data;
- To do hypothesis testing. For example, we want to check if the data follow some basic rule of economics or some basic model of heat transfer or fluid mechanics.

A bird's eye view of the general curve fitting problem is shown in Fig. 3.1.

(a) Exact fit: Example, enthalpy h = f(T,P). When we want to write a program to simulate a thermodynamic system,[1] it is very cumbersome to use tables of temperature and pressure, and it is easier to use functions. For property data like the above, or calibration data like thermocouple emf versus temperature, exact fits are possible and desired.

---

[1] System in this book refers to a collection of components. It is different from the concept of system in thermodynamics wherein no mass can enter or leave a system.

Fig. 3.1 A bird's eye view
of the general curve fitting
problem

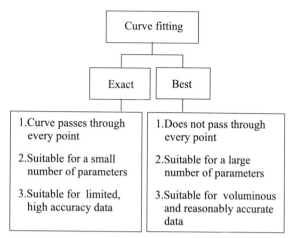

(b) Best fit: Any regression in engineering problems like a Nusselt number correlation. For example, consider flow over a heated cylinder. Under a steady state,
we measure the total amount of heat which is dissipated in order to maintain
the cylinder temperature a constant. We increase the velocity for every Reynolds
number, and we determine the heat transfer rate. Using Newton's law of cooling,
the heat transfer rate is converted to a heat transfer coefficient. From this, we
get the dimensionless heat transfer coefficient called the Nusselt number. The
Nusselt number goes with the correlating variables as follows:

$$Nu = aRe^b Pr^c \tag{3.1}$$

It is then possible for us to get a, b, and c from experiments and curve fitting.

The basis for all the above ideas comes from calculations or experiments. Regardless
of what we do, we get values of the dependent variables only at a discrete number
of points. Our idea of doing calculations or experiments is not to just get the values
at discrete points. As a scientist, we know that if the experiment is done carefully,
we will get 5 values of the Nusselt number for 5 values of Reynolds number. We are
not going to stop there. We would rather want to have a predictive relationship like
the one given in Eq. 3.1, which can be used for first predicting Nu at various values
of Re and Pr for which we did not do experiments and further, possibly simulate the
whole system of which the heated cylinder may be a part.

So with the data we have, before we can do system simulation and optimization,
there is an intermediate step involved which requires that all the data be converted
into equation forms, which can be played around with.

3.2   Exact Fit and Its Types

Important exact fits are listed below:

1. Polynomial interpolation,
2. Lagrange interpolation,
3. Newton's divided difference polynomial(s),
4. Spline fit.

The exact fit can also be called interpolation.

3.2.1   Polynomial Interpolation

Let us propose that y varies quadratically with x. So we write the relation as

$$y = a_o + a_1 x + a_2 x^2 \qquad (3.2)$$

where a_o, a_1, and a_2 are constants.
So in order to get the 3 constants, we need the values of y at 3 values of x.

$$y_o = f(x_o) \qquad (3.3)$$
$$y_1 = f(x_1) \qquad (3.4)$$
$$y_2 = f(x_2) \qquad (3.5)$$

So we have 3 equations in 3 unknowns which can be solved to obtain a_o, a_1, and a_2. There will be no error associated with the proposed equation if it is applied at x_0, x_1, and x_2. There will be some error when we apply it in between x_0, x_1, and x_2. This is because we are approximating the physics by a quadratic. With 3 points, we can make a quadratic pass through. We can only do that much.

Example 3.1 Evaluate the natural logarithm of 5 using quadratic interpolation, given

$$ln(1) = 0$$
$$ln(3) = 1.099$$
$$ln(7) = 1.946$$

Report the error by comparing your answer with the value obtained from your calculator.

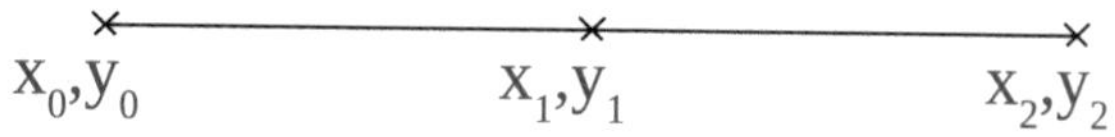

Fig. 3.2 Lagrange interpolation for three points

$$X_0, Y_0 \qquad X_1, Y_1 \qquad X_2, Y_2$$

Solution

$$\text{when } x = 1, \quad 0 = a_o + a_1 + a_2 \tag{3.6}$$

$$\text{when } x = 3, \quad 1.099 = a_o + 3a_1 + 9a_2 \tag{3.7}$$

$$\text{when } x = 7, \quad 1.946 = a_o + 7a_1 + 49a_2 \tag{3.8}$$

Solving for a_o, a_1, and a_2

$$a_o = -0.719$$
$$a_1 = 0.775$$
$$a_2 = -0.056$$

$$y = -0.719 + 0.775x - 0.056x^2 \tag{3.9}$$

when $x = 5$, $y = 1.757$; $y_{actual} = \ln(5) = 1.608$ (from calculator).

Hence the percentage error is 9.

This demonstrates how the polynomial interpolation works. We can complicate this by increasing the degree of the polynomial (and taking more points), but then the polynomial becomes very oscillatory.

3.2.2 Lagrange Interpolating Polynomial

Consider three points x_0, x_1, and x_2, and the corresponding y values are y_0, y_1, and y_2, respectively. A depiction of this is given in Fig. 3.2.

We can use the polynomial interpolation we just saw. However, the Lagrange interpolating polynomial is much more potent as it can easily be extended to any number of points. It is easy for us to take the first derivative, the second derivative, and so on with Lagrange interpolation. The post-processing work of CFD software, after we get the velocities and temperatures, generally uses the Lagrange interpolating polynomial to obtain the gradients of velocity and temperature. Once we get the gradient of temperature, we can directly get the Nusselt number and then correlate for the Nusselt number if we have data at a few values of velocity (please recall Eq. 3.1).

In Fig. 3.2, x_o, x_1, and x_2 need not be equally spaced. For the above situation, the Lagrange interpolating polynomial is given by

$$y = \frac{(x - x_1)(x - x_2)}{(x_0 - x_1)(x_0 - x_2)} y_0 + \frac{(x - x_0)(x - x_2)}{(x_1 - x_0)(x_1 - x_2)} y_1 + \frac{(x - x_0)(x - x_1)}{(x_2 - x_0)(x_2 - x_1)} y_2 \tag{3.10}$$

This is called the **Lagrange interpolating polynomial of order 2**. When written in compact mathematical form, the Lagrangian interpolating polynomial of order N is given by

$$y = \sum_{i=0}^{N} y_i \prod_{\substack{j=0 \\ j \neq i}}^{N} \frac{(x - x_j)}{(x_i - x_j)} \tag{3.11}$$

It is possible for us to get the first, second, and higher order derivatives. The first derivative will still be a function of x. The second derivative will be a constant for a second-order polynomial. For a third-order polynomial, we need the function y at four points and the second derivative will be linear in x.

Example 3.2 A long heat-generating wall is bathed by a cold fluid on both sides. The thickness of the wall is 10 cm and the thermal conductivity of the wall material k is 45 W/mK. The temperature of the cold fluid is 30 °C. The problem geometry, with the boundary conditions, is shown in Fig. 3.3. Thermocouples located inside the wall show the following temperatures (see Table 3.1), under a steady state.

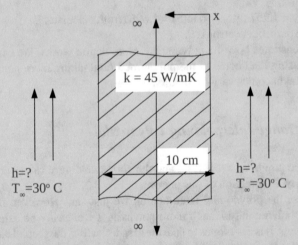

Fig. 3.3 Problem geometry with boundary conditions for Example 3.2

Table 3.1 Temperature of the heat generating wall at various locations (Example 3.2)

x, m	T(x), °C
0	80
0.01	82.1
0.03	84.9
0.05	87.6

The temperature distribution is symmetric about the midplane and the origin is indicated in Fig. 3.3. Using the second-order Lagrange interpolation formula, determine the heat transfer coefficient "h" at the surface. If the cross-sectional area of the wall is $1m^2$, determine the volumetric heat generation rate in the wall for steady-state conditions.

Solution

What is it we are trying to do here? We are trying to get T as a function of x and then write as q = −k dT/dx. q that comes from the wall is due to conduction. Therefore $q_{cond} = q_{conv}$.

The next logical question is instead of doing all this, why cannot we directly insert thermocouples in the flow? That will affect the flow and probably may even make a laminar flow turbulent and hence is not a good idea. We do such measurements with instruments like the hot wire anemometer, but the boundary layer itself is so small and thin that it is very difficult to do those measurements. Therefore, we would much rather prefer to do the measurements on the wall where it is easy to insert thermocouples and then use the conduction–convection coupling and Fourier's law, and then we can determine the heat transfer coefficient. This is the so-called "standard operating procedure" in convection heat transfer!

The problem at hand is typically called an **inverse problem**. We have a mathematical model for this, which is

$$\frac{d^2 T}{dx^2} + \frac{q_v}{k} = 0 \tag{3.12}$$

We have some measurements of temperature at x_o, x_1, and x_2. So if we marry the mathematical model with these measurements, we are able to get much more information about the system. What is the information we are getting? We are getting two more parameters, which are the heat transfer coefficient and the volumetric heat generation rate. The straight problem is very simple; there is a wall and there is convection at the boundary. The heat transfer coefficient and the volumetric heat generation rate are given. What is the surface temperature? Or what is the temperature 1 cm away from the surface? These are easy to answer!

But here, we are making some measurements and inferring some properties of the system. This is what is known as an inverse problem.

Here, we have 4 readings, but we need to choose only 3 for the Lagrange method. Which 3 do we choose? We choose something closest to the wall, so we are going to employ the first 3 readings.

$$y = \frac{(x - x_1)(x - x_2)}{(x_0 - x_1)(x_0 - x_2)} T_0 + \frac{(x - x_0)(x - x_2)}{(x_1 - x_0)(x_1 - x_2)} T_1$$

$$+ \frac{(x - x_0)(x - x_1)}{(x_2 - x_1)(x_2 - x_1)} T_2 \qquad (3.13)$$

$$\frac{dT}{dx} = \frac{[2x - (x_1 + x_2)]}{[(x_0 - x_1)(x_0 - x_2)]} T_0 + \frac{[2x - (x_0 + x_2)]}{[(x_1 - x_0)(x_1 - x_2)]} T_1$$

$$+ \frac{[2x - (x_0 + x_1)]}{[(x_2 - x_0)(x_2 - x_1)]} T_2 \qquad (3.14)$$

Determining dT/dx at x_o, we get

$$\left.\frac{dT}{dx}\right|_{x_0} = \frac{2(0) - (0.01 + 0.03)}{(0 - 0.01)(0 - 0.03)} 80 + \frac{2(0) - (0 + 0.03)}{(0.01 - 0)(0.01 - 0.03)} 82.1$$

$$+ \frac{2(0) - (0 + 0.01)}{(0.03 - 0)(0.03 - 0.01)} 84.9$$

$$\left.\frac{dT}{dx}\right|_{x_o} = 233.33 \text{K/m}$$

$$k\left.\frac{dT}{dx}\right|_{x_o} = h(T(x_0) - T_\infty)$$

$$45 \times 233.33 = h(80 - 30)$$

$$h = 210 \text{W/m}^{2K}$$

Figure 3.4 shows the qualitative variation of the temperature across the wall while the arrow indicates the direction of heat flow.

Fig. 3.4 Qualitative temperature distribution and direction of heat flow for Example 3.2

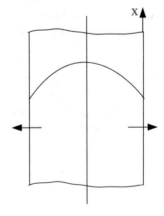

The last part of the problem is the determination of the volumetric heat generation rate. There are several ways of obtaining this. The heat transfer from the right side is $hA\Delta T$ and that from the left side is also $hA\Delta T$. So that makes a total of $2hA\Delta T$. Where does all this heat come from? From the heat generated. Therefore, equating the two, we have

$$q_v V = 2hA\Delta T \tag{3.15}$$

So, we determine q_v from this. The above procedure does not take recourse to the governing equation. (This is strictly not correct as the governing equation is itself a differential form of this energy balance with the constitutive relation (Fourier's law) embedded in it).

When we are employing 3 temperatures, we can consider any 3 of the 4. But the pair $(0, 80\,^\circ\text{C})$ is very crucial because we are evaluating the gradient at the solid–fluid interface. So, using the overall energy balance,

$$q_v V = 2(hA\Delta T)$$
$$q_v \cancel{A} L = 2 \times 210 \cancel{A} \times 50$$
$$q_v = \frac{2 \times 210 \times 50}{0.10} = 2.1 \times 10^5 \text{ W/m}^3$$

There is a heat-generating wall that has a thickness of 10 cm whose thermal conductivity is given by 45 W/mK, and the heat generation rate is estimated to be 2.1×10^5 W/m^3. Now it is possible for us to apply our knowledge of heat transfer and obtain the maximum temperature and see whether it is the same as what is measured at 0.05 m. The governing equation for the problem under consideration is

$$\frac{d^2 T}{dx^2} + \frac{q_v}{k} = 0 \tag{3.16}$$

The solution to the above equation is

$$\frac{dT}{dx} = -\frac{q_v x}{k} + A \tag{3.17}$$

$$T = -\frac{q_v x^2}{2k} + Ax + B \tag{3.18}$$

$T = 80\,^\circ\text{C}$ at $x = 0$, which gives $B = 80\,^\circ\text{C}$.
At $x = +0.05$ m, $dT/dx = 0$, hence $A = \frac{q_v (0.05)}{k}$.
Substituting for A and B in Eq. 3.18, at $x = 0.05$m, we get $T = 85.8\,^\circ\text{C}$.
This value is quite close to the data given in Table 3.1.

This is a very simple presentation of inverse heat transfer. Inverse heat transfer can be quite powerful. For example, in the international terminals at an airport during the Swine flu seasons of 2009 and 2010 and during the COVID-19 in 2020 and 2021, when people were arriving, thermal infrared images were taken to evaluate if

Fig. 3.5 Lagrangian interpolation polynomial of order 2 with three points

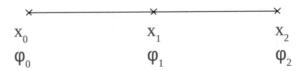

any of the passengers show symptoms of flu. Here, we are basically working out an inverse problem. When the imager sees that the radiation from the nose and other parts exceeds a certain value, it is a sign that the passenger could be infected.

Thermal infrared radiation can also be used for cancer detection, say breast tumor, for example. If we can use an infrared camera and if we obtain the surface temperature image, if there is a tumor inside the breast, the metabolism will be high. This will cause the volumetric thermal heat generation to be high compared to the tissue which is noncancerous. Therefore, this will show up as a signature on the surface temperature image of the breast. This is called a breast thermogram. Now, from these temperatures, one solves an inverse problem to determine the size and location of the tumor.

3.2.2.1   Lagrange Interpolation Polynomial: Higher Order Derivatives

The Lagrangian interpolating polynomial of order 2 for $\phi = f(x)$ (refer to Fig. 3.5) is given by

$$\phi = \frac{(x-x_1)(x-x_2)}{(x_0-x_1)(x_0-x_2)}\phi_0 + \frac{(x-x_0)(x-x_2)}{(x_1-x_0)(x_1-x_2)}\phi_1 + \frac{(x-x_0)(x-x_1)}{(x_2-x_0)(x_2-x_1)}\phi_2$$
$$(3.19)$$

Equation 3.19 is exactly satisfied at $x = x_o$, $x = x_1$ and $x = x_2$, and there will be no error at these three points. But at all intermediate points, there will be an error as we are approximating the functional form for ϕ by a quadratic. Given that we have 3 points and we want the polynomial to pass through all the 3 points, the best fit is a polynomial of order 2.

Let us now determine the first and second derivative of the polynomial ϕ with respect to x.

$$\frac{d\phi}{dx} = \frac{[2x - (x_1 + x_2)]}{[(x_0 - x_1)(x_0 - x_2)]}\phi_0 + \frac{[2x - (x_0 + x_2)]}{[(x_1 - x_0)(x_1 - x_2)]}\phi_1 +$$
$$\frac{[2x - (x_0 + x_1)]}{[(x_2 - x_0)(x_2 - x_1)]}\phi_2$$
$$(3.20)$$

$$\frac{d^2\phi}{dx^2} = \frac{2}{[(x_0 - x_1)(x_0 - x_2)]}\phi_0 + \frac{2}{[(x_1 - x_0)(x_1 - x_2)]}\phi_1$$
$$+ \frac{2}{[(x_2 - x_0)(x_2 - x_1)]}\phi_2$$
$$(3.21)$$

Fig. 3.6 Depiction of the central difference scheme

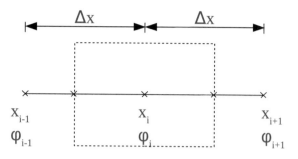

For equispaced intervals such that $x_1 - x_o = x_2 - x_1 = \Delta x$, we have

$$
\begin{aligned}
\frac{d^2\phi}{dx^2}\Big|_{x=x_1} &= \frac{2}{2\Delta x^2}\phi_0 - \frac{2\phi_1}{\Delta x^2} + \frac{2\phi_2}{2\Delta x^2} \\
&= \frac{\phi_0 - 2\phi_1 + \phi_2}{\Delta x^2}
\end{aligned}
\tag{3.22}
$$

Purely by using the concept of an interpolating polynomial, we are able to get the second derivative of the variable ϕ, where this variable could be anything like temperature, stream function, and potential difference. The above is akin to the central difference method in finite differences, which is one way of getting a discrete form for derivatives. The central difference method basically uses the Taylor series approximation. Consider Fig. 3.6, where ϕ is the variable of interest. Using the finite difference method, $d^2\phi/dx^2$ can be written as

$$
\frac{d^2\phi}{dx^2}\Big|_{x_i} = \frac{\frac{d\phi}{dx}\big|_{x+\frac{1}{2}} - \frac{d\phi}{dx}\big|_{x-\frac{1}{2}}}{\Delta x}
\tag{3.23}
$$

$$
\begin{aligned}
\frac{d^2\phi}{dx^2}\Big|_{x_i} &= \frac{\frac{\phi_{i+1}-\phi_i}{\Delta x} - \frac{\phi_i-\phi_{i-1}}{\Delta x}}{\Delta x} \\
&= \frac{\phi_{i+1} - 2\phi_i + \phi_{i-1}}{\Delta x^2}
\end{aligned}
\tag{3.24}
$$

This is the way the central difference is worked out in the finite difference method. Now we can see that the results obtained using the finite difference method are the same as what we obtained using the Lagrange interpolating polynomial.

When we use Lagrange polynomials of order 3 and 4, it leads to results that are similar to ones obtained with higher order schemes in the finite difference method.

The right-hand side term of Eq. 3.24 can be written as

$$
\frac{\phi_E - 2\phi_P + \phi_W}{\Delta x^2}
$$

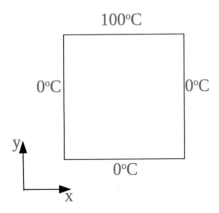

Fig. 3.7 Two-dimensional steady-state heat conduction across a slab without heat generation

Let us now look at an example to make these ideas more clear. Consider two-dimensional steady-state heat conduction in a slab as given in Fig. 3.7. The governing equation is

$$\nabla^2 T = 0 \tag{3.25}$$

$$\text{or} \quad \frac{\partial^2 T}{\partial x^2} + \frac{\partial^2 T}{\partial y^2} = 0 \tag{3.26}$$

The above equation is called the Laplace equation, in which ∇^2 is referred to as the Laplacian operator.

We can solve the above equation using finite differences. We already have a representation for $\frac{d^2\phi}{dx^2}$, which is

$$\frac{d^2\phi}{dx^2} = \frac{\phi_E - 2\phi_P + \phi_W}{\Delta x^2} \tag{3.27}$$

Similarly

$$\frac{d^2\phi}{dy^2} = \frac{\phi_N - 2\phi_W + \phi_S}{\Delta x^2} \tag{3.28}$$

when $\Delta x = \Delta y$,

$$\frac{\partial^2\phi}{\partial x^2} + \frac{\partial^2\phi}{\partial y^2} = 0$$

$$\frac{\phi_E - 2\phi_P + \phi_W}{\Delta x^2} + \frac{\phi_N - 2\phi_P + \phi_S}{\Delta x^2} = 0 \tag{3.29}$$

$$\text{which reduces to } \phi_P = \frac{\phi_E + \phi_W + \phi_N + \phi_S}{4} \tag{3.30}$$

Please note that Eq. 3.30 can also be obtained by using Lagrange interpolating polynomials for $\phi(x, y)$ for one variable at a time and getting the second derivative at node P.

If we simplify further, we see that the value of the temperature at a particular node turns out to be the algebraic average of the temperatures at its surrounding nodes. Looks reasonable, provided, this is also the commonsensical answer we would have obtained. For the problem under question, if we apply this formula at the center point, it would be $(100 + 0 + 0 + 0)/4 = 25$ °C. Regardless of the sophistication of the mathematical technique we use, the center temperature must be 25 °C. This serves as basic validation of the approximation we have done to the governing equation.

3.2.3 Newton's Divided Difference Method

A general polynomial can be represented as

$$
\begin{aligned}
y = a_0 &+ a_1(x - x_0) + a_2(x - x_0)(x - x_1) \\
&+ a_3(x - x_0)(x - x_1)(x - x_2) + \cdots \\
&+ a_n(x - x_0)(x - x_1) \cdots (x - x_{n-1})
\end{aligned}
\tag{3.31}
$$

Here, $x_0, x_1, x_2, \ldots, x_n$ all need not be equispaced. We are trying to fit a polynomial for y in terms of x. We have to get all the coefficients $a_0, a_1, a_2, \ldots, a_n$. If we substitute $x = x_0$, we get $y_0 = a_0$ because all the other terms become 0. This way, we get a_0. Now we substitute for $x = x_1$. We get $y = a_0 + a_1(x_1 - x_0)$. All the other terms will vanish. We already know a_0. We know $(x_1 - x_0)$ and hence we can get a_1. We then thus obtain all the "a"s in this way in a recursive fashion.

Suppose we have a cubic polynomial, we have four equations with four unknowns and the system can be solved. Here all the unknowns are determined sequentially and not simultaneously. There is no need to solve the simultaneous equations. It is a lot simpler than the polynomial interpolation. Even the Lagrange interpolation is a lot simpler compared to the polynomial interpolation we did earlier because the coefficients need not have to be simultaneously determined.

3.2.4 Spline Approximation

If f(x) varies with x as shown in Fig. 3.8, we would like to join the 4 points by a smooth curve. The key is to fit lower order polynomials for subsets of the points, and mix and match all of them to get a "nice" smooth curve. But all of them do not follow the same equation. Locally for every 3 points, there is a parabola or for every 4 points there is a cube. But in order to avoid discontinuities at the intermediate points, we have to match the functions as well as the slopes at these intermediate points. We

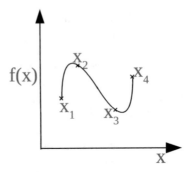

Fig. 3.8 General representation of a spline approximation

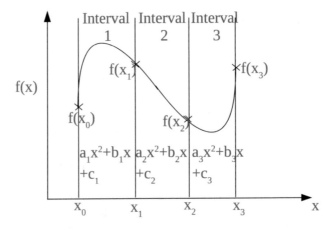

Fig. 3.9 Spline approximation with a quadratic polynomial

will again get a set of simultaneous equations, which when solved simultaneously will get us the values of the coefficients and this is how spline fitting is done.

Consider a spline approximation with quadratic fit in each subinterval as depicted in Fig. 3.9. We can divide it into 3 intervals.

The equations for the three intervals are as follows:

$$\text{Interval 1:} f(x) = a_1 x^2 + b_1 x + c_1 \tag{3.32}$$

$$\text{Interval 2:} f(x) = a_2 x^2 + b_2 x + c_2 \tag{3.33}$$

$$\text{Interval 3:} f(x) = a_3 x^2 + b_3 x + c_3 \tag{3.34}$$

We have the value of the function only at 4 points corresponding to x_0, x_1, x_2, and x_3. There are 9 unknowns to be determined—$a_1, a_2, a_3, b_1, b_2, b_3, c_1, c_2, c_3$. We require 9 equations for solving for these. At the intermediate points, there is only one value of f(x) whether we use the polynomial in the interval to the left or to the right. So for every point, we have two equations either coming from behind or from ahead. If n is the number of intervals, there are $n - 1$ intermediate points. The total number of intervals here is 3, i.e., $n = 3$. Intermediate points are 2. For every point, we have 2 equations and hence we have a total of $2(n - 1)$ equations.

At the end points, we have 2 equations. So the total number of equations we have is $2n - 2 + 2 = 2n$. The number of constants on the other hand is $3n$.

There should be continuity of the slope at the intermediate points. Hence, $f'(x)$ should be the same whether we come from the left or right. That means

$$2a_1x_1 + b_1 = 2a_2x_2 + b_2 \tag{3.35}$$

Like this, we can generate equations at the intermediate points. We have $(n - 1)$ such equations. So now we have $2n + n - 1 = 3n - 1$ equations totally. We are still short by one!

The second derivative at either of the two end points may be assumed to be 0. This means that either $2a_1$ or $2a_3$ is 0, depending on the end point chosen, which means that either a_1 or a_3 is 0. So the last condition is $f''(x) = 0$ at one end.

Normally, we use the cubic spline when there are hundreds of points. When we do this, our eyes will not notice that this juggling has been done somewhere. There is no other simpler way to mathematically close it. This is what a graphing software typically does when we use an approximating polynomial. So much mathematics goes on in the background! (See Chapra Steven and Canale Raymond 2009, for a fuller discussion on spline fits.)

This brings us to the end of exact fits.

3.3 Best Fit

All of the above methods discussed in Sect. 3.2 are applicable for properly determined systems where the number of points is equal to the order of the polynomial +1 and so on. For example, if we have 3 points, we have a second-order polynomial. Oftentimes, we have overdetermined systems. For example, if we perform the experiments for determining the heat transfer coefficient and obtain results for 10 Reynolds numbers, we know that $Nu = aRe^bPr$ and when the Prandtl number, Pr, is fixed, we have two constants. To determine these, 2 data points are sufficient. But we have 10 data points. So we have to get the best fit. What is our criterion of "best"? Whether we want to minimize the difference, or the difference in the modular form, or we want to minimize the maximum deviation from any point, or we want to minimize the square of the error, or higher order power of the error?

Oftentimes, we only get data points which are discrete. (For example, the performance of equipment like a turbomachine.) However, we are interested in getting a functional relationship between the independent and the dependent variables, so that we are able to do system simulation and eventually optimization. If the goal is calibration, we want to have an exact fit, where the curve should pass through all the points. We discussed various strategies like Newton's divided difference polynomial,

Lagrange interpolation polynomial, and so on. These are basically exact fits where we have very few points, whose measurements are very accurate. However, there are several cases where the measurements are error prone. There are also far too many points, and it is very unwise on our part to come up with a ninth-degree or a twelfth-degree polynomial that will generally show a highly oscillatory behavior. These are **overdetermined systems** as already mentioned. What do we exactly mean by this? **These are systems that have far too many data points compared to the number of equations required to regress the particular form of the equation.**

For example, if we know that the relationship between enthalpy and temperature is linear, we can state it as $h = a + bT$.

So if we have enthalpy at two values of temperature, we can get both a and b. But suppose we have 25 or 30 values of temperature, each of which has an enthalpy and all of which also have an error associated with it. Now, this is an overdetermined system as two pairs of temperature-enthalpy can be used to determine a and b. Among these values of a and b, which is the most desirable, we do not know. Therefore, we have to come up with a **strategy of how to handle the overdetermined system**. We do not want the curve to pass through all the points. So what form of the curve do we want? Polynomial or exponential or power law or any other form? Some of the possible representations are given below.

$$\text{Polynomial: } y = a + bx + cx^2 \tag{3.36}$$

$$\text{Exponential: } y = ae^{bt} \tag{3.37}$$

$$\text{Power law: } y = ax^b \tag{3.38}$$

The polynomial given here is a general depiction; we may have higher order polynomials also, or we can have the exponential form $y = ae^{bt}$ which is typically the case when we have an initial value problem, for example, concentration decreases with time or the population is changing with time; or we can have the power law form, for example, the Nusselt number or skin friction coefficient. Who will tell us the best way to get the values for a, b, and c if it is a polynomial, or how to get a and b, if it is exponential or the power law? We have chosen the form. But what is the strategy to be used, since it is an overdetermined system? So we now have to discuss the strategies for the best fit.

3.4   Strategies for Best Fit

Let us take a straight line. $y = ax + b$. We have the data as shown in Table 3.2.

We want (a, b). We have the table containing the values of y corresponding to different values of x. We require the values of a and b to be determined from this. How do we get the table of values? We could perform experiments in the laboratory or calculations on the computer. What could be this x and y? x could be the Reynolds number while y could be skin friction coefficient, x could be temperature difference while y could be heat flux. We are looking at a simple one-variable problem.

Table 3.2 Simple tabulation of y for various values of x

x	y
x_1	y_1
x_2	y_2
.	.
.	.
.	.
x_n	y_n

Fig. 3.10 Regression with minimizing differences

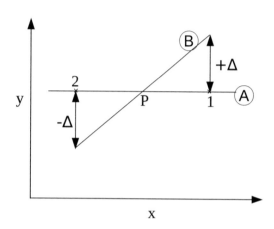

Strategy 1:

So the first strategy is to minimize R, or rather

$$\text{Minimize } S = \sum_{i=1}^{n} R_i = \sum_{i=1}^{n} [y_i - (ax_i + b)] \tag{3.39}$$

where n refers to the number of points.

This is the simplest possibility. But it has a problem. Let us highlight this difficulty with an example. Consider two points 1 and 2 as shown in Fig. 3.10.

Common sense tells us that we can join these two by a straight line. Let this be called line A. Suppose we take a point P on this line, now we pass another line through it, namely line B. So as seen in Fig. 3.10, line B has a deviation of $-\Delta$ on one side of line A, while the deviation is $+\Delta$ on the other side from line A. These two will cancel out and this will give exactly the same value of S or the sum of residues, as opposed to the correct line (which is A in our case). Any line other than the vertical will reduce the sum of the residues S to 0. So we will not get a unique line. Hence, from a common sense point of view, it is possible that large negative errors are compensated by large positive errors. In view of this, this is not the best strategy for the best fit of a curve.

Fig. 3.11 Regression with
minimizing modulus of the
differences

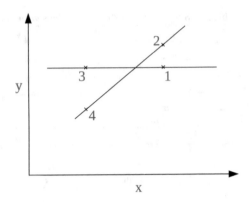

Strategy 2:

Another strategy would be to minimize the modulus of the difference between the
data and the fit.

$$\text{Minimize } S = \sum_{i=1}^{n} |R_i| = \sum_{i=1}^{n} |y_i - (ax_i + b)| \tag{3.40}$$

Here, we minimize the modulus of the difference between y_{data} and y_{model}, where
a large negative error cannot cancel out a large positive error. It now raises a lot of
hope in us. Let us take a hypothetical example to see whether this strategy works.
There are 4 points as seen from Fig. 3.11. Let us join points 1 and 3 by a line and
points 2 and 4 by another line. Any line between these two lines will try to minimize
S. There could be so many lines which will satisfy this. However, the original goal
was to obtain unique values of a and b which will give the best fit. So this too does
not work!

Strategy 3:

Consider a set of data points 1, 2, 3, 4, and 5. Common sense tells us that a line passing
through points 1, 2, 3, and 4 (or very close to them) is a good way of representing the
data. But point 5 gets neglected in this process. In order to accommodate point 5, let
us choose the dashed line as shown in Fig. 3.12 as the "best" fit. On what basis was
this chosen? The dashed line (of Fig. 3.12) minimizes the maximum deviation from
any point. So it satisfies what is known as the **minimax criterion**. This basically
comes from the decision theory. Minimax is based on maximizing the minimum
gain and minimizing the maximum loss.

As far as we are concerned, we are trying to minimize the maximum distance from
any point. But point 5 is a rank outsider, and there is something fundamentally wrong
with this value. Statistically, this point 5 is called an **outlier**; it is an **"outstanding
point"**! Minimax criterion unnecessarily tries to give undue importance to an outlier.
Sometimes often in meetings too, the most vocal person will be heard and given

Fig. 3.12 Regression with minimax strategy

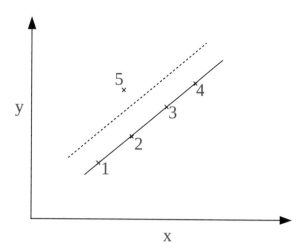

importance though his/her view may not be the most reasonable! For the problem at hand, common sense suggests that we better remove the point (here, point 5) and repeat the simulations/experiments. Suddenly, we cannot get a new physics or heat transfer that goes up and down. There is something wrong with that point. Maybe an error in taking the reading, or not allowing a steady state to be reached or a fluctuation in the operating conditions like voltage and so on.

With all these strategies having failed, we go on to the least square regression (LSR), which is Strategy 4. Here, we try to see if by minimizing the square of the differences between the data and the fit, we can get a good handle on the problem (an excellent treatment on regression is available in Chapra and Canale, 2009).

3.4.1 Least Square Regression (LSR)

Let $y = ax + b$. If we have two points, we can get a and b right away. If we have 200 points, then we can have so many combinations of a and b. Now we are trying to fit one global value for a and b which best fits the data. Whenever we have a procedure by which we determine a and b, we substitute a value of x in ax + b from the resulting y of the data and $(y_{data} - y_{fit})$ is called the residue. The residue in general will be nonzero. We take the square of this and try to minimize the sum of these residues.

$$S = \sum_{i=1}^{N} R_i^2 = \sum_{i=1}^{N} [y_i - (ax_i + b_i)]^2 \qquad (3.41)$$

in which y_i refers to y_{data}, and $(ax_i + b_i)$ refers to y_{fit}. The square takes care of all the negative and positive errors and also helps give us the best value of a and b that fits all data. This is called the **L2 norm or the Euclidean norm**. In order to obtain a and b, we differentiate the norm as follows.

$$\frac{\partial S}{\partial a} = 0 = 2 \sum (y_i - ax_i - b_i)(-x_i) \tag{3.42}$$

$$\frac{\partial S}{\partial b} = 0 = 2 \sum (y_i - ax_i - b_i)(-1) \tag{3.43}$$

On rearranging the above two equations, we have

$$-\sum x_i y_i + a \sum x_i^2 + b \sum x_i = 0 \tag{3.44}$$

$$-\sum y_i + a \sum x_i + nb = 0 \tag{3.45}$$

The above equations can be solved for a and b as

$$nb = \sum y_i - a \sum x_i$$

$$b = \frac{\sum y_i - a \sum x_i}{n} \tag{3.46}$$

Substituting for b, we get

$$-\sum x_i y_i + a \sum x_i^2 + \frac{\sum y_i - a \sum x_i}{n} \sum x_i = 0$$

$$-n \sum x_i y_i + na \sum x_i^2 + \sum y_i \sum x_i - a(\sum x_i)^2 = 0$$

Please note that the two terms $\sum x_i^2$ and $(\sum x_i)^2$ are not the same.

$$a = \frac{n \sum x_i y_i - \sum x_i \sum y_i}{n \sum x_i^2 - (\sum x_i)^2} \tag{3.47}$$

$$b = \frac{\sum y_i \sum x_i^2 - \sum x_i y_i \sum x_i}{n \sum x_i^2 - (\sum x_i)^2} \tag{3.48}$$

Now we have the unique values of a and b that best represent the data. What these mean and whether there is substantial improvement in our understanding by trying to force $y = ax + b$ need to be discussed.

Let us start with a simple example. Let us say we have fully developed turbulent flow in a pipe. We send water in at the rate of $\dot{m}$. Now, we wrap the pipe with strip heaters, whose power can be varied. We can measure the temperature at the inlet and the outlet. The heaters can also be individually controlled so that we can maintain a constant temperature at the outside. We send cold water into the pipe, the outside of which is maintained at one particular temperature, or is input with electrical heating so that it is essentially a heat exchanger that is trying to heat the water. A schematic of this situation is shown in Fig. 3.13. It is possible for us to change the velocity and when we do that, in order to maintain the temperature of the outer wall a constant,

Fig. 3.13 Fully developed
turbulent flow in a pipe

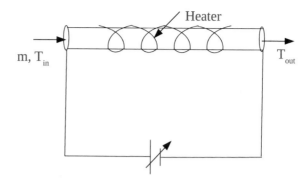

Fig. 3.14 T-x diagram for
fully developed turbulent
flow in pipe

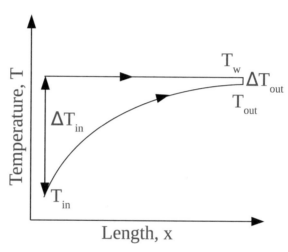

we can change the power settings of the heater. For every change in velocity, we can change the heater setting and keep the outer temperature a constant. When we change the velocity, we change the Reynolds number and the heat transfer rate. Hence, we are studying how the heat transfer rate changes with fluid velocity. This is essentially what we do in convective heat transfer. If the pipe were assumed to be at a particular temperature, then the wall temperature of the pipe is a constant while the water temperature goes up like what is shown in Fig. 3.14. The temperature difference at the inlet is ΔT_{in} while at the other end it is ΔT_{out}.

Because one is a straight line and the temperature distribution is exponential, we cannot use the arithmetic mean temperature difference. Therefore, it becomes contingent upon us to use the log mean temperature difference. The log mean temperature difference is defined as the temperature difference at one end of the heat exchanger (the pipe in this case) minus the temperature difference at the other end of the heat exchanger divided by the logarithm of the ratio of these two temperature differences.

$$LMTD = \frac{\Delta T_{out} - \Delta T_{in}}{ln(\Delta T_{out}/\Delta T_{in})} \quad (3.49)$$

Energy balance also needs to be satisfied thus

$$\dot{m}Cp\Delta T = UA\Delta T_{LMTD} \tag{3.50}$$

where $\dot{m}$ is the mass flow rate of the fluid, Cp is the specific heat capacity of the fluid at constant pressure, U is the overall heat transfer coefficient, and A is the surface area. The outside temperature is fixed, so we do not have to use the U because we know what the temperature on the outside is, and so we will replace U by h.

$$Q = hA\Delta T_{LMTD} \tag{3.51}$$

$$h = \frac{Q}{A\Delta T_{LMTD}} = \frac{Nu.k_f}{d} = \frac{aRe^b Pr^c k_f}{d} \tag{3.52}$$

In the above equations, k_f is the thermal conductivity of the fluid, and Nu is the dimensionless heat transfer coefficient or the Nusselt number (in the above formulation, conduction resistance across the pipe wall is neglected). Let us work this out numerically.

Example 3.3 Experiments on a fully developed turbulent flow in a pipe under steady-state conditions have been carried out with water as the medium. The variation of the Nusselt number with Reynolds number is given in Table 3.3. Assuming a power law for the Nusselt number, of the form $Nu = CRe^m$, using the least square regression, gets the best fit for C and m.

Table 3.3 Nusselt number for various Reynolds numbers (Example 3.3)

S.No.	Re	Nu
1	2500	24
2	4000	36
3	7000	55
4	12000	84
5	18000	119

Solution

We have straightaway introduced a complication. A little while ago, we elucidated the procedure to regress a straight line but the problem says there is a power law variation. What should we do as the first step? We take natural logarithms on both sides and reduce the power law form to a linear equation.

Table 3.4 LSR for Example 3.3

S.No.	Re	Nu	ln(Re),x	ln(Nu),y	x^2	xy
1	2500	24	7.82	3.178	61.15	24.85
2	4000	36	8.29	3.586	68.72	29.7
3	7000	55	8.85	4.01	78.32	35.4
4	12000	84	9.39	4.432	88.17	41.61
5	18000	119	9.798	4.78	96	46.83
Σ			44.16	19.986	392.36	178.39

$$Nu = CRe^m \tag{3.53}$$
$$ln(Nu) = ln(C) + mln(Re) \tag{3.54}$$
$$Y = aX + b \tag{3.55}$$

$$Y = ln(Nu) \tag{3.56}$$
$$a = m \tag{3.57}$$
$$X = ln(Re) \tag{3.58}$$
$$b = ln(C) \tag{3.59}$$

The dimensionless Nusselt number can be written as $Nu = aRe^m Pr^n$. Suppose we are considering experiments over water; the fluid is fixed, and so is the Prandtl number. Hence, aPr^n becomes a fixed quantity for the problem under question. So we get the expression $Nu = cRe^m$.

We can solve the problem using a spreadsheet like MS Excel or the open office spreadsheet, but it is better that we do hand calculations for the present.

$$\bar{x} = 8.83, \text{ and } \bar{y} = 3.994$$

Using Eqs. 3.47 and 3.48, and values from Table 3.4, we get the following.

$$a = \frac{(5 \times 178.39 - 44.16 \times 19.986)}{(5 \times 392.36 - (44.16)^2)}$$
$$a = 0.801$$
$$b = \frac{(19.986 \times 392.36 - 178.39 \times 44.16)}{(5 \times 392.36 - (44.16)^2)}$$
$$b = -3.31$$
$$C = 0.0464$$
$$Nu = 0.0464Re^{0.801} \tag{3.60}$$

We now proceed to determine the goodness of the fit. In order to do this, first we calculate Y_{fit} by substituting the value of the Reynolds number, raising it to the power

Table 3.5 Calculation of goodness of fit for Example 3.3

S.No.	Re	Nu	$\bar{Y}$	Y_{fit}	$(Y - \bar{Y})^2$	$(Y - Y_{fit})^2$
1	2500	24	3.9944	3.06	0.65	3.6e-4
2	4000	36	3.9944	3.535	0.17	2.3e-3
3	7000	55	3.9944	3.988	4e-3	3.2e-4
4	12000	84	3.9944	4.41	0.2	2.9e-4
5	18000	119	3.9944	4.73	0.62	1.7e-3
Σ					1.65	0.05

of 0.801, multiplying it by 0.0464, and taking its natural logarithm. (This Reynolds exponent of 0.8 is very typical of turbulent flows.)

It is instructive to look at the last two columns of Table 3.5, where $(Y - \bar{Y})^2$ and $(Y - Y_{fit})^2$ are tabulated.

$$\sum (Y - \bar{Y})^2 = S_t \text{ and } \sum (Y - Y_{fit})^2 = S_r.$$

There may be a dilemma as to whether we want to check in the ax + b equation or directly in the equation for the Nusselt number! But it really does not matter to evaluate the goodness of the correlation, though the correct procedure would be to compare the natural logarithm of the Nusselt number.

Now we introduce a new term r^2, which is the coefficient of determination, as

$$r^2 = \frac{S_t - S_r}{S_t} = \frac{1.65 - 0.05}{1.65}$$

$$r^2 = 0.97$$

r^2 is called as **coefficient of determination**.
$\sqrt{r^2} = r$ is known as **correlation coefficient**. $r = \sqrt{0.97} = 0.98$.

3.4.1.1   Meaning and Uses of Correlation Coefficient

We are able to determine the constants a and b, and we are able to calculate some statistical quantities like r which is close to 1. So we believe that the correlation is good. We did some heat transfer experiments. We varied the Reynolds number and got the Nusselt number. We believe that these two are related by a power law form and hence went ahead and did a least square regression and we got some values. Has all this really helped us or not? We need some statistical measures to answer this question. Suppose we had no heat transfer knowledge, and just have the Reynolds number and the Nusselt number, what we will have possibly done as the first step is to look at the Nusselt number (y) and get its mean. Then we will tell the outside world that we do not know any functional relationship but the mean value is likely to be like this. We get the mean value of ln y as 3.9944. But we do not stop with the mean of the natural logarithm of the Nusselt number. If we say that the mean is like this, then it is expected to have a variance which is like $(Y - \bar{Y})^2$. Therefore, the total residual which is given by the sum of the square of the difference between the ln(Nusselt number)-ln(Nusselt number)$_{average}$ will be 1.65, and the average of this will be 0.33.

However, if we get a little smart and we say that instead of qualifying only by the mean, if we can have a functional correlation ax + b, we can regress a and b, and see whether doing this really helps us. For the problem under consideration, we are saying that with respect to the fit, we are able to reduce the sum of the residuals S_t from 1.65 to $S_r = 0.05$. So, of the 1.65, 1.60 is explained by the fact that the Nusselt number goes as aRe^m. Therefore, 97% variance in the data is explained by this correlation. Therefore, it is a good correlation. In the absence of a correlation, we will report only the mean and the standard deviation. But apart from this, we go deeper and find out some physics behind this and propose a power law relationship and we are able to explain almost all of the variance in the data with very high r^2.

Sometimes, we may get a correlation coefficient which is negative (with a minimum of -1), which may arise in some correlation where, when x increases, y decreases. Suppose we have a correlation that gives us only 60%, either our experiments are erroneous or there are additional variables that we have not taken into account. So when we do a regression, it tells us many things. It is just not the experiments or the simulations that will help us understand science! **Regression itself can be a good teacher!**

Any correlation that has a high r^2 is not necessarily a good correlation. It is a good correlation purely from a statistical perspective. We do not know if there is a causal relationship between x and y. For example, let us calculate the ratio of the number of rainy days in India in a year to the total number of days (365). Let us look at the ratio of the number of one-day international cricket matches won by India to the total number of matches played in a year. Let us say we look at the data for the last 5 years. The two may have a beautiful correlation. But we are just trying to plot two irrelevant things. Such a correlation is known as a "spurious correlation"!

So, first we have to really know whether there exists a physical relationship between variables under question. This can come when we perform experiments or when we are looking at the nondimensional form of the equations or by using the Buckingham Pi theorem.

Outside of the above, we can also draw a **parity plot**. If we plot Y_{data} and Y_{fit}, the center line is a 45° line about which the points will be found. The points must be equally spread on both sides of the 45° line, called the parity line. A typical parity plot is given in Fig. 3.15. If all the points lie on the parity line, it is incredible. This may also lead to suspicion! Furthermore, any experiment will have a natural variation. Approximately, 50% of the points should be above the parity line, and 50% of the points should be below the line. When all the points are bunched together, it means there is some higher order physics. For example, when the Nusselt number is increasing and we get more error, the correlation suggests there are some additional effects like the effect of variation of viscosity that we have not taken into account. This parity plot is also called a **scattergram** or scatter plot. On the same parity plot, we can have red, blue, and green colors to indicate fluids of different Prandtl

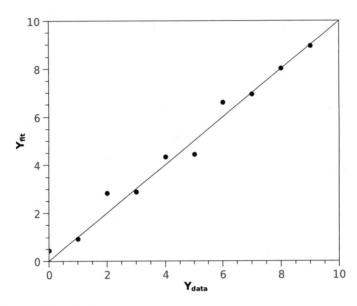

Fig. 3.15 A typical parity plot

numbers. Beyond a certain point, it is just creativity, when it comes to how we present the results and how we put the plots together!

Suppose we want to get the height of a student in this class, we measure the height of each student and take the average. We then report the average height and the variance S_r. If the class is large, the heights are likely to follow a Gaussian or a normal distribution. Now we can go one step further and attempt a correlation. Suppose we know the date of birth of each person, the height is possibly directly proportional to the date of birth, such that y = ax + b. With the date of birth, we get a much better fit than just taking all the heights alone and getting the mean. So the goodness of the fit $y = ax + b$ is just a measure of whether with the date of birth, we are able to predict heights, much better than just reporting the mean of the population.

3.4.1.2 The Concept of Maximum Likelihood Estimate (MLE)

Let us take 4 points (1, 2, 3, and 4), as shown in Fig. 3.16. We want to draw a straight line through this, which is the LSR fit for example. Assuming that all the measurements are made using the same instruments, we can assume that the distribution of errors in the measurements follows a Gaussian with a standard deviation of σ. If $Y_{fit} = ax + b$, the probability for getting Y_1 is given by

$$P(Y_1|(b)) = \frac{1}{\sqrt{2\pi\sigma^2}} e^{\frac{-[Y_1-(ax_1+b)]^2}{2\sigma^2}} \tag{3.61}$$

Fig. 3.16 Relationship of LSR with probability

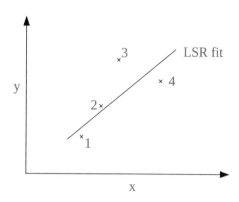

This comes from the normal or Gaussian distribution. Similarly, the probability of getting y_2 for the same a and b is given by

$$P(Y_2|(a,b)) = \frac{1}{\sqrt{2\pi\sigma^2}}e^{\frac{-[Y_2-(ax_2+b)]^2}{2\sigma^2}} \tag{3.62}$$

Therefore, the probability of getting Y_1, Y_2, and so on is given by the product of the individual probabilities. These are independent events. The total probability L is given by

$$L = P(Y_1, Y_2, \ldots |(a,b)) = \prod_{i=1}^{N} P_i$$

$$L = \frac{1}{(2\pi\sigma^2)^{N/2}}e^{\frac{-\sum_{i=1}^{N}[Y_i-(ax_i+b)]^2}{2\sigma^2}} \tag{3.63}$$

$$-2ln(L) = Nln(2\pi\sigma^2) + \frac{\sum_{i=1}^{N} R^2}{\sigma^2} \tag{3.64}$$

Now we want to determine the best values of a and b that maximize P. This procedure is known as the **Maximum Likelihood Estimation**. The standard deviation σ is not a variable in the problem; it is known (in fact, for the maximum likelihood estimation procedure to work, we do not even have to know σ. If we know that *sigma* is a constant across measurements, that would suffice). If $-2ln(L) = P$, we can make the first derivative stationary as follows:

$$\frac{\partial P}{\partial a} = 0 = \frac{\partial P}{\partial b} \tag{3.65}$$

and get a and b. The resulting equation (3.65) is exactly the same as what we obtained earlier in the derivation of LSR for a straight line (Eqs. 3.41–3.43).

3.4.2 Performance Metrics of LSR

1. Before the correlation, we have the data Y_i and the mean $\bar{Y}$ and hence S_t is the sum of the deviation of each value from the mean $S_t = \sum_{i=1}^{N}(Y - \bar{Y})^2$.
2. S_r is the residue given by $S_r = \sum_{i=1}^{N}(Y_i - Y_{fit,i})^2$.
3. Now the coefficient of determination is basically what is the percentage or how much reduction is brought about in S_t by proposing a correlation wherein we fit Y as a function of x. The square root of this coefficient of determination is the correlation coefficient.
4. There is one more quantity of interest, the standard error of estimate. It is something like the Root Mean Square error (RMSE).

$$\text{Std. error} = \sqrt{\frac{\sum_{i=1}^{N}(Y_i - Y_{fit,i})^2}{(n-2)}} \tag{3.66}$$

The numerator in this is actually S_r. We divide by $(n-2)$ and not n because if there are 2 points and it is a straight line, there is no question of an error. It is an exact fit. Therefore, this formula breaks down or this formula is meaningless if we try to do a linear fit for 2 points. So when n = 2, the denominator becomes zero and the expression becomes infinity, which does not mean that the standard error is infinity. It simply tells us not to use this formula for n = 2. This is good when n is much greater than 2 like 20 or 30 data points. For example, if we develop a correlation between the Nusselt number and the Reynolds number, and we have 40 or 50 data points, instead of dividing by 48 if we divide by 46, it will not make much of a difference. We will get more or less the same standard error. The n and $(n-2)$ look so different only when n is very small!

Let us look at another example in linear regression. Consider the cooling of a first-order system—say, mercury in a glass thermometer kept in a patient's mouth.

The system has a surface area A, mass m and on the outside, we have a heat transfer coefficient of "h" and an ambient temperature of T_∞. The governing equation for the temperature of the volume of mercury is

$$mCp\frac{dT}{dt} = -hA(T - T_\infty) \tag{3.67}$$

$$\theta = T - T_\infty \tag{3.68}$$

$$mCp\frac{d\theta}{dt} = -hA\theta \tag{3.69}$$

$$\theta = \theta_i e^{-t/\tau} \tag{3.70}$$

This is called a first-order system. There are some inherent assumptions.

1. The heat transfer coefficient is constant and does not depend on the temperature of the body (which is a questionable assumption). The mass and specific heat are constant; these are no questionable assumptions. But is the heat transfer coefficient really a constant? Under natural convection, this is not so. The heat transfer coefficient is really variable, but we keep it a constant here.
2. Because the body is cooling and giving heat to the surroundings, the temperature of the surroundings is not increasing. That is why it is a first-order system. But we can complicate it by having a second-order system. For example, consider an infant crying for milk. Let us say that the mother is trying to prepare baby food. She puts in milk powder and hot water, and mixes them in a glass bottle. If the milk is too hot, the bottle with milk is immersed in a basin of water. Unfortunately here, T_∞ is the temperature of the surrounding water. This starts increasing as it gets the heat from the milk in the bottle. We have got to model this as a second-order system.

For the problem under consideration, we assume that T_∞ is a constant. The goal of this exercise is to estimate the time constant if we have temperature versus time history. Basically, we can generate different curves for τ_1, τ_2, and so on. There will be one particular value of τ at which $\sum_{i=1}^{N}(\theta_i - \theta_{i,fit})^2$ is minimum. That is the best estimate of the time constant for this system. From τ, if we know m, Cp, and A, it gives us a very simple method to determine the heat transfer coefficient by means of an unsteady heat transfer experiment.

Example 3.4 Consider the cooling of a first-order system of mass m, specific heat C_p, and surface area A, at an initial temperature T_i°C. Such a system is placed in "cold" quiescent air or still air at $T_\infty = 30^\circ$C with the heat transfer coefficient of h. From the experimental data given in Table 3.6, determine the time constant τ and the initial temperature T_i. Evaluate the standard error of the estimate of the temperature and the correlation coefficient.

Table 3.6 Temperature–time history for Example 3.4

S. No.	t, s (x)	T, °C
1	10	93.3
2	30	82.2
3	60	68.1
4	90	57.9
5	130	49.2
6	180	41.4
7	250	36.3
8	300	32.9

Table 3.7 LSR for Example 3.4

S. No.	t, s (x)	$\theta, °C$	$ln(\theta), y$	x^2	xy
1	10	63.3	4.1479	100	41.4789
2	30	52.2	3.9551	900	118.6525
3	60	38.1	3.6402	3600	218.4129
4	90	27.9	3.3286	8100	299.5764
5	130	19.2	2.9549	16900	384.1383
6	180	11.4	2.4336	32400	438.0504
7	250	6.3	1.8405	62500	460.1374
8	300	2.9	1.0647	90000	319.4132
Σ	1050		23.3656	214500	2279.8600

Solution

We first reduce the solution to the standard equation of a straight line.

$$\theta = \theta_i e^{-t/\tau} \tag{3.71}$$

$$ln(\theta) = ln(\theta_i) - t/\tau \tag{3.72}$$

$$Y = ax + b$$

$$Y = ln(\theta)$$

$$x = t$$

$$a = -1/\tau$$

$$b = ln(\theta_i)$$

Now we can regress for a and b (see Table 3.7).

$$a = \frac{n\sum x_i y_i - \sum x_i \sum y_i}{n\sum x_i^2 - (\sum x_i)^2}$$

$$a = -0.0103$$

$$b = \frac{\sum y_i \sum x_i^2 - \sum x_i y_i \sum x_i}{n\sum x_i^2 - (\sum x_i)^2}$$

$$b = 4.2674$$

$$\theta = \theta_i e^{-t/100}$$

$$\theta = 71.3379 e^{-t/100}$$

T at t = 0 is 101.7 °C. This is the initial temperature of the system. We can now evaluate the goodness of the fit (see Table 3.8).

Table 3.8 Calculation of goodness of fit for Example 3.4

$\bar{y}$	y_{fit}	$(y - y_{fit})^2$	$(y - \bar{y})^2$
2.9207	4.1700	0.0003	1.5060
2.9207	3.9700	0.0000	1.0699
2.9207	3.6700	0.0001	0.5177
2.9207	3.3700	0.0002	0.1664
2.9207	2.9700	0.0005	0.0012
2.9207	2.4700	0.0002	0.2373
2.9207	1.7700	0.0191	1.1667
2.9207	1.2700	0.0155	3.4447
$\sum$	23.66	0.0359	8.1099

$$r^2 = \frac{S_t - S_r}{S_t} = \frac{8.1099 - 0.0359}{8.1099}$$

$$r = 0.9978 \text{ or } 99.78\%$$

This is a terrific correlation. The correlation coefficient is very high and very good. So it is a very accurate representation of the data. That means that if we are able to regress the data (i.e., the data can be represented as $e^{-t/\tau}$), then 99.7% of the variance of the data can be explained by proposing this curve. The standard error of the estimate is given by

$$\sqrt{\frac{S_r}{n-2}} = 0.0931$$

This standard error of the estimate is acceptable. That means at any time t, if we are using the correlation and obtaining the value of ln (θ), there is a 99% chance that this ln (θ) will lie between the mean and $\pm 3 \times 0.08 = \pm 0.24$. It gives us the confidence with which we can predict ln (θ). This is an independent measure of the validity of the estimate. So when ln (θ) is high, 0.08 is small and it is fine. But when this ln (θ) is going down, it becomes comparable. Therefore, when we do the experiment, we must try to use the maximum data when the system is hot. When the system is approaching the temperature of the surroundings and ln (θ) is very small, when sufficient time has elapsed and the driving force for heat transfer itself is small, the estimates are more prone to error. Other effects may also come into play in this example. In fact, we can sub-divide the data into 3 sets as early, middle, and final phases and for each of these phases, we may get a different value of τ.

MATLAB code for Example 3.4

```matlab
1   clear;
2   clc;
3
4   % Time data
5   t=[10;30;60;90;130;180;250;300];
6
7   % Temperature data
8   T=[93.3;82.2;68.1;57.9;49.2;41.4;36.3;32.9];
9
10
11  % Ambient temperature
12  T_inf = 30;
13
14  % Calculating theta
15  theta=T-T_inf;
16
17  % Number of data points
18  n=length(t);
19
20  y=log(theta);
21
22  % Calculating a and b
23  a=(n*sum(t.*y)-sum(t)*sum(y))/(n*sum(t.^2)-(sum(t))^2);
24  b=(sum(y)*sum(t.^2)-sum(t.*y)*sum(t))/(n*sum(t.^2)-(sum(t))^2);
25
26  % linear fit equation
27  y_fit=a.*t+b;
28
29  % time constant
30  time=-1/a;
31
32  theta0=exp(b);
33
34  % initial temperature
35  T0=theta0 + T_inf;
36
37  S_r=sum((y-y_fit).^2);
38  S_t=sum((y-mean(y)).^2);
39
40  % correlation coefficient
41  r=sqrt((S_t-S_r)/S_t);
42  % standard error
43  error=sqrt(S_r/(n-2));
44
45  % Print
46  prt = ['a = ',num2str(a)...
47      ', b = ',num2str(b)...
48      ', Correlation coefficient (r) = ',num2str(r)...
49      ', Standard error = ',num2str(error)];
50  disp(prt)
```

The output of the program is

$a = -0.010261, b = 4.2674$
$Correlation\ coefficient\ (r) = 0.99778$
$Standard\ error = 0.077388$

3.4.3 Linear Least Squares in Two Variables

There are some other forms that are amenable to linear regression. Let us look at functions of more than one variable $Y = f(x_1, x_2)$. Now suppose we propose that

$$Y = C_0 + C_1 x_1 + C_2 x_2 \tag{3.73}$$

which is the simplest linear form in two variables

$$S = \sum (Y_i - [C_0 + C_1 x_1 + C_2 x_2])^2 \tag{3.74}$$

$$\frac{\partial S}{\partial C_0} = 0, \frac{\partial S}{\partial C_1} = 0, \frac{\partial S}{\partial C_2} = 0 \tag{3.75}$$

This S is the same as $\sum (Y_i - Y_{fit})^2$ which has to be minimized. What is the immediate motivation for trying to understand a form like this? Many thermal sciences problems are two-variable problems. For example, convective heat transfer across a tube is a function of the length and diameter of the tube. Another example is the flow over a cylinder, where the Nusselt number goes as $Nu = aRe^m Pr^n$. The expression is the same for turbulent flow over a pipe. When we take the logarithm on both sides, we get

$$ln(Nu) = ln(a) + m\,ln(Re) + n\,ln(Pr) \tag{3.76}$$

This is of the form $Y = C_0 + C_1 x_1 + C_2 x_2$ where x_1 is the Reynolds number and x_2 is the Prandtl number. Like this, we have several systems in heat transfer and thermal engineering where the power law form can be reduced to the linear form. This is called **multiple linear regression**. For certain kinds of problems, we can do all these tricks and solve but for some, beyond a certain point, we have to take up nonlinear regression. More on this later.

Some other forms that can be linearized:

1. $Y = ax^b$ can be linearised as $ln(Y) = ln(a) + b\,ln(x)$
2. $Y = ae^{bx}$ can be linearised as $ln(Y) = ln(a) + bx$
3. $Y = \frac{ax}{b+x}$ can be linearised as $\frac{1}{Y} = \frac{b+x}{ax}, \frac{1}{Y} = \frac{b}{a}\frac{1}{x} + \frac{1}{a}$

Table 3.9 Linear least square regression with matrix algebra

S.No	x	y	x^2	xy
1	0	3.2	0	0
2	1	6.7	1	67
3	2	11.5	4	23
4	3	14.4	9	432
Σ	6	35.8	14	

3.4.4   Linear Least Squares with Matrix Algebra

Consider a straight line fit $y = ax + b$. To make matters simple, let us take only 4 data points, as shown in Table 3.9.

$$S = \sum_{i=1}^{N} [y_i - (ax_i + b)]^2 \tag{3.77}$$

$$\frac{\partial S}{\partial a} = 0 \tag{3.78}$$

$$\frac{\partial S}{\partial b} = 0 \tag{3.79}$$

$$a = \frac{n \sum x_i y_i - \sum x_i \sum y_i}{n \sum x_i^2 - (\sum x_i)^2} = 3.84$$

$$b = \frac{\sum y_i - a \sum x_i}{n} = 3.19$$

$$y = 3.84x + 3.19$$

We can write the above in a compact form as follows.

$$[Z^T Z]\{A\} = [Z^T]\{Y\} \tag{3.80}$$

$$[A] = \begin{bmatrix} a \\ b \end{bmatrix}$$

$$[Z] = \begin{bmatrix} 0 & 1 \\ 1 & 1 \\ 2 & 1 \\ 3 & 1 \end{bmatrix}$$

$$[Z^T] = \begin{bmatrix} 0 & 1 & 2 & 3 \\ 1 & 1 & 1 & 1 \end{bmatrix}$$

$$[Z^T Z] = \begin{bmatrix} 0 & 1 & 2 & 3 \\ 1 & 1 & 1 & 1 \end{bmatrix} \begin{bmatrix} 0 & 1 \\ 1 & 1 \\ 2 & 1 \\ 3 & 1 \end{bmatrix} = \begin{bmatrix} 14 & 6 \\ 6 & 4 \end{bmatrix}$$

The RHS is given by

$$\begin{bmatrix} 0 & 1 & 2 & 3 \\ 1 & 1 & 1 & 1 \end{bmatrix} \begin{bmatrix} 3.2 \\ 6.7 \\ 11.5 \\ 14.4 \end{bmatrix} = \begin{bmatrix} 72.9 \\ 35.8 \end{bmatrix}$$

$$\begin{bmatrix} \sum x_i^2 & \sum x_i \\ \sum x_i & n \end{bmatrix} \begin{bmatrix} a \\ b \end{bmatrix}$$

When we propose $[Z^T Z][A] = [Z^T][Y]$, we have very concisely written the linear least squares formulation. Now all the tools available for matrix inversion can be used. There is no need to simultaneously solve two equations like what we did before. We will see, how by matrix inversion, we can get the same answers. The right side is called the **forcing vector**. So with the matrix formulation, we have

$$[Z^T Z] = \begin{bmatrix} 14 & 6 \\ 6 & 4 \end{bmatrix}$$

$$[Z^T Y] = \begin{bmatrix} 72.9 \\ 35.8 \end{bmatrix}$$

$$\begin{bmatrix} 14 & 6 \\ 6 & 4 \end{bmatrix} \begin{bmatrix} a \\ b \end{bmatrix} = \begin{bmatrix} 72.9 \\ 35.8 \end{bmatrix}$$

$$\begin{bmatrix} a \\ b \end{bmatrix} = \frac{1}{20} \begin{bmatrix} 14 & 6 \\ 6 & 4 \end{bmatrix}^{-1} \begin{bmatrix} 72.9 \\ 35.8 \end{bmatrix}$$

$$\begin{bmatrix} a \\ b \end{bmatrix} = \frac{1}{20} \begin{bmatrix} 4 & -6 \\ -6 & 14 \end{bmatrix} \begin{bmatrix} 72.9 \\ 35.8 \end{bmatrix}$$

$$\begin{bmatrix} a \\ b \end{bmatrix} = \frac{1}{20} \begin{bmatrix} 4 \times 72.9 - 6 \times 35.8 \\ -6 \times 72.9 + 14 \times 35.8 \end{bmatrix} = \begin{bmatrix} 3.84 \\ 3.19 \end{bmatrix}$$

The values of a and b are the same as those obtained without using matrix algebra. This is a very smart way of doing it. When there are many equations and unknowns, we can use the power of matrix algebra to solve the system of equations elegantly.

3.5   Nonlinear Least Squares

3.5.1   *Introduction*

We need to first establish the need for nonlinear regression in thermal systems. If we disassemble the chassis of our desktop, we will see the processor and an aluminum sink on top of it. There may be more than one such sink. There are at least two fans, one near the outlet, the other fan is dedicated to the CPU. When we boot, the second fan will not turn on. Only when we run several applications, the second fan turns on when the CPU exceeds a certain temperature. Basically, we have a heat sink like what is shown in Fig. 3.17.

Suppose the whole heat sink is such that it is made of a highly conducting material and can be assumed to be at the same temperature, we can treat it as a lumped capacitance system. So when the system is turned on, we want to determine its temperature response. There is a processor that is generating heat at the rate of Q Watts, has a mass m, specific heat C_p and the heat transfer coefficient afforded is h, and the surface area is A; if we treat the heat sink and the processor to be at the same temperature, which is a reasonably good assumption to start with, the governing equation will be

$$mC_p\frac{dT}{dt} = Q - hA(T - T_\infty) \tag{3.81}$$

It is to be noted that in Eq. 3.81, m is the combined mass of the CPU and heat sink, and C_p is the mass averaged specific heat.

Initially, when we start, the processor is constantly generating heat at the rate of Q Watts. At $t = 0$, $T = T_\infty$. So when $T = T_\infty$, even though hA is available, the ΔT is so less that it is not able to compensate the Q, hence $(Q - hA\Delta T)$ is positive which forces the system to a higher temperature. The system temperature keeps climbing up. While this is happening, the heat generation from the processor, Q, remains the same. As the temperature difference ΔT keeps increasing, a time will come when $Q = hA\Delta T$ and the left side is switched off, which means that $mC_p\, dT/d\tau$ becomes 0 and the system approaches a steady state.

Fig. 3.17 Depiction of a heat sink used in a desktop PC

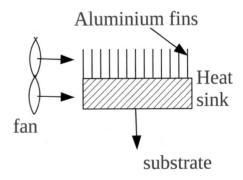

Fig. 3.18 Heating, steady-state, and cooling regimes of a first-order system

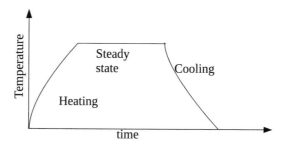

After the system approaches a steady state, when we are trying to shut it down, $Q = 0$. Then $mC_p dT/d\tau = -hA\Delta T$, so it will follow a cooling curve that is analogous to what we studied previously. Therefore, if we plot the temperature–time history for this problem, when it is starting up, operating under a steady state for a long time and then shutting down, the temperature profile will look like what is shown in Fig. 3.18.

If we are actually a chip designer, before we start to worry about the unsteady state, we will worry about the steady-state temperature. So for the steady-state temperature, we do not have to solve all these equations. When equilibrium is established

$$Q = hA(T_s - T_\infty) \tag{3.82}$$

$$T_s = T_\infty + \frac{Q}{hA} \tag{3.83}$$

Usually, a limit is set by the manufacturer on the maximum permissible temperature of the chip, say $T_s = 80\,°C$. The ambient $T_\infty = 30\,°C$ is also fixed. We know Q, which is the heat the processor is going to dissipate. Therefore we do not have much choice. If there is a fan, and if we know that for forced convection $h = 100\,W/m^2K$, then what we are left with is to calculate the surface area of the heat sink. This is one aspect of chip design. But if we need information on how long it will take for the system to reach this temperature, how long will it take for the system to cool to a particular temperature, we have to solve the differential equation. The solution to the differential equation is interesting.

$$mCp\frac{dT}{dt} = Q - hA(T - T_\infty) \tag{3.84}$$

$$\text{Let } T - T_\infty = \theta \tag{3.85}$$

We first set the RHS to zero and write

$$mC_p\frac{d\theta}{dt} + hA\theta = 0 \tag{3.86}$$

The complementary function is given by

$$\theta = Ae^{-t/\tau} \tag{3.87}$$

$$\text{where } \tau = \frac{mCp}{hA} \tag{3.88}$$

The particular integral will be $\theta = Q/hA$. The particular integral is one that satisfies the differential equation exactly. If we take $\theta = Q/hA$ and substitute it in the differential equation (Eq. 3.84), which is not a function of time, the first term vanishes. The complete solution to the problem will be the sum of the complementary function and the particular integral. The general solution will then be

$$\theta = \frac{Q}{hA} + Be^{-t/\tau} \tag{3.89}$$

There is a constant B that has to be figured out from the initial conditions. When we start, at time t = 0, $\theta = 0$. Substituting this in the solution, we get B = $-Q/hA$. So the final solution is

$$\theta = \frac{Q}{hA}[1 - e^{-t/\tau}]$$
$$\theta = a[1 - e^{-bt}] \tag{3.90}$$

The above form cannot be reduced to a linear regression model regardless of the number and intensity of algebraic manipulations. We have come to the stage where a and b cannot be determined using linear least squares. In nonlinear least squares, we first assume initial values for a and b, determine the residual, and then keep changing a and b, and see whether the residual goes down. **In one shot, we cannot get the solution because it cannot be linearized.**

3.5.2   Gauss–Newton Algorithm

Let us look at the following equation:

$$Y = a(1 - e^{-bt}) \tag{3.91}$$

This cannot be reduced to a linear form in spite of all possible manipulations, as already mentioned. Therefore we take recourse to nonlinear regression. The above can be rewritten as

$$Y_i = f(x_i, a, b) + e_i \tag{3.92}$$

What does this mean? If we happen to obtain a function $f(x_i)$, for which the parameters a and b are known, if we substitute the values of a and b and for a known value of x_i, we can get the value of Y_i. But that value of $f(x_i)$ need not agree with the Y_{data}. After all, the $f(x_i)$ is a form that we are trying to fit. So the difference between Y_i and the $f(x_i)$ with the best values of a and b will result in some error called e_i.

For simplicity, we consider this as a 2-parameter problem. The function need not be only in the form given above, it can be of any nonlinear form with two parameters a and b. So for a general 2-parameter problem, where linear regression will not hold, we are proposing this.

We have to assume a value for a and b and then expand $f(x_i)$ from the jth iteration to the $j+1$th iteration using the Taylors series. We truncate the series after a certain number of terms and determine the difference $(Y_{data} - Y_{fit})$ for all the data points. Then in a least square sense, we try to minimize this difference. We compare the old values of a and b with the new values of a and b. But that alone will not do. With the old values of a and b, we find out the values of Y_{fit}. We then take the square of $Y_{data} - Y_{fit}$ for each of the i values. We sum it up and get $S_0 = \sum R^2$. Now at iteration 0, for assumed values of a and b, we will get S_0, which is the sum of the residues. Now going through this procedure, we will get new values of a and b, which are called a_{new} and b_{new}. Using these new values, we will get the new value of Y_{fit}. We find out $(Y_{fit} - Y_{data})^2$ which for this iteration is S, and see how "S" is going down in the iterations. After the sum of residues, S, has reached a sufficiently small value, we can take a call and say that this is enough.

So basically, the key is that we are still using linear least squares, which means that we have to approximate the nonlinear form by an appropriate linear form. Expanding $f(x_i)$ in the vicinity of j using Taylors series, we have

$$f(x_i)_{j+1} = f(x_i)_j + \frac{\partial f(x_i)_j}{\partial a}\Delta a + \frac{\partial f(x_i)_j}{\partial b}\Delta b + \text{Higher order terms} \qquad (3.93)$$

When we try to regress, starting with this approach, if we are truncating after the first-order terms, then we are linearizing the nonlinear form using the Taylors series and this is the Gauss Newton algorithm or GNA. Now substituting for $f(x_i)$ using the expression for $f(x_i)_{j+1}$ from the Taylors series expansion in Eqs. 3.91 or 3.92, we have

$$Y_i = f(x_i)_j + \frac{\partial f(x_i)_j}{\partial a}\Delta a + \frac{\partial f(x_i)_j}{\partial b}\Delta b + e_i \qquad (3.94)$$

$$[Y_i - f(x_i)_j] = \frac{\partial f(x_i)_j}{\partial a}\Delta a + \frac{\partial f(x_i)_j}{\partial b}\Delta b + e_i \qquad (3.95)$$

$$\{D\} = [Z_j]\{\Delta A\} + [E] \qquad (3.96)$$

where

$$[Z_j] = \begin{bmatrix} \frac{\partial f_1}{\partial a} & \frac{\partial f_1}{\partial b} \\ \frac{\partial f_2}{\partial a} & \frac{\partial f_2}{\partial b} \\ \vdots & \vdots \\ \frac{\partial f_n}{\partial a} & \frac{\partial f_n}{\partial b} \end{bmatrix} \qquad (3.97)$$

$[Z_j]$ is called the Jacobian or the sensitivity matrix. We cannot evaluate the Jacobian matrix unless we have the values of a and b, which we are actually seeking. One cannot go to the next step unless we assume a and b. This was not the case for the linear least squares. We have to assume values for a and b, work out all these, and get the new values of a and b in nonlinear least squares.

If D is given like this, can we write the representation using linear least square as follows?

$$\{D\} = \begin{bmatrix} Y_1 - f(x_1) \\ Y_2 - f(x_2) \\ \vdots \\ Y_n - f(x_n) \end{bmatrix} \tag{3.98}$$

The D matrix is in fact called the **forcing vector**. We call Z_j column vector signifying that for the jth value of iteration, it is not constant here. Now using linear least squares,

$$[Z_j^T Z]\{\Delta A\} = [Z_j^T]\{D\} \tag{3.99}$$

On solving the above equation, we get $\{\Delta A\}$ which is nothing but $\left\{ \begin{smallmatrix} \Delta a \\ \Delta b \end{smallmatrix} \right\}$
Starting with $\{a_j \quad b_j\}^T$, we get

$$a_{j+1} = a_j + \Delta a \tag{3.100}$$
$$b_{j+1} = b_j + \Delta b \tag{3.101}$$

We continue with our iterations, till the stopping criterion is satisfied.

$$\left| \frac{a_{j+1} - a_j}{a_j} \right| \times 100 \le \epsilon_1$$

$$\left| \frac{b_{j+1} - b_j}{b_j} \right| \times 100 \le \epsilon_2$$

ϵ_1 and ϵ_2 are decided by us. They can both be equal or different.

3.5.2.1 Key Points in Nonlinear Regression

(i) Basically, nonlinear least squares is an iterative and often cumbersome procedure. The best way to implement this in a multivariable problem is to write a MATLAB script or use a nonlinear regression tool.

(ii) As is the case with any iterative technique, successful solutions are critically dependent on our initial guess. If the initial guess is way off, the solution may converge very slowly, oscillate or diverge. So the Gauss–Newton algorithm (GNA) is not a universal cure; it cannot solve all nonlinear regression problems. However, it is a strategy which is worth trying. Suppose we go through this algorithm and get stuck in between, there are some powerful tools that are available for us to correct.

(iii) In GNA, we do not use the second-order terms. So there is a chance that the scheme can falter along the way.

Example 3.5 Consider the heating of a first-order system whose temperature excess is known to vary as follows. $\theta = T - T_\infty$; $\theta = a(1 - e^{-bt})$, where θ is in °C and t is in seconds. Using the data given in Table 3.10, with an initial guess of $a = 40\,°C$ and $b = 5 \times 10^{-3}s^{-1}$, perform three iterations of the Gauss–Newton algorithm. Compare the residuals before and after the iteration.

Table 3.10 Data for Example 3.5

S. No.	t, s	θ_i,°C
1	10	3.1
2	41	11.8
3	79	21.1
4	139	29.8
5	202	37.4
6	298	42.5

Solution

$$\theta = 40[1 - e^{-t/200}]$$

We get a high value for the residue here (see Table 3.11) which tells us that the values for a and b are not correct.

1st iteration:

$$[Z_0] = \begin{bmatrix} 0.0488 & 380.49 \\ 0.1854 & 1336.02 \\ 0.3263 & 2128.83 \\ 0.5009 & 2774.85 \\ 0.6358 & 2942.89 \\ 0.7746 & 2686.44 \end{bmatrix}$$

$$[Z_0^T][Z_0] = \begin{bmatrix} 1.40 & 6302.89 \\ 6302.89 & 3.00 \times 10^7 \end{bmatrix}$$

Table 3.11 Values from the initial guess and first iteration

t, s	θ_i, °C	θ_{fit}	$(\theta_i - \theta_{fit})^2$	$\theta_{fit,new}$	$(\theta - \theta_{fit,new})^2$
10	3.1	1.95	1.32	3.38	0.0793
41	11.8	7.41	19.24	12.36	0.3087
79	21.1	13.05	64.76	20.80	0.0878
139	29.8	20.04	95.32	30.01	0.0421
202	37.4	25.43	143.25	36.03	1.8688
298	42.5	30.99	132.59	41.08	2.0083
Σ			456.4742		4.3951

Now let us get the forcing vector D.

$$[Z_0]^T [D] = [Z_0]^T \begin{bmatrix} 3.10 - 1.95 \\ 11.90 - 7.41 \\ 21.00 - 13.05 \\ 29.90 - 20.04 \\ 37.30 - 25.43 \\ 42.70 - 30.99 \end{bmatrix} = \begin{bmatrix} 24.91 \\ 1.17 \times 10^5 \end{bmatrix}$$

$$\begin{bmatrix} \Delta a \\ \Delta b \end{bmatrix} = \begin{bmatrix} 1.40 & 6302.89 \\ 6302.89 & 3.00 \times 10^7 \end{bmatrix}^{-1} \begin{bmatrix} 24.91 \\ 1.17 \times 10^5 \end{bmatrix}$$

Solve the above to get the values of Δa and Δb:

$$\begin{bmatrix} \Delta a \\ \Delta b \end{bmatrix} = \begin{bmatrix} 5.71 \\ 2.69 \times 10^{-3} \end{bmatrix}$$

$$a_{new} = 40 + 5.71 = 45.71$$
$$b_{new} = 5 \times 10^{-3} + 2.69 \times 10^{-3} = 7.69 \times 10^{-3}$$
$$\theta_{new} = 45.71[1 - e^{-t/130.10}]$$

2nd iteration:

$$[Z_0] = \begin{bmatrix} 0.0753 & 380.4918 \\ 0.2745 & 1339.9448 \\ 0.4611 & 1917.7100 \\ 0.6630 & 2109.8617 \\ 0.7942 & 1872.7426 \\ 0.9029 & 1303.3960 \end{bmatrix}$$

$$[Z_0^T] [Z_0] = \begin{bmatrix} 2.1792 & 5346.4147 \\ 5346.4147 & 1.53 \times 10^7 \end{bmatrix}$$

Now let us get the forcing vector D.

$$[Z_0]^T [D] = [Z_0]^T \begin{bmatrix} 3.10 - 3.39 \\ 11.90 - 12.36 \\ 21.00 - 20.77 \\ 29.90 - 29.87 \\ 37.30 - 35.77 \\ 42.70 - 40.67 \end{bmatrix} = \begin{bmatrix} 2.18 \\ 3.88 \times 10^3 \end{bmatrix}$$

$$\begin{bmatrix} \Delta a \\ \Delta b \end{bmatrix} = \begin{bmatrix} 2.1458 & 5509.4278 \\ 5509.4278 & 1.64 \times 10^7 \end{bmatrix}^{-1} \begin{bmatrix} 2.18 \\ 3.88 \times 10^3 \end{bmatrix}$$

Solve the above to get the values of Δa and Δb:

$$\begin{bmatrix} \Delta a \\ \Delta b \end{bmatrix} = \begin{bmatrix} 2.96 \\ -7.59 \times 10^{-4} \end{bmatrix}$$

$a_{new} = 45.7095 + 2.96 = 48.6743$

$b_{new} = 7.69 \times 10^{-3} - 7.59 \times 10^{-4} = 6.927 \times 10^{-3}$

$\theta_{new} = 48.6743[1 - e^{-t/144.36}]$

3nd iteration:

$$[Z_0] = \begin{bmatrix} 0.0669 & 454.1666 \\ 0.2472 & 1502.2336 \\ 0.4215 & 2224.6416 \\ 0.6182 & 2583.1014 \\ 0.7532 & 2426.3106 \\ 0.8731 & 1840.7771 \end{bmatrix}$$

$$[Z_0^T][Z_0] = \begin{bmatrix} 1.9551 & 6371.041256 \\ 6371.0413 & 2.34 \times 10^7 \end{bmatrix}$$

Now let us get the forcing vector D.

$$[Z_0]^T[D] = [Z_0]^T \begin{bmatrix} 3.10 - 3.26 \\ 11.90 - 12.03 \\ 21.00 - 20.51 \\ 29.90 - 30.09 \\ 37.30 - 36.66 \\ 42.70 - 42.50 \end{bmatrix} = \begin{bmatrix} 0.56 \\ 1.92 \times 10^3 \end{bmatrix}$$

$$\begin{bmatrix} \Delta a \\ \Delta b \end{bmatrix} = \begin{bmatrix} 1.9551 & 6371.041256 \\ 6371.0413 & 2.34 \times 10^7 \end{bmatrix}^{-1} \begin{bmatrix} 0.56 \\ 1.92 \times 10^3 \end{bmatrix}$$

Solve the above to get the values of Δa and Δb:

$$\begin{bmatrix} \Delta a \\ \Delta b \end{bmatrix} = \begin{bmatrix} 0.1477 \\ 4.20 \times 10^{-5} \end{bmatrix}$$

$a_{new} = 48.6743 + 0.1477 = 48.8220$

$b_{new} = 6.93 \times 10^{-3} + 4.20 \times 10^{-5} = 6.97 \times 10^{-3}$

$\theta_{new} = 48.8220[1 - e^{-t/143.49}]$

We now look at the power of the Gauss–Newton algorithm. In the first iteration, when θ was 3.1, the θ_{fit} was 1.95. Immediately after one iteration, the $\theta_{fit,new}$ is now 3.38, which is very close to the θ. If one sees the second row, $\theta_{fit,new}$ has increased from 7.41 to 12.36 in just one iteration where the actual value is 11.8. $\sum(\theta - \theta_{fit})^2$ has dropped from 456.47 to 4.39 in just after one iteration (refer to Table 3.11). This is the result after just one iteration of the Gauss–Newton algorithm. There will be a significant improvement in the residuals if the problem is well conditioned. At the end of the third iteration from the MATLAB output below, $\sum(\theta - \theta_{fit})^2$ has become even smaller and the solution has almost approached the true solution.

MATLAB code for Example 3.5

```
1   clc;
2   clear;
3
4   % Input Data
5   xt = [10,41,79,139,202,298];              % input X data
6   yt = [3.1,11.8,21.1,29.8,37.4,42.5];      % input Y data
7   x = xt';
8   y = yt';
9
10  % Intial guess
11  a = 40;                                    % initial guess for a and b
12  b = 0.005;
13
14  % Pre-allocation
15  z = zeros(length(xt),2);                   % Jacobian matrix
16  D = zeros(length(xt),1);                   % forcing vector matrix
17  y_f = zeros(length(xt),1);                 % y fit matrix
18
19  errTol = 10^-10;                           % error tolerance
20  countmax=10;                               % maximum iterations
21
22  count = 0;
23  label = 0;
24  while label==0
25
26
27      syms av bv xv;
28      y_ff=av*(1-exp(-1*bv*xv));             % function y=a[1-exp(-bx)]
29      dyda=diff(y_ff,av);                    % derivative of y w.r.t a
30      dydb=diff(y_ff,bv);                    % derivative of y w.r.t b
31
32      y_f=single(subs(y_ff,{av,bv,xv},{a,b,x}));
33      z(:,1)=single(subs(dyda,{av,bv,xv},{a,b,x}));
34      z(:,2)=single(subs(dydb,{av,bv,xv},{a,b,x}));
35      D=y-y_f;
36
37      if count>0
38
39          ztz=(z'*z);
40          ztd=z'*D;
41          DV=inv(ztz)*ztd;
```

```
42              a=a+DV(1);
43              b=b+DV(2);
44
45          end
46
47          y_fn=single(subs(y_ff,{av,bv,xv},{a,b,x}));
48
49          err = sum((y - y_fn).^2);
50
51          if err < errTol || count==countmax
52              label =1;
53          end
54
55          % Print
56          prt = ['Itr = ',num2str(count),...
57              ', a = ',num2str(a),...
58              ', b = ',num2str(b),...
59              ', 1/b = ',num2str(1/b),...
60              ', err = ',num2str(err),'\\'];
61          disp(prt)
62
63          count=count+1;
64
65      end
```

The output of the program is

```
Itr = 0, a = 40, b = 0.005, err = 456.4742
Itr = 1, a = 45.7095, b = 0.0076861, err = 4.3951
Itr = 2, a = 48.6743, b = 0.0069272, err = 1.0509
Itr = 3, a = 48.822, b = 0.0069692, err = 0.88806
Itr = 4, a = 48.8239, b = 0.0069686, err = 0.88806
Itr = 5, a = 48.8239, b = 0.0069686, err = 0.88806
```

3.5.2.2 What Are the Difficulties with the Gauss–Newton Algorithm?

1. Sometimes, it is not possible to calculate the partial derivatives. It is difficult because analytically we may encounter a very complex mathematical expression. In those cases, we calculate them numerically using numerical differentiation.
2. If the initial guess is not chosen properly or the system is not well conditioned, it may diverge, may not converge, or may converge slowly. Some modifications can be done in such a case. One possibility is to add an identity matrix in the formula, which is a nonnegative damping factor. This is the contribution by Levenberg who said it will condition the matrix properly and will lead to better results (see Eq. 3.103).

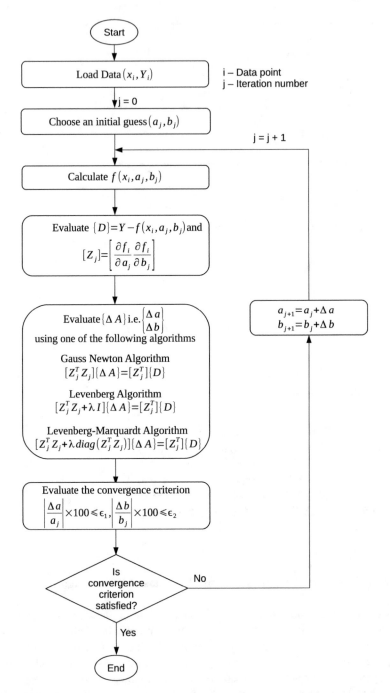

Fig. 3.19 Flowchart for solution to nonlinear regression problems

$$[Z^T Z + \lambda I]\{\Delta A\} = [Z^T]\{D\} \tag{3.102}$$

in which λ is the nonnegative damping factor.

3. An additional improvement is the use of only the diagonal elements of $Z^T Z$ and for all the other elements, use 0. This was the modification proposed by Marquardt (see Eq. 3.103). He employed the idea that a correction to the algorithm is required, but instead of the identity matrix, he chose to use the diagonal of $Z^T Z$. The resulting method is known as the Levenberg–Marquardt algorithm.

$$[Z^T Z + \lambda diag(Z^T Z)]\{\Delta A\} = [Z^T]\{D\} \tag{3.103}$$

A flowchart for solving nonlinear problems (with specific reference to Example 3.5) using GNA, Levenberg and Levenberg-Marquardt Algorithm is shown in Fig. 3.19.

The value of λ in Eqs. 3.102 and 3.103, which is the nonnegative damping factor, can be chosen only by experience. So we start with one value of λ and find out the results. One can employ the product of $\alpha\lambda$ as the new λ, where α is a factor and see how it progresses. If the convergence is very fast, we can sort of "cool down" λ and it can be reduced. If convergence is slow, we can assign higher values of λ. If $\lambda = 0$, it becomes the Gauss–Newton algorithm. If the λ is very high, it approaches the steepest descent or the steepest ascent (ascent—for a maximization problem).

For the sake of simplicity, the value of λ is taken as constant in the examples 3.6 and 3.7.

Example 3.6 Revisit Example 3.5 and solve it using the Levenberg method. Perform at least 3 iterations.

Solution

Let us solve this problem using the Levenberg method (Table 3.12).

Table 3.12 Values from the initial guess and first iteration

t, s	θ_i, °C	θ_{fit}	$(\theta_i - \theta_{fit})^2$	$\theta_{fit,new}$	$(\theta - \theta_{fit,new})^2$
10	3.1	1.95	1.32	3.39	0.0845
41	11.8	7.41	19.24	12.36	0.3174
79	21.1	13.05	64.76	20.77	0.1088
139	29.8	20.04	95.32	29.87	0.0044
202	37.4	25.43	143.25	35.77	2.6441
298	42.5	30.99	132.59	40.67	3.3448
Σ			456.4742		6.5038

We get a high value for the residue here (see Table 3.13) which tells us that the values for a and b are not correct.

1st iteration:

$$[Z_0] = \begin{bmatrix} 0.0488 & 380.49 \\ 0.1854 & 1336.02 \\ 0.3263 & 2128.83 \\ 0.5009 & 2774.85 \\ 0.6358 & 2942.89 \\ 0.7746 & 2686.44 \end{bmatrix}$$

$$[Z_0^T][Z_0] + \lambda I = \begin{bmatrix} 1.40 & 6302.89 \\ 6302.89 & 3.00 \times 10^7 \end{bmatrix} + 0.01 \times \begin{bmatrix} 1 & 0 \\ 0 & 1 \end{bmatrix}$$

Now let us get the forcing vector D.

$$[Z_0]^T [D] = [Z_0]^T \begin{bmatrix} 3.10 - 1.95 \\ 11.90 - 7.41 \\ 21.00 - 13.05 \\ 29.90 - 20.04 \\ 37.30 - 25.43 \\ 42.70 - 30.99 \end{bmatrix} = \begin{bmatrix} 24.91 \\ 1.17 \times 10^5 \end{bmatrix}$$

$$\begin{bmatrix} \Delta a \\ \Delta b \end{bmatrix} = \begin{bmatrix} 1.41 & 6302.89 \\ 6302.89 & 3.00 \times 10^7 \end{bmatrix}^{-1} \begin{bmatrix} 24.91 \\ 1.17 \times 10^5 \end{bmatrix}$$

Solve the above to get the values of Δa and Δb:

$$\begin{bmatrix} \Delta a \\ \Delta b \end{bmatrix} = \begin{bmatrix} 5.04 \\ 2.83 \times 10^{-3} \end{bmatrix}$$

$$a_{new} = 40 + 5.04 = 45.04$$
$$b_{new} = 5 \times 10^{-3} + 2.83 \times 10^{-3} = 7.83 \times 10^{-3}$$
$$\theta_{new} = 45.04[1 - e^{-t/127.79}]$$

2nd iteration:

$$[Z_0] = \begin{bmatrix} 0.0753 & 416.5431 \\ 0.2745 & 1339.9448 \\ 0.4611 & 1917.7100 \\ 0.6630 & 2109.8617 \\ 0.7942 & 1872.7426 \\ 0.9029 & 1303.3960 \end{bmatrix}$$

$$[Z_0^T][Z_0] + \lambda I = \begin{bmatrix} 2.1792 & 5346.4147 \\ 5346.4147 & 1.53 \times 10^7 \end{bmatrix} + 0.01 \times \begin{bmatrix} 1 & 0 \\ 0 & 1 \end{bmatrix}$$

Now let us get the forcing vector D.

$$[Z_0]^T [D] = [Z_0]^T \begin{bmatrix} 3.10 - 3.39 \\ 11.90 - 12.36 \\ 21.00 - 20.77 \\ 29.90 - 29.87 \\ 37.30 - 35.77 \\ 42.70 - 40.67 \end{bmatrix} = \begin{bmatrix} 2.87 \\ 5.05 \times 10^3 \end{bmatrix}$$

$$\begin{bmatrix} \Delta a \\ \Delta b \end{bmatrix} = \begin{bmatrix} 2.1892 & 5346.4147 \\ 5346.4147 & 1.53 \times 10^7 \end{bmatrix}^{-1} \begin{bmatrix} 2.18 \\ 3.88 \times 10^3 \end{bmatrix}$$

Solve the above to get the values of Δa and Δb:

$$\begin{bmatrix} \Delta a \\ \Delta b \end{bmatrix} = \begin{bmatrix} 3.46 \\ -8.78 \times 10^{-4} \end{bmatrix}$$

$$a_{new} = 45.04 + 3.46 = 48.50$$
$$b_{new} = 7.83 \times 10^{-3} - 8.78 \times 10^{-4} = 6.95 \times 10^{-3}$$
$$\theta_{new} = 48.50[1 - e^{-t/143.94}]$$

3nd iteration:

$$[Z_0] = \begin{bmatrix} 0.0671 & 452.4807 \\ 0.2479 & 1495.7306 \\ 0.4224 & 2213.3307 \\ 0.6193 & 2566.8899 \\ 0.7542 & 2408.0510 \\ 0.8738 & 1823.4242 \end{bmatrix}$$

$$[Z_0^T][Z_0] + \lambda I = \begin{bmatrix} 1.9603 & 6335.1352 \\ 6335.1352 & 2.31 \times 10^7 \end{bmatrix} + 0.01 \times \begin{bmatrix} 1 & 0 \\ 0 & 1 \end{bmatrix}$$

Now let us get the forcing vector D.

$$[Z_0]^T [D] = [Z_0]^T \begin{bmatrix} 3.10 - 3.26 \\ 11.90 - 12.02 \\ 21.00 - 20.49 \\ 29.90 - 30.04 \\ 37.30 - 36.58 \\ 42.70 - 42.38 \end{bmatrix} = \begin{bmatrix} 0.77 \\ 2.53 \times 10^3 \end{bmatrix}$$

$$\begin{bmatrix} \Delta a \\ \Delta b \end{bmatrix} = \begin{bmatrix} 1.9703 & 6335.1352 \\ 6335.1352 & 2.31 \times 10^7 \end{bmatrix}^{-1} \begin{bmatrix} 0.77 \\ 2.53 \times 10^3 \end{bmatrix}$$

Solve the above to get the values of Δa and Δb:

$$\begin{bmatrix} \Delta a \\ \Delta b \end{bmatrix} = \begin{bmatrix} 0.3059 \\ 2.56 \times 10^{-5} \end{bmatrix}$$

$$a_{new} = 48.5032 + 0.3059 = 48.8091$$
$$b_{new} = 6.95 \times 10^{-3} + 2.56 \times 10^{-5} = 6.97 \times 10^{-3}$$
$$\theta_{new} = 48.8091[1 - e^{-t/143.41}]$$

MATLAB code for Example 3.6

```
1   clc;
2   clear;
3
4   % Input Data
5   xt = [10,41,79,139,202,298];                    % input X data
6   yt = [3.1,11.8,21.1,29.8,37.4,42.5];            % input Y data
7   L = 0.01;       %Hyperperameter lambda value
8   x = xt';
9   y = yt';
10
11  % Intial guess
12  a = 40;                                         % initial guess for a and b
13  b = 0.005;
14
15  % Pre-allocation
16  z = zeros(length(xt),2);                        % Jacobian matrix
17  D = zeros(length(xt),1);                        % forcing vector matrix
18  y_f = zeros(length(xt),1);                      % y fit matrix
19
20  errTol = 10^-10;                                % error tolerance
21  countmax=10;                                    % maximum iterations
22
23  count = 0;
24  label = 0;
25  while label==0
26
27
28      syms av bv xv;
29      y_ff=av*(1-exp(-1*bv*xv));                  % function y=a[1-exp(-bx)]
30      dyda=diff(y_ff,av);                         % derivative of y w.r.t a
31      dydb=diff(y_ff,bv);                         % derivative of y w.r.t b
32
33      y_f=single(subs(y_ff,{av,bv,xv},{a,b,x}));
34      z(:,1)=single(subs(dyda,{av,bv,xv},{a,b,x}));
35      z(:,2)=single(subs(dydb,{av,bv,xv},{a,b,x}));
36      D=y-y_f;
37
38      if count>0
39
40          zt=(z'*z);
41          ztz=(zt+(L*eye(size(zt))));
```

```
42          ztd=z'*D;
43          DV=inv(ztz)*ztd;
44          a=a+DV(1);
45          b=b+DV(2);
46
47     end
48
49     y_fn=single(subs(y_ff,{av,bv,xv},{a,b,x}));
50
51     err = sum((y - y_fn).^2);
52
53     if err < errTol || count==countmax
54         label =1;
55     end
56
57     % Print
58     prt = ['Itr = ',num2str(count),...
59         ', a = ',num2str(a),...
60         ', b = ',num2str(b),...
61         ', err = ',num2str(err),'\\'];
62     disp(prt)
63
64     count=count+1;
65
66  end
```

The output of the program is

> *Itr = 0, a = 40, b = 0.005, err = 456.4742*
> *Itr = 1, a = 45.0449, b = 0.0078256, err = 6.5038*
> *Itr = 2, a = 48.5033, b = 0.0069472, err = 1.1879*
> *Itr = 3, a = 48.8092, b = 0.0069728, err = 0.88811*
> *Itr = 4, a = 48.8234, b = 0.0069687, err = 0.88806*
> *Itr = 5, a = 48.8239, b = 0.0069686, err = 0.88806*

Example 3.7 Revisit Example 3.5 and solve it using the Levenberg–Marquardt method. Perform at least 3 iterations.

Solution

Let us solve this problem using the Levenberg–Marquardt method.

We get a high value for the residue here (see Table 3.13) which tells us that the values for a and b are not correct.

Table 3.13 Values from the initial guess and first iteration

t, s	θ_i, °C	θ_{fit}	$(\theta_i - \theta_{fit})^2$	$\theta_{fit,new}$	$(\theta - \theta_{fit,new})^2$
10	3.1	1.95	1.32	3.39	0.0660
41	11.8	7.41	19.24	12.30	0.2511
79	21.1	13.05	64.76	20.78	0.1023
139	29.8	20.04	95.32	30.11	0.0933
202	37.4	25.43	143.25	36.29	1.2301
298	42.5	30.99	132.59	41.55	0.8949
Σ			456.4742		2.6377

1st iteration:

$$[Z_0] = \begin{bmatrix} 0.0488 & 380.49 \\ 0.1854 & 1336.02 \\ 0.3263 & 2128.83 \\ 0.5009 & 2774.85 \\ 0.6358 & 2942.89 \\ 0.7746 & 2686.44 \end{bmatrix}$$

$$[Z_0^T][Z_0] + \lambda\, diag(Z^T Z) = \begin{bmatrix} 1.40 & 6302.89 \\ 6302.89 & 3.00 \times 10^7 \end{bmatrix}$$

$$+ 0.01 \times \begin{bmatrix} 1.40 & 0 \\ 0 & 3.00 \times 10^7 \end{bmatrix}$$

Now let us get the forcing vector D.

$$[Z_0]^T[D] = [Z_0]^T \begin{bmatrix} 3.10 - 1.95 \\ 11.90 - 7.41 \\ 21.00 - 13.05 \\ 29.90 - 20.04 \\ 37.30 - 25.43 \\ 42.70 - 30.99 \end{bmatrix} = \begin{bmatrix} 24.91 \\ 1.17 \times 10^5 \end{bmatrix}$$

$$\begin{bmatrix} \Delta a \\ \Delta b \end{bmatrix} = \begin{bmatrix} 1.41 & 6302.89 \\ 6302.89 & 3.03 \times 10^7 \end{bmatrix}^{-1} \begin{bmatrix} 24.91 \\ 1.17 \times 10^5 \end{bmatrix}$$

Solve the above to get the values of Δa and Δb:

$$\begin{bmatrix} \Delta a \\ \Delta b \end{bmatrix} = \begin{bmatrix} 6.56 \\ 2.48 \times 10^{-3} \end{bmatrix}$$

$$a_{new} = 40 + 6.56 = 46.56$$
$$b_{new} = 5 \times 10^{-3} + 2.48 \times 10^{-3} = 7.48 \times 10^{-3}$$
$$\theta_{new} = 46.56[1 - e^{-t/133.65}]$$

2nd iteration:

$$[Z_0] = \begin{bmatrix} 0.0721 & 432.0501 \\ 0.2642 & 1404.6893 \\ 0.4463 & 2036.7543 \\ 0.6466 & 2287.4426 \\ 0.7794 & 2074.7312 \\ 0.8924 & 1492.3301 \end{bmatrix}$$

$$[Z_0^T][Z_0] + \lambda \, diag(Z^T Z) = \begin{bmatrix} 2.0962 & 5739.1245 \\ 5739.1245 & 1.81 \times 10^7 \end{bmatrix}$$
$$+ \, 0.01 \times \begin{bmatrix} 2.0962 & 0 \\ 0 & 1.81 \times 10^7 \end{bmatrix}$$

Now let us get the forcing vector D.

$$[Z_0]^T [D] = [Z_0]^T \begin{bmatrix} 3.10 - 3.36 \\ 11.90 - 12.30 \\ 21.00 - 20.78 \\ 29.90 - 30.11 \\ 37.30 - 36.29 \\ 42.70 - 41.55 \end{bmatrix} = \begin{bmatrix} 2.85 \\ 1.50 \times 10^3 \end{bmatrix}$$

$$\begin{bmatrix} \Delta a \\ \Delta b \end{bmatrix} = \begin{bmatrix} 2.1171 & 5739.1245 \\ 5346.4147 & 1.83 \times 10^7 \end{bmatrix}^{-1} \begin{bmatrix} 2.87 \\ 5.05 \times 10^3 \end{bmatrix}$$

Solve the above to get the values of Δa and Δb:

$$\begin{bmatrix} \Delta a \\ \Delta b \end{bmatrix} = \begin{bmatrix} 1.94 \\ -4.54 \times 10^{-4} \end{bmatrix}$$

$$a_{new} = 46.56 + 1.94 = 48.50$$
$$b_{new} = 7.48 \times 10^{-3} - 4.54 \times 10^{-4} = 7.03 \times 10^{-3}$$
$$\theta_{new} = 48.50[1 - e^{-t/142.28}]$$

3nd iteration:

$$[Z_0] = \begin{bmatrix} 0.0679 & 452.1065 \\ 0.2504 & 1490.7325 \\ 0.4261 & 2199.1311 \\ 0.6235 & 2538.0128 \\ 0.7582 & 2368.7984 \\ 0.8769 & 1779.7580 \end{bmatrix}$$

$$[Z_0^T][Z_0] + \lambda \, diag(Z^T Z) = \begin{bmatrix} 1.9814 & 6280.1462 \\ 6280.1462 & 2.25 \times 10^7 \end{bmatrix}$$

$$+ 0.01 \times \begin{bmatrix} 1.9814 & 0 \\ 0 & 2.25 \times 10^7 \end{bmatrix}$$

Now let us get the forcing vector D.

$$[Z_0]^T [D] = [Z_0]^T \begin{bmatrix} 3.10 - 3.29 \\ 11.90 - 12.14 \\ 21.00 - 20.67 \\ 29.90 - 30.24 \\ 37.30 - 36.78 \\ 42.70 - 42.53 \end{bmatrix} = \begin{bmatrix} 0.26 \\ 6.56 \times 10^2 \end{bmatrix}$$

$$\begin{bmatrix} \Delta a \\ \Delta b \end{bmatrix} = \begin{bmatrix} 2.0012 & 6280.1462 \\ 6280.1462 & 2.27 \times 10^7 \end{bmatrix}^{-1} \begin{bmatrix} 0.26 \\ 6.56 \times 10^2 \end{bmatrix}$$

Solve the above to get the values of Δa and Δb:

$$\begin{bmatrix} \Delta a \\ \Delta b \end{bmatrix} = \begin{bmatrix} 0.28 \\ -4.94 \times 10^{-5} \end{bmatrix}$$

$$a_{new} = 48.50 + 0.28 = 48.78$$
$$b_{new} = 7.03 \times 10^{-3} - 4.94 \times 10^{-5} = 6.98 \times 10^{-3}$$
$$\theta_{new} = 48.78[1 - e^{-t/143.29}]$$

MATLAB code for Example 3.7

```
1  clc;
2  clear;
3
4  % Input Data
5  xt = [10,41,79,139,202,298];        % input X data
6  yt = [3.1,11.8,21.1,29.8,37.4,42.5];  % input Y data
7  L = 0.01;     %Hyperperameter lambda value
8  x = xt';
```

```matlab
9    y = yt';
10
11   % Intial guess
12   a = 40;                              % initial guess for a and b
13   b = 0.005;
14
15   % Pre-allocation
16   z = zeros(length(xt),2);             % Jacobian matrix
17   D = zeros(length(xt),1);             % forcing vector matrix
18   y_f = zeros(length(xt),1);           % y fit matrix
19
20   errTol = 10^-10;                     % error tolerance
21   countmax=10;                         % maximum iterations
22
23   count = 0;
24   label = 0;
25   while label==0
26
27
28       syms av bv xv;
29       y_ff=av*(1-exp(-1*bv*xv));       % function y=a[1-exp(-bx)]
30       dyda=diff(y_ff,av);              % derivative of y w.r.t a
31       dydb=diff(y_ff,bv);              % derivative of y w.r.t b
32
33       y_f=single(subs(y_ff,{av,bv,xv},{a,b,x}));
34       z(:,1)=single(subs(dyda,{av,bv,xv},{a,b,x}));
35       z(:,2)=single(subs(dydb,{av,bv,xv},{a,b,x}));
36       D=y-y_f;
37
38       if count>0
39
40           zt=(z'*z);
41           ztz=(zt+(L*diag(diag((zt)))));
42           ztd=z'*D;
43           DV=inv(ztz)*ztd;
44           a=a+DV(1);
45           b=b+DV(2);
46
47       end
48
49       y_fn=single(subs(y_ff,{av,bv,xv},{a,b,x}));
50
51       err = sum((y - y_fn).^2);
52
53       if err < errTol || count==countmax
54           label =1;
55       end
56
57       % Print
58       prt = ['Itr = ',num2str(count),...
59           ', a = ',num2str(a),...
```

```
60              ',  b  =  ',num2str(b) ,...
61              ',  err  =  ',num2str(err),'\\'];
62        disp(prt)
63
64        count=count+1;
65
66   end
```

The output of the program is

Itr = 0, a = 40, b = 0.005, err = 456.4742
Itr = 1, a = 46.5618, b = 0.0074825, err = 2.6377
Itr = 2, a = 48.5026, b = 0.0070284, err = 0.93062
Itr = 3, a = 48.7858, b = 0.006979, err = 0.88839
Itr = 4, a = 48.8187, b = 0.0069701, err = 0.88807
Itr = 5, a = 48.8232, b = 0.0069688, err = 0.88806

Problems

3.1 The dynamic viscosity of liquid water in Ns/m^2 varies with temperature (T in K) as given in Table 3.14.

   (a) Using Newton's divided difference method, appropriate for 4 data points, obtain an exact fit to the viscosity as a function of temperature.
   (b) Using the fit obtained in (a), estimate the viscosity at 295 K.
   (c) Compare the result obtained in (b) with a linear interpolation.

3.2 The specific volume of saturated water vapor (v_g) varies with temperature (T) as shown in Table 3.15.
   Using Lagrange interpolation polynomial of order 2, determine the specific volume of saturated water vapor at 60°C. Compare this with the estimate obtained using a linear interpolation and comment on the results.

3.3 The saturation pressure of water (P) in kPa varies with temperature (T in K) as given in Table 3.16. Using Lagrange interpolation, determine the value of saturation pressure at 305 K.

Table 3.14 Variation of viscosity with temperature (problem 3.1)

S. No.	Temperature, K	Viscosity $\times 10^4$, Ns/m^2
1	275	16.52
2	290	10.8
3	300	8.55
4	315	6.31

Table 3.15 Specific volume for various temperatures for problem 3.2

T, °C	$v_g m^3$/kg
30	32.90
50	12.04
80	3.409

Table 3.16 Saturation pressure against temperature for problem 3.3

S. No.	Temperature, K	Pressure, kPa
1	273	0.611
2	293	2.34
3	313	7.38

Table 3.17 Variation of thermal conductivity with temperature for problem 3.4

Temperature, T in K	400	450	500
k, W/mK	0.0339	0.0372	0.0405

3.4 The thermal conductivity of dry air (k) for various temperatures is given in Table 3.17.

Obtain "k" at 470K using Lagrange interpolation and compare it with a linear interpolation. Comment on your result.

3.5 In a forced convection heat transfer experiment, the dimensionless heat transfer coefficient or the Nusselt number is known to vary with the Reynolds number in a power law fashion, as $Nu = aRe^b$, where a and b are constants. The experimental results are tabulated in Table 3.18.

(a) Using Least Squares Regression, estimate the parameters a and b.

(b) Determine the standard error of the estimate and the correlation coefficient.

3.6 Consider the cooling of an aluminum plate of dimensions $150 \times 150 \times 3$ (all in mm). The plate loses heat by convection from all its faces to still air. Radiation from the plate can be neglected. The plate can be assumed to be spatially isothermal. The ambient temperature is constant at $T_\infty = 30°C$. The temperature–time response of the plate, based on experiments is given in Table 3.19.

The temperature excess $\theta = T - T_\infty$ is known to vary as $\theta/\theta_i = exp(-t/\tau)$ where θ_i is the initial temperature excess given by $T_i - T_\infty$ and $\tau = mC_p/hA$, the time constant.

Table 3.18 Nusselt number data for problem 3.5

S. No.	Re	Nu
1	5000	50
2	1.7×10^4	80
3	5.5×10^4	161
4	1×10^5	202
5	2.25×10^5	300

Table 3.19 Transient temperature history of the aluminum plate for problem 3.6

S. No.	Time t, s	Temperature T, °C
1	10	98.5
2	40	96.1
3	80	90.2
4	120	85.9
5	180	82.8
6	240	75.1

(a) With the data given above, perform a linear least squares regression and obtain the best estimates of θ_i and τ. Additionally, determine the correlation coefficient and the standard error of the estimate of the temperature.

(b) Given $C_p = 940$ J/kg K and $\rho = 2700 kg/m^3$ for aluminum, from the results obtained in part (a), determine the heat transfer coefficient (h) for the situation.

3.7 An experiment was designed to examine the effect of load, x (in appropriate units) on the probability of failure of specimens of a certain industrial component. The following results were obtained from the experiment (see Table 3.20).
The regression model suitable for this problem is of the following form (also known as the logistic regression model):

$$p = \frac{1}{1 + e^{-(a+bx)}}$$

where "p" is the probability of failure of the component.

(a) Using the above model with the data given in Table 3.20, obtain the "best" estimates of a and b.

(b) Estimate the standard error of the estimate of "p" from the regression equation and the correlation coefficient.

Table 3.20 Variation of number of failures with load for problem 3.7

Load, x	Number of specimens	Number of failures
10	400	17
30	500	49
60	450	112
85	600	275
95	550	310
110	350	253

Table 3.21 Temperature–time history of the device for problem 3.8

S. No.	t, s	T, K
1	200	365
2	400	339
3	600	320
4	800	304
5	1000	290

3.8 A device operating in outer space, undergoing cooling, can be treated as a first-order lumped capacitance system, losing heat to outer space at $0\,K$ purely by radiation (no convection). The hemispherical emissivity of the device is ϵ, the surface area of the device is A, the mass of the device is m, and its specific heat C_p. The initial temperature of the device is T_i and there is no heat generation in the system.

(a) State the governing equation for the system temperature (T) as a function of time.

(b) Prove that the general solution to (a) can be given by

$$T = \left[\frac{1}{T_i^3} + \frac{\epsilon \sigma A t}{m c_p}\right]^{-1/3}$$

where σ is the Stefan–Boltzmann constant $(5.67 \times 10^{-8}\ W/m^2 K^4)$ and "t" is the time in seconds.

(c) Consider such a system made of aluminum for which $m = 0.1$ kg, $A = 0.08$ m^2, and $c_p = 900 J/kgK$. Experimental data of T versus t is obtained and given in Table 3.21. The challenge is to determine T_i and ϵ from the data and from the solution given in part b, by treating this as a nonlinear regression problem. Perform two iterations of the Gauss–Newton algorithm (GNA) with an initial guess of $T_i = 450K$ and $\epsilon = 0.3$ and compare the $\sum R^2$ values before and after performing the iterations. [You may choose to perform additional iterations on the computer by writing your own code or using a program like MATLAB].

Reference

Chapra Steven, C., & Canale Raymond, P. (2009). *Numerical methods for engineers*. New York, USA: Mc Graw Hill.

Chapter 4
Optimization—Basic Ideas and Formulation

4.1 Introduction

We now come to a very important part of the book, namely optimization. Before we could do this, we needed to know system simulation, regression, and we assume some prior knowledge of modeling.

What is optimization? Mathematically, ***optimization is the process of finding conditions that give us the maximum or minimum values of a function***. We want to find an extremum that may be a maximum or a minimum. As far as engineering is concerned, it is expected that every engineer will optimize. Whatever be the design, after proposing, we want to optimize it. Therefore, we can say that optimization is always expected from engineers, but is not "always" done. It seems perfectly logical that once we design, say a power plant, we want to optimize it. We want to optimize the clutch, the brake, our time, our resources, and so on. Everyone wants to optimize something or the other. Even so, there are many instances where it is not done. This is usually because of something called the **cost benefit**.

- In very large engineering projects on the other hand, sometimes it may not be possible for us to bring them down to a mathematical form and write the set of constraints and objective functions. Even if we are able to do so, it may be mathematically so complex that it is not possible for us to optimize.
- The second thing is that there are situations where it is possible for us to optimize. But because of the time and the effort involved in optimizing it, we do not want that. More often than not, we are satisfied with a design that works reasonably well and satisfies all the performance requirements. One such example is the pump and piping system for the apartment complex we saw earlier.
- The third reason may be an absolute lack of knowledge on our part about optimization and the techniques available. We sometimes do not know and hence say it is not required!
- There are other projects in which it is not worth doing optimization. For example, an engineering system may have reached saturation. An electrical motor efficiency

© The Author(s) 2021
C. Balaji, *Thermal System Design and Optimization*,
https://doi.org/10.1007/978-3-030-59046-8_4

is already 98 or 99%. A copper or aluminum fin efficiency is 98 or 99%. There is not much scope for optimization here (generally).
- In certain cases, the additional time, effort, and money we put into the optimization effort is not justified.

So for various reasons, optimization is not done. But since this is a book on optimization, one can argue that optimization should be done for everything. But that is far from the truth. Optimization is not required for every design. Therefore, when we are talking about optimization, we are talking about only those problems for which optimization is meaningful.

There is a large class of problems for which it is meaningful to do it. For example, we want to design an aircraft and certainly would want to reduce the weight. The Boeing 787 Dreamliner comes with a lot of plastic. The 747-400 Jumbo, with 4 engines, is no longer the benchmark. It is now possible to fly 16 h nonstop on 2 engines, from Mumbai to New York. So the redundancy of 4 engines is not required. Commercial transoceanic aviation may eventually settle down for 2 engines. The 2-engine craft saves so much fuel and is a highly optimized design compared to the Jumbo. Design of a satellite is yet another example where the objective is to minimize the weight. Optimization is a must here. Likewise, there is a wide range of engineering problems where optimization is nonnegotiable and is definitely required. Let us look at some examples.

Examples in Mechanical engineering: Some possible objectives of optimization

- Maximum efficiency of a power plant
- Maximum efficiency of an IC engine
- Maximum COP of an air conditioning system or refrigerator
- Minimizing the cost of manufacturing
- Optimal scheduling for a machine on which several jobs can be done.

Examples in Chemical engineering

- Optimum product mix in a petroleum refining fractionating column.

Examples in Electrical engineering

- Minimizing transmission losses (T&D loss)
- Minimizing communication time between nodes in a network.

Examples in Computer engineering

- Maximizing processor speed
- Maximizing memory storage.

There is no dearth of optimization problems. In a lighter vein, "We can easily make a career out of it", so to speak. Before we go on to writing the objective function and defining the constraints, let us look at an example.

Example 4.1 An ethylene refining plant receives 500 kg/hr of 50% pure ethylene. It refines it into two types of outputs 1 and 2. Type 1 has 90% purity. Type 2 has 70% purity. The raw material cost is Rs 40/kg. The selling price of Type 1 is Rs 200/kg and that of Type 2 is Rs 120/kg. Packaging facilities allow a maximum of 200 kg/hr of Type 1 and 225 kg/hr of Type 2. The transportation cost is Rs 8/kg for Type 1 and Rs 16/kg for Type 2, and the total transportation cost should not exceed Rs 4000/hr. Set up the optimization problem to maximize the profit and state all the constraints. Do not try to solve the optimization problem.

Solution

Let us first get the expression for profit.
Profit = sales − (investment + transportation cost)

$$\text{Profit} = (200x_1 + 120x_2) - (40 \times 500) - (8x_1 + 16x_2) = 192x_1 + 104x_2 - 20000 \tag{4.1}$$

We want to maximize the profit. Without constraints, we can imagine having x_1 and x_2 as infinity, which is not possible (just like they say, economics is all about limited resources and unlimited wants).

Constraints:

1. Mass balance: $0.9x_1 + 0.7x_2 \leq 0.5 \times 500$
   $0.9x_1 + 0.7x_2 \leq 250$.
2. Transportation cost: $8x_1 + 16x_2 \leq 4000$ *or* $x_1 + 2x_2 \leq 500$.
3. Bounds for variables:
   There are some constraints arising outbounds for variables. We have to set limits on variables. Everything cannot be allowed to freely float.
   $x_1 \leq 200$
   $x_2 \leq 225$.
4. There are some nonnegativity constraints. For example, we cannot produce -100 kgs of Type 1. So,
   $x_1 \geq 0$
   $x_2 \geq 0$.

The above is called the **formulation of the optimization problem**.
Can we plot the constraints as a graph? Yes. First we plot $x_1 = 200$ and $x_2 = 225$, and these represent the bounds for variables (Fig. 4.1).

Then we plot the transportation constraint. When $x_2 = 0$, $x_1 = 500$. When $x_1 = 0$, $x_2 = 250$. We connect the 2 points by a line and shade the inner region enclosed by it (Fig. 4.2).

Finally, we plot the mass balance constraint. When $x_2 = 0$, $x_1 = 277$. When $x_1 = 0$, $x_2 = 357$.

Fig. 4.1 Example 4.1 with bounds plotted

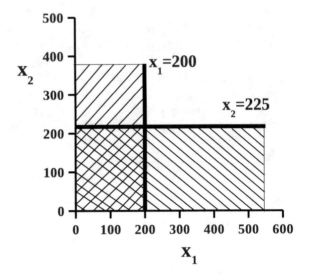

Fig. 4.2 Example 4.1 with transportation constraint plotted

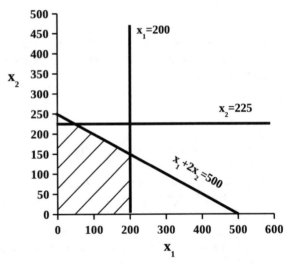

So what is the final feasible region? It is the portion shaded in Fig. 4.3. What is the revelation here? The objective function is in the background thus far. The feasible region is completely decided by the constraints. In this region, we have to determine the point or points that will maximize the objective function.

In certain cases, it is possible that the feasible region itself reduces to a point. The constraints decide the final solution. We will then have a set of simultaneous equations and we solve them and get a solution and that is it. Whether it is the optimum or not, we do not know; that is the only solution available. And far worse, if the number of constraints is more than the variables, we have an **_overdetermined system_** that cannot be solved at all.

Fig. 4.3 Example 4.1 with transportation and mass balance constraints plotted

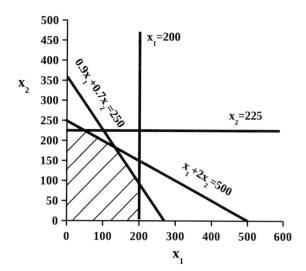

This begs the question "What holds the key to the optimization?". The answer is "the constraints".

Each of the points in the feasible region is a valid solution and will not violate any of the constraints. But each of these points is not equally desirable because each will result in a different value of the profit. Therefore, we will now have to develop an optimization algorithm which will help us find out which of these points in the feasible region maximizes the objective function y.

We are so fortunate that the objective function and the constraints are all linear. But this is a highly nonlinear world. As thermal engineers, we know that we frequently encounter the term $(T^4 - T_\infty^4)$ in radiation, which is highly nonlinear. Certain problems involving manufacturing, job allocation, and some applications where the objective function and the constraints are linear, can be solved by a special technique called Linear Programming (LP). We look at LP in detail in Chap. 7.

4.2 General Representation of an Optimization Problem

$$Y = Y[x_1, x_2, x_3 \ldots x_n] \tag{4.2}$$

subject to

$$\phi_1 = \phi_1[x_1, x_2, x_3 \ldots x_n] = 0 \tag{4.3}$$
$$\phi_2 = \phi_2[x_1, x_2, x_3 \ldots x_n] = 0 \tag{4.4}$$
$$\vdots$$
$$\phi_m = \phi_m[x_1, x_2, x_2 \ldots x_n] = 0 \tag{4.5}$$

$$\psi_1 = \psi_1[x_1, x_2, x_2 \ldots x_n] \leq r_1 \qquad (4.6)$$
$$\psi_2 = \psi_2[x_1, x_2, x_2 \ldots x_n] \leq r_2 \qquad (4.7)$$
$$\vdots$$
$$\psi_k = \psi_k[x_1, x_2, x_2 \ldots x_n] \leq r_k \qquad (4.8)$$

Equation 4.2 is the objective function and also called *figure of merit* in many places.

Equations 4.3–4.5 represent the *equality constraints*. They arise when we have to satisfy the basic laws of nature like the law of conservation of mass or Newton's second law of motion. For instance, suppose we say that the mass balance has to be satisfied exactly and no mass can be stored within a system, the mass balance constraint of Example 4.1 will have to be represented as an equality as $0.9x_1 + 0.7x_2 = 250$.

However, when we say the maximum temperature in a laptop, for example, should not exceed 70 °C, we write the constraint as $T \leq 70$. *Inequality constraints* like this arise when restrictions or limitations are set on the possible values of certain parameters, because of safety and other considerations. Equations 4.6–4.8 represent the inequality constraints.

Sometimes it is possible to treat equality constraints also as inequality constraints. If x = 5, one can represent it as a set of inequality constraints x < 6 and x > 4. If we have a powerful solver that works very well for problems with inequality constraints, such a strategy can be gainfully employed.

We also have *nonnegativity constraints*, wherein we declare that all x's are greater than or equal to 0. Bounds for variables are also prescribed under normal circumstances.

The objective function need not be linear, but can be hyperbolic, logarithmic, exponential, and so on. Solving the objective function with the constraints may be very formidable for some problems. For example, to get one solution, we may have to do a finite element or a CFD simulation.

4.2.1 Properties of Objective Functions

$$Max[a + y_1(x_1 \ldots x_n)] = a + Max[y_1(x_1 \ldots x_n)] \qquad (4.9)$$

The above is possible when "a" is a constant. So we just optimize y and add the constant to it, after the optimization is accomplished.

$$Max(y_1) = Min(-y_1) \qquad (4.10)$$

This is called the duality principle.

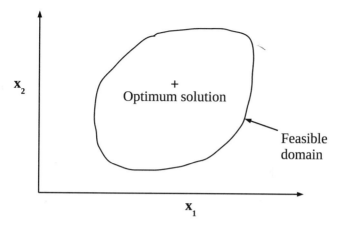

Fig. 4.4 Depiction of the feasible domain and the optimum for a two-variable problem

4.2.2   Cardinal Ideas in Optimization

1. The constraints allow us to explore the solution space; they do not decide the final solution themselves. They do not decide the values of all the variables. If they decide the values of all the variables, there is nothing we can do. Though the constraints bind the values, they give us the freedom to play around with the variables.
2. Where does the smartness lie? It lies in getting the global optimum without working out the value of y for each of the points in the feasible domain. First, there is the feasible domain in which we have got the choice of so many solutions of which all are not equally desirable (see Fig. 4.4). The constraints allow us some breathing space. But in doing so, we do not unimaginatively and exhaustively search for all the points in the feasible domain. Therefore, there is a need to develop appropriate optimization techniques to obtain the optimum without having to exhaustively search for all the points that satisfy all the constraints. Therein lies the key to developing a successful optimization technique.

4.2.3   Flowchart for Solving an Optimization Problem

A typical flowchart for solving an optimization problem is given in Fig. 4.5. First, we have to establish the need for optimization, i.e., get convinced that the problem is worth optimizing. Having decided that optimization is required, the next step is to set up the objective function. Following this will be setting up of the constraints. We need to set limits on the variables, called bounds. The next step is to choose the method that we are going to use. Then comes a decision box, which checks if the

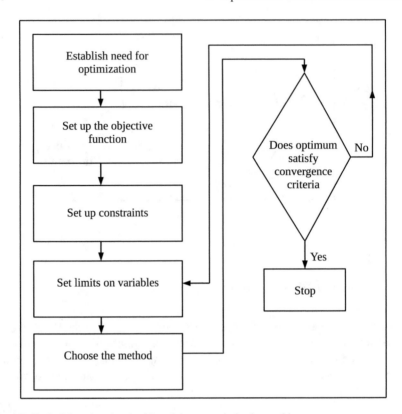

Fig. 4.5 Typical flowchart involved in solving an optimization problem

optimum satisfies the convergence criteria? If it is a yes, we stop, else we go back. There can be several types to this, several variants of the flowchart but this gives us a broad idea of how we go about solving an optimization problem.

So now, we have an idea of the pictorial description of optimization of y as a function of x_1 and x_2, where the opportunity to optimize arises, where the smartness lies, and how we go about solving it. This recipe can be applied if we have an idea of which technique works for a particular kind of problem and then one employs the most appropriate technique for the problem under consideration.

Even before starting all this, for the pump and piping problem, we adopted a crude method, wherein we plotted the total cost as a function of the pressure rise ΔP and found out where the cost becomes a minimum. This is a possible **helter skelter approach** to optimization problems, wherein many variables and/or constraints are not involved.

For a two-dimensional problem, we can start by choosing one value of x_1 and x_2 each and determining the value of y. We can take points along the east, west, north, and south of the initial guess. We find in which way y is moving and move the solution in that direction and again repeat the same procedure. This is called the **Lattice search**.

A helter skelter approach, for example, will be to choose some combinations of variables and get y and then find out where y is a maximum or a minimum and declare that as the optimum. It is dangerous but maybe better than not doing any optimization at all. Such a procedure does not guarantee that we have attained a global optimum in a real-life engineering problem. But there may be some limitations in some engineering problems. For example, there may be only a handful of combinations or there may be a set of combinations that can be manufactured and so, if we work out y for this set of variables, it may be enough. In this book, we are not talking about optimization where such techniques would work. We are talking about situations where there are many variables and the constraints give us the freedom to search the feasible domain resulting in a large number of feasible solutions such that a "quick scientific guessing" of the optimum is not possible.

As aforesaid, there could be some crude approach where we can get an idea of which is better than others. But the global optimum is not guaranteed. So whenever we optimize, we start from step 2 of Fig. 4.5, only when there is a need for optimization and we are convinced that the helter skelter methods will not work. If either because of money and effort, we do not want to optimize and we do not need it or our problem is such that some person working in the industry tells us that some 10 or 20 combinations are enough to be evaluated, we have to just "close the shop and go home". However, in this book we are talking about serious optimization problems such as minimizing the weight of an aircraft, increasing the specific fuel consumption of a street car, and minimizing the payload of a satellite where there are many variables and equally many possible design solutions.

4.2.4   Optimization Techniques

There are several ways of classifying optimization techniques and one such is given in Fig. 4.6. Optimization methods are broadly classified into calculus methods and search methods. Under each method we can have a single-variable or a multivariable problem. Again, a single or a multivariable problem can be a constrained or an unconstrained variable optimization problem.

In calculus methods, we use the information on derivatives to determine the optimum. In search methods, we use objective function information mostly and start with an initial point and progressively improve the objective function.

In calculus methods, however, we completely ignore the objective function and just determine the values of $x_1, x_2 \ldots x_n$ at which y becomes an extremum. We do not worry about the value of the optimum. The value of the function y at the optimum is a post-processed quantity in calculus methods. So the calculation of the objective function is pushed to the end. However, in search methods, it is the objective function that we are always comparing.

The important requirement for a calculus method is that the constraints must be differentiable and must be equalities. If they are inequalities, then it is a lot of trouble.

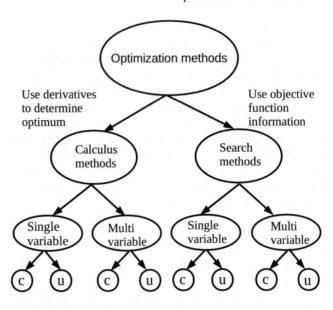

- Constraints must be equalities   c- constrained
- Constraints must be differentiable   u-unconstrained

Fig. 4.6 Classification of optimization problems

Please note that the first division into calculus and search methods is based on the type of method used. Further down, the divisions are based on the problem. We have indicated both types of optimization problems we normally encounter and the methods used for solving this. We can have a multivariable constrained optimization problem which can be solved by a search or a calculus method. Or we can have a single-variable unconstrained optimization problem that can be solved using an appropriate method.

Problems

4.1 In a steam power plant, 5 kg/s steam enters the turbine. Bleeding occurs at two stages as shown in Fig. 4.7. The bled steam is used for preheating. The prices are, Rs.4/kWh for electricity, Rs. 0.15/kg for low-pressure steam, and Rs. 0.25/kg for high-pressure steam. Assume that each kg/s into the generator can produce 0.025 kWh electricity.

To prevent overheating of the generator, the mass flow into it should be less than 3 kg/s. To prevent unequal loading on the shaft, the extraction rates should be such that $2x_1 + 3x_2 \leq 10$. The design of the bleed outlets allows the constraint $6x_1 + 5x_2 \leq 20$. Formulate the optimization problem for maximizing the profit from the plant.

Fig. 4.7 Schematic for
problem 4.1

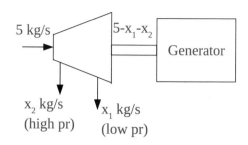

4.2 A one-dimensional pin fin losing heat by natural convection has length L and diameter d (both in m). The volume of the fin is fixed at V m^3. The base temperature of the fin, T_b, the ambient temperature, T_∞, and the heat transfer coefficient, h, are known and are constant. The fin is adiabatic at the tip. It is desired to maximize the heat transfer from the fin for a given volume, V.

(a) Formulate this as a two-variable, one-constraint optimization problem in L and D.
(b) By substituting for L or D from the constraint, convert the problem into a single-variable optimization problem in either L or D.
(c) Do you feel that an optimum L or D exists? Justify your answer.

4.3 Consider a power plant on a spacecraft working on an internally reversible Carnot cycle on an organic fluid between two temperatures T_H and T_L, where T_H is the temperature of evaporation of the organic fluid and T_L is the condensation temperature. The heat rejection from the condenser has to be accomplished by radiation to the outer space at a temperature of T_∞ K. The emissivity of the condenser surface is ϵ, and the total surface area available is A m^2.

(a) Formulate the optimization problem for maximizing the work output from the plant, with T_H being fixed and T_L being the variable.
(b) Solve the above problem for the special case of $T_\infty = 0$ K, using calculus.
(c) For the case of $T_H = 350K$, $T_\infty = 3$ K, $\epsilon = 0.9$, and A = $10m^2$, determine the optimal value of T_L and the corresponding work output and efficiency of the space power plant . You may use the successive substitution method for solving any nonlinear equation you may encounter.

4.4 Computer-based exercise[†]

(a) An overhead tank of a big apartment complex has a capacity of 12000 liters. It is desired to select a pump and piping system to transport water from the sump to the tank. The distance between the two is 270 m, and the tank is at a level 22 m above the sump. For operational convenience, the time to fill the tank shall be 60 min. Losses in the expansions, contractions, bends, and elbows have to be calculated appropriately. Design a workable system for the above and sketch the layout.

4.5 Using data for PVC pipes from local market sources, assign realistic values for the cost of the pipe (PVC), pump, and running cost including maintenance costs. With the help of any method known to you, obtain the value of ΔP developed by the pump at which the total cost will be minimum. The pump is expected to work every day and the average daily consumption of water is 24000 l. The cost of electricity may be assumed to be Rs.[1] 5.50 per unit and invariant with respect to time. The life of the system may be assumed to be 20 years. Let x the increase in electricity cost per year in percentage. Consider two values of x — 6% and 7%.

Output expected:

1. Setting up of the optimization problem with data
2. Sample calculations,
3. Plots on Excel/MATLAB ,
4. Final configuration and sketch.

†(Solution to the problem 4.4 is given in Appendix)

---

[1](1 USD ≈ Rs. 70 (as of April 2019)).

Chapter 5
Lagrange Multipliers

5.1 Introduction

The method of Lagrange multipliers is one of the most powerful optimization techniques. This can be used to solve both unconstrained and constrained problems with multiple variables. So, is it a cure for all as it can solve all kinds of problems? No! Because (i) the constraints must be equalities, (ii) the number of constraints must be less than the number of variables, and (iii) the objective function and constraints must be differentiable. These restrictions notwithstanding, there is a wide class of problems including those in thermal engineering which can be solved using Lagrange multipliers.

5.2 The Algorithm

Consider a general optimization problem.

$$\text{Maximize or Minimize } y = y[x_1, x_2, x_3.....x_n] \tag{5.1}$$
$$\text{subject to } \phi_1 = \phi_1[x_1, x_2, x_3....x_n] = 0 \tag{5.2}$$
$$\phi_2 = \phi_2[x_1, x_2, x_3....x_n] = 0 \tag{5.3}$$
$$\vdots$$
$$\phi_m = \phi_m[x_1, x_2, x_2....x_n] = 0 \tag{5.4}$$

For the Lagrange multiplier method to work, "m" should be less than equal to "n". If m = n, the constraints themselves will fix the solution and no optimum exists. If m < n, there will be a feasible domain, where we can possibly explore. If m > n, we have an over constrained optimization problem that cannot be solved. The Lagrange multiplier method to optimize an "m" constraint "n" variable optimization problem is akin to solving a set of equations that are generally written as follows:

© The Author(s) 2021
C. Balaji, *Thermal System Design and Optimization*,
https://doi.org/10.1007/978-3-030-59046-8_5

$$\nabla y - \lambda \nabla \phi = 0 \qquad (5.5)$$

$$\frac{\partial y}{\partial x_1} - \lambda_1 \frac{\partial \phi_1}{\partial x_1} - \lambda_2 \frac{\partial \phi_2}{\partial x_1} \dots - \lambda_m \frac{\partial \phi_m}{\partial x_1} = 0 \qquad (5.6)$$

$$\vdots$$

$$\frac{\partial y}{\partial x_n} - \lambda_1 \frac{\partial \phi_1}{\partial x_n} - \lambda_2 \frac{\partial \phi_2}{\partial x_n} \dots - \lambda_m \frac{\partial \phi_m}{\partial x_n} = 0 \qquad (5.7)$$

Where λs are scalars. We have (m + n) equations in total. The total number of variables is also (m + n). So if they are simultaneously solved, we can get the values of $x_1, x_2 .. x_n$ at which y becomes stationary. There is an added bonus. We also obtain the m values of λ, λ_1 to λ_m, which are called the Lagrange multipliers. It is too premature at this stage to discuss what the λ's are, but we will look at a physical interpretation of λ after we work through a few examples.

5.2.1 Unconstrained Optimization Problems

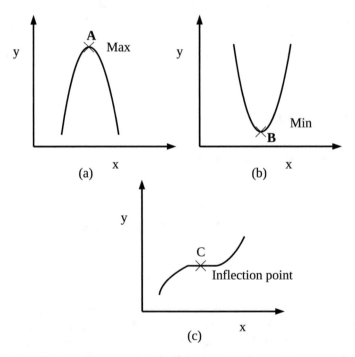

Fig. 5.1 Typical depiction of **a** maximum, **b** minimum and **c** inflection point

Here, there are no constraints (ϕ's do not exist). So, there are no Lagrange multipliers. Hence this method reduces to $\nabla y = 0$. For a one-variable problem, we have $dy/dx = 0$. Typically, we can have three situations as seen in Fig. 5.1. Here, A is a maximum, B is a minimum while C is an inflection point, where the second derivative is also 0. The Lagrange multiplier method, therefore, will give us only the values of the independent variables at which the function becomes stationary. It helps us to locate the extremum. However, necessary and sufficient second-order conditions are required to determine whether the optimum is a maximum or a minimum or an inflection point. A typical depiction of minimum for a two-variable unconstrained optimization problem is given in Fig. 5.2.

Having said that, we must also add that in many (not all !) engineering problems, once we have made something stationary it is possible for us to figure out intuitively whether we are heading toward a maximum or minimum.

Fig. 5.2 Depiction of minimum for a two-variable problem

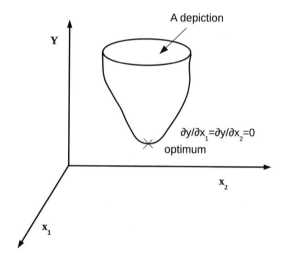

Example 5.1 Determine the minimum of the following function using the Lagrange multiplier method.

$$y = (x_1 - 8)^2 + (x_2 - 6)^2 \tag{5.8}$$

Solution

First we realize that the above is a two-variable unconstrained optimization problem. Equation 5.8 represents the equation of a circle with center at (8, 6).

Fig. 5.3 Iso-objective
contours for Example 5.1

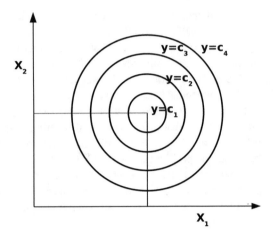

We first set $\frac{\partial y}{\partial x_1} = 0$, $\frac{\partial y}{\partial x_2} = 0$ thus get the value of x_1 and x_2 at the optimum. We substitute for x_1 and x_2 in Eq. 5.8 and obtain y.

We will denote the optimum by (x_1^+, x_2^+) to distinguish it from general x_1 and x_2. The Lagrange multiplier equations for this problem reduce to $\nabla y = 0$

$$\frac{\partial y}{\partial x_1} = 0 \text{ and } \frac{\partial y}{\partial x_2} = 0 \qquad (5.9)$$

The solution is $(x_1^+, x_2^+) = (8, 6)$ and $y^+ = 0$

Let us now proceed to a little more involved unconstrained optimization problem.

Example 5.2 Maximize the following function using the Lagrange multiplier method.

$$y = 6 - [x_1^2 + (1/(x_2 x_3)) + (x_2/x_1) + (x_3/8)] \qquad (5.10)$$

$$x_1, x_2, x_3 \geq 0$$

Solution

The above problem is an unconstrained optimization problem. We say that x_1, x_2 and x_3 should be greater than 0 because they are physical variables. We cannot produce -4 tables and -6 chairs for example. Maximization of $a + y$ is equal to $a +$ maximization of y and minimization of $(a - y)$ is equal to (a-max(y)). We can make use of this for solving this unconstrained optimization problem.

Applying the method of Lagrange multipliers, we have

$$\frac{\partial y}{\partial x_1} = 0 = \frac{\partial y}{\partial x_2} = \frac{\partial y}{\partial x_3} \qquad (5.11)$$

$$\frac{\partial y}{\partial x_1} = -[2x_1 - (x_2/x_1^2)] = 0 \tag{5.12}$$

$$\frac{\partial y}{\partial x_2} = -[-1/(x_2^2 x_3) + (1/x_1)] = 0 \tag{5.13}$$

$$\frac{\partial y}{\partial x_3} = -[-1/(x_2 x_3^2) + (1/8)] = 0 \tag{5.14}$$

$$2x_1^3 = x_2 \tag{5.15}$$

$$x_1 = x_2^2 x_3 \tag{5.16}$$

$$x_2 x_3^2 = 8 \tag{5.17}$$

$$x_2 = (8/x_3^2), \quad x_1 = (64/x_3^3) \tag{5.18}$$

$$x_3^+ = 4.87, \ x_2^+ = 0.337, \ x_1^+ = 0.55 \tag{5.19}$$

$$y^+ = 3.86 \tag{5.20}$$

We now take on a more realistic problem.

Example 5.3 Consider a solar thermal application where the hot water produced by a solar collector is kept in a cylindrical storage tank and its use
regulated so that it is also available during night time. The storage tank has
a capacity of 4000 l. Convective losses from the tank have to be minimized.
Radiative losses can be neglected. Ambient temperature T_∞ and convection
coefficient h are constant. The hot water temperature may be assumed constant
in the analysis. Solve this as an unconstrained optimization problem in r and
h, where r is the radius of the tank and h is the height of the tank using the
Lagrange multiplier method.

Solution

We want to solve this as an unconstrained optimization problem. However, there is
a constraint in this problem. So first the constraint is substituted into the objective
function to convert the latter into an unconstrained problem. The losses from both
the top and the bottom walls also need to be accounted. We have to just minimize A.

$$Q = hA\Delta T \tag{5.21}$$

We want to minimize Q. Minimizing A is akin to minimizing Q for this problem
(Area A = lateral surface area + top surface area + bottom surface area).
Hence, the minimization problem can be posed as

$$\text{Min } A = 2\pi r^2 + 2\pi rh \tag{5.22}$$

$$\text{Constraint: } \pi r^2 h = 4 \ (volume) \tag{5.23}$$

$$A = 2\pi r^2 + 2\pi r . 4/(\pi r^2) \tag{5.24}$$

$$A = 2\pi r^2 + (8/r) \tag{5.25}$$

Differentiating A with respect to r and equating it to 0, we can get the value of r. Hence, the minimization problem can be posed as

$$\frac{\partial A}{\partial r} = 4\pi r - (8/r^2) = 0 \tag{5.26}$$

$$4\pi r = 8/r^2 \tag{5.27}$$

$$r^3 = 2/\pi \tag{5.28}$$

$$r^+ = 0.860 \; m \tag{5.29}$$

$$\pi r^2 h = 4 \tag{5.30}$$

$$h^+ = 1.72 \; m \tag{5.31}$$

$$A^+ = 13.94 \; m^2 \tag{5.32}$$

$$\frac{\partial A}{\partial r} = 4\pi r - (8/r^2) \tag{5.33}$$

$$\frac{\partial^2 A}{\partial r^2} = 4\pi + (16/r^3) \tag{5.34}$$

Since $\frac{\partial^2 A}{\partial r^2}$ is +ve. A^+ is a minimum. If the temperature and the convection coefficient are constant, the height of a cylindrical tank must be twice its radius or equal to its diameter.

5.2.2 Constrained Optimization Problems

Example 5.4 Revisit Example 5.3 and solve it as a two-variable, one-constraint optimization problem.

$$\text{Min } A = 2\pi r^2 + 2\pi r h \tag{5.35}$$

$$\text{Constraint: } \pi r^2 h = 4 \tag{5.36}$$

$$\phi = \pi r^2 h - 4 = 0 \tag{5.37}$$

Solution

The Lagrange multiplier equations for this problem are

$$\frac{\partial A}{\partial r} - \lambda \frac{\partial \phi}{\partial r} = 0 \tag{5.38}$$

$$\frac{\partial A}{\partial h} - \lambda \frac{\partial \phi}{\partial h} = 0 \tag{5.39}$$

$$\phi = 0 \tag{5.40}$$

We have 3 equations and 3 unknowns r, h, and λ

$$(4\pi r + 2\pi h) - \lambda \, 2\pi r h = 0 \tag{5.41}$$

$$2\pi r - \lambda \pi r^2 = 0 \tag{5.42}$$

$$\pi r^2 h = 4 \tag{5.43}$$

$$4r + 2h = 2\lambda r h \tag{5.44}$$

$$\lambda = 2/r \tag{5.45}$$

$$\pi r^2 h = 4 \tag{5.46}$$

$$4r + 2h = 2 \, (2/r) \, r \, h \tag{5.47}$$

$$4r = 2h \tag{5.48}$$

$$h = 2r \tag{5.49}$$

$$\pi r^2 \, 2r = 4 \tag{5.50}$$

$$r^+ = 0.86 \, m \tag{5.51}$$

$$h^+ = 2r^+ = 1.72 \, m \tag{5.52}$$

$$\lambda = 2.335 \, m^{-1} \tag{5.53}$$

Now we have the optimum values r^+, h^+, A^+ along with a new parameter called λ, which has the units m^{-1}. We have to now interpret what this λ means. In order to do this, let us undertake a small exercise. Suppose we change the volume of the tank from 4000 to 4500 l, we want to see what happens to the solution. For this case, it can be shown that $r^+ = 0.89$ m. The other quantities of interest (for v = 4500 l) are shown below.

$$r^+ = 0.89 \, m, \quad h^+ = 1.79 \, m, \quad A^+ = 15.09 \, m^2 \tag{5.54}$$

We evaluate the change in area to the change in volume. When the volume changes from 4000 to 4500 liters, the surface area has changed from 13.94 to $15.09 m^2$.

$$\frac{\Delta A}{\Delta V} = \frac{(15.09 - 13.94)}{0.5} = 2.3 m^{-1} \tag{5.55}$$

What is the value of λ here? 2.3. What is $\Delta A/\Delta V$? 2.3. So λ is nothing but the change in objective function with respect to a change in the constraint. λ is called the shadow price. So λ is the *Lagrange multiplier, or the sensitivity coefficient, or the* **shadow price**. If we relax the constraint from 4000 to $4500 \, m^3$, how much additional area can we get? What is the sensitivity? The answer to these questions is λ!. In fact, it comes from the governing equation itself.

$$\nabla y - \lambda \nabla \phi = 0 \tag{5.56}$$

$$\lambda = \frac{\nabla y}{\nabla \phi} \approx \frac{\Delta y^+}{\Delta \phi^+} \text{ (will be shown in due course)} \tag{5.57}$$

(+ indicates at the optimum)

Hence, λ is the change in the objective function to the change in constraint. If there are m constraints, each of these λ's represents the sensitivity of the objective function to that particular constraint. In Example 5.4, we had only one constraint and hence λ is the sensitivity of the area to the constraint, which is the volume. λ is a very important physical parameter. When we work with the Lagrange multiplier method, outside of the optimum values, we also get to calculate the sensitivity coefficients which is helpful to evaluate the effect of a relaxation of a constraint on the optimal solution.

Example 5.5 Determine the shortest distance from the point (0, 1) to the parabola $x^2 = 4y$ by (a) eliminating x (b) Lagrange multiplier technique. Explain why approach (a) fails to solve the problem while approach (b) does not fail. (This problem is adapted from Engineering Optimization by Ravindran et al. 2006).

Solution

A plot of the parabola is given. The shortest distance from the point (0, 1) is so obvious. The Lagrange multiplier should also give the same answer. We will start with approach (b). The first step is to minimize z given by

$$\text{Minimize } z = \sqrt{(x - 0)^2 + (y - 1)^2} \tag{5.58}$$

Fig. 5.4 Plot of $x^2 = 4y$ (Example 5.5)

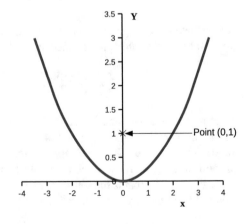

$$R = z^2 = x^2 + (y - 1)^2 \tag{5.59}$$

$$\text{subject to: } \phi = x^2 - 4y = 0 \tag{5.60}$$

The Lagrange multipliers equations are

$$\frac{\partial R}{\partial x} - \lambda \frac{\partial \phi}{\partial x} = 0 \tag{5.61}$$

$$\frac{\partial R}{\partial y} - \lambda \frac{\partial \phi}{\partial y} = 0 \tag{5.62}$$

$$\phi = 0 \tag{5.63}$$

x, y and λ are the 3 unknowns, there are 3 equations. It is possible for us to solve the 3 equations to determine the 3 unknowns.

$$2x - \lambda 2x = 0 \tag{5.64}$$

$$2(y - 1) + \lambda 4 = 0 \tag{5.65}$$

$$\phi = x^2 - 4y = 0 \tag{5.66}$$

From Eq. 5.64, we have $\lambda = 1$, substituting for $\lambda = 1$ in Eq. 5.65, we have

$$2(y - 1) + 4 = 0 \tag{5.67}$$

$$(y - 1) = -2 \tag{5.68}$$

$$y = -1 \tag{5.69}$$

Substituting for $y = -1$ in the Eq. 5.66, we get $x^2 = -4!!!$ which is absurd. Why does this happen? When we cancel the x in Eq. 5.64, we are assuming that x is not 0. But, unfortunately $x = 0$ is the solution!. That is the catch in this problem. We cannot cancel x from both sides of Eq. 5.64. So, even the Lagrange multiplier technique fails and we have to be alert! Therefore from Eq. 5.64, we get $x = 0$. From Eq. 5.66 $\phi = 0$ and therefore $y = 0$.

$$z = \sqrt{(x - 0)^2 + (y - 1)^2} = 1$$

This problem cautions us to exercise restraint, whenever we have the urge to convert a constrained optimization problem into an unconstrained one.
What about the other method (approach(a))?

$$z^2 = x^2 + (y - 1)^2 \tag{5.70}$$

$$x^2 = 4y \tag{5.71}$$

$$z^2 = 4y + y^2 - 2y + 1 \tag{5.72}$$

$$z^2 = (y + 1)^2 \tag{5.73}$$

$$z = y + 1 \tag{5.74}$$

We are stuck here and cannot proceed further with the solution.

The Lagrange method, without the intervention of the analyst, leads us to the solution automatically because when we solve Eq. 5.64, x = 0 is a distinct possibility. But in method (a), we have to give additional arguments to get the answer. So method "a"is quite inferior compared to method "b". z = y + 1 is correct, but we do not know the value of y$^+$ to get the minimum distance z$^+$.

5.3 Graphical Interpretation of the Lagrange Multiplier Method

The methodology we would like to use for illustrating this is as follows. We first take a two-variable, one-constraint problem. Using the regular Lagrange multiplier method, we first obtain the solution to convince ourselves that we do not have a fictitious problem in hand. Then using graph sheets, we plot what is required and try to interpret from the solution of the Lagrange multiplier and the plot, if there is a correlation between the two.

Example 5.6 Minimize $y = 4x_1 + 3x_2$, subject to $(x_1 - 8)^2 + (x_2 - 6)^2 = 25$

Solution

The above problem could be minimization of some cost, subject to some criterion. We solve it first as a constrained problem using the Lagrange multiplier method. Using the Lagrange multiplier equations, we have

$$\frac{\partial y}{\partial x_1} - \lambda \frac{\partial \phi}{\partial x_1} = 0 \tag{5.75}$$

$$\frac{\partial y}{\partial x_2} - \lambda \frac{\partial \phi}{\partial x_2} = 0 \tag{5.76}$$

$$\phi = (x_1 - 8)^2 + (x_2 - 6)^2 - 25 = 0 \tag{5.77}$$

$$4 - \lambda\, 2(x_1 - 8) = 0 \tag{5.78}$$

$$3 - \lambda\, 2(x_2 - 6) = 0 \tag{5.79}$$

$$(x_1 - 8)^2 + (x_2 - 6)^2 = 25 \tag{5.80}$$

$$2\lambda(x_1 - 8) = 4 \tag{5.81}$$

$$2\lambda(x_2 - 6) = 3 \tag{5.82}$$

$$\frac{x_1 - 8}{x_2 - 6} = \frac{4}{3} \tag{5.83}$$

$$3x_1 - 24 = 4x_2 - 24 \tag{5.84}$$

$$3x_1 - 4x_2 = 0 \tag{5.85}$$

$$x_1 = (4/3)x_2 \tag{5.86}$$

$$((4/3)x_2 - 8)^2 + (x_2 - 6)^2 = 25 \tag{5.87}$$

$$2.77x_2^2 - 33.33x_2 + 75 = 0 \tag{5.88}$$

$$x_2^2 - 12.03x_2 + 27.08 = 0 \tag{5.89}$$

$$x_2 = \frac{12.03 \pm \sqrt{(12.03)^2 - 4 \times 27.08}}{2} \tag{5.90}$$

$$x_2 = \frac{12 \pm 6}{2} = 9 \text{ or } 3 \tag{5.91}$$

$$x_1 = 12 \text{ or } 4 \tag{5.92}$$

$$y^+ = 4x_1 + 3x_2 \tag{5.93}$$

$$y^+ = 25 \tag{5.94}$$

But we do not know if y^+ is a minimum or a maximum. One possibility is to convert the problem into an unconstrained one and take the second derivative.

Now taking a graph sheet with x_1 in the x-axis and x_2 along the y-axis, we need to plot the constraint.

Then we will have to draw several iso-objective lines now, each of which represents different values of y. For example $y = 10 = 4x_1 + 3x_2$ can be first plotted. Now we can plot one more line $4x_1 + 3x_2 = 20$. If we take a scale and move these lines close to the constraints, one particular line will just touch the constraint. That is where we get the solution. This happens when $y = 4x_1 + 3x_2 = 25$.

Let us first plot the line $y = 10$ and then $y = 20$. 2 points each are enough to plot the line (shown below).

y=10

x_1	x_2
0	3.33
2.5	0

y=20

x_1	x_2
0	6.67
5	0

The two lines $y = 10$ and $y = 20$ are seen plotted here in Fig. 5.5 and the actual solution obtained using the Lagrange multiplier is also indicated. When we move from $y = 10$ to $y = 20$, we are approaching the curve and moving closer to it. It means that we are near the answer but not quite there. We draw iso-objective lines (lines representing y = constant where this constant keeps on changing) and move these lines (which are basically parallel) such that one of those lines becomes a tangent to the constraint. When one such iso-objective line touches the constraint, at that point, we get the solution because the iso-objective line is meeting the constraint.

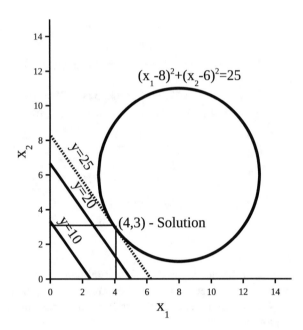

Fig. 5.5 Plot of $(x_1 - 8)^2 + (x_2 - 6)^2 = 25$ of Example 5.6 with a few iso-objective lines

Furthermore, any point on the constraint is a valid solution to the problem because the constraint cannot be violated. Now if we move further, again the constraint will be violated. So there may be other $y = c$ lines that may cut the constraint at one or more points, but none can have a value of y lower than what is obtained when $y = c$ becomes a tangent to the constraints.

When we are to the left of the constraint curve, we can get values of y that are very low. However, the constraint $(x_1 - 8)^2 + (x_2 - 6)^2 = 25$ will not be met.

The final solution to this problem is $x_1^+ = 4$, $x_2^+ = 3$ and $y^+ = 25$.

What does the Lagrange multiplier method do? The tangent to the constraint equation at the optimal point and the iso-objective line are parallel to each other. An alternative way of saying this is the gradient vectors will be parallel to each other. If the gradient vectors are parallel to each other, we are not saying that they will have the same magnitude. But what we are saying here is, the gradient vectors have to be collinear. They can even be pointing in the opposite direction. If we want to say that the tangent to this curve and the iso-objective line are parallel to each other, it is analogous to saying that ∇y and $\nabla \phi$ must be parallel.

However, their magnitudes need not be the same. This is possible only when

$$\nabla y - \lambda \nabla \phi = 0 \tag{5.95}$$

or

$$\nabla y + \lambda \nabla \phi = 0 \tag{5.96}$$

Therefore we say that ∇y and $\nabla \phi$ must be collinear vectors. But is that all? The solution must lie on the constraint curve so that the constraint equation is also satisfied. Therefore, the above condition must be satisfied in conjunction with $\phi = 0$, because the final solution we are seeking is a point on the constraint curve. Therefore, if we come up with the above set of equations and solve this set of scalar equations in conjunction with $\phi = 0$, then we will get a solution to the original optimization problem. This is the graphical interpretation of the Lagrange multiplier.

The important point to remember is ∇y and $\nabla \phi$ are collinear vectors. The Lagrange multiplier λ has to be involved because the magnitudes of y and ϕ need not be the same. λ is only a scalar, so it does matter whether we use $\nabla y - \lambda \nabla \phi = 0$ or $\nabla y + \lambda \nabla \phi = 0$. Please remember that finally the solution must lie on the constraint.

5.4 Mathematical Proof of the Lagrange Multiplier Method

Now for a quick mathematical proof. Let us consider a two-variable, one-constraint optimization problem.

$$\text{Max/Min} \; : \; y(x_1, x_2) \tag{5.97}$$

$$\text{subject to} \; : \; \phi(x_1, x_2) = 0 \tag{5.98}$$

This is standard formulation of the two-variable, one-constraint optimization problem. We are seeking a solution to this. If we want to write out the expression for the differential of ϕ, we have

$$d\phi = \frac{\partial \phi}{\partial x_1} dx_1 + \frac{\partial \phi}{\partial x_2} dx_2 \tag{5.99}$$

When we are seeking a solution to the optimization problem, the constraint has to be necessarily satisfied. For this, it must satisfy $\phi = 0$; Therefore $d\phi = 0$.

$$d\phi = 0 \tag{5.100}$$

$$d\phi = (\frac{\partial \phi}{\partial x_1}) dx_1 + (\frac{\partial \phi}{\partial x_2}) dx_2 \tag{5.101}$$

$$\therefore dx_1 = -\frac{(\frac{\partial \phi}{\partial x_2}) dx_2}{(\frac{\partial \phi}{\partial x_1})} \tag{5.102}$$

Let us start working on dy now.

$$dy = (\frac{\partial y}{\partial x_1}) dx_1 + (\frac{\partial y}{\partial x_2}) dx_2 \tag{5.103}$$

It is possible for us to substitute for dx_1 in Eq. (5.103) from the expression obtained in Eq. (5.102). Substituting for dx_1 from Eq. 5.102, we have

$$dy = -\frac{\frac{\partial y}{\partial x_1}\frac{\partial \phi}{\partial x_2}dx_2}{\frac{\partial \phi}{\partial x_1}} + \frac{\partial y}{\partial x_2}dx_2 \tag{5.104}$$

Let us say

$$\lambda = -\frac{\frac{\partial y}{\partial x_1}}{\frac{\partial \phi}{\partial x_1}} \tag{5.105}$$

substituting for λ in Eq. 5.104

$$dy = [\frac{\partial y}{\partial x_2} - \lambda\frac{\partial \phi}{\partial x_2}]dx_2 \tag{5.106}$$

If we are seeking a solution to an optimization problem, regardless of the values of dx_1 and dx_2, dy has to necessarily be 0. Therefore, the term within the bracket in Eq. 5.106 has to be 0 (as there is no point making dx_2 equal to 0).

$$\frac{\partial y}{\partial x_2} - \lambda\frac{\partial \phi}{\partial x_2} = 0 \tag{5.107}$$

From the definition of λ (Eq. 5.105), we already have

$$\frac{\partial y}{\partial x_1} - \lambda\frac{\partial \phi}{\partial x_1} = 0 \tag{5.108}$$

and finally

$$\phi = 0 \tag{5.109}$$

$$\therefore \nabla y - \lambda\nabla\phi = 0 \text{ from Eqs. 5.107 and 5.108} \tag{5.110}$$

These are the Lagrange multiplier equations stated at the beginning of the chapter.

5.5 Economic Significance of the Lagrange Multipliers

From the definition of λ, we see that

$$\lambda = [\frac{\nabla y}{\nabla \phi}]^+ \tag{5.111}$$

+ indicates λ is evaluated at the optimal point

$$\lambda = \left[\frac{\frac{\partial y}{\partial x_1} \cdot i + \frac{\partial y}{\partial x_2} \cdot j}{\frac{\partial \phi}{\partial x_1} \cdot i + \frac{\partial \phi}{\partial x_2} \cdot j} \right] \times \frac{(dx_1 i + dx_2 j)}{(dx_1 i + dx_2 j)} \tag{5.112}$$

$$\lambda = \left[\frac{\Delta Y^+}{\Delta \phi^+} \right] \tag{5.113}$$

Therefore, λ is the ratio of the change in the objective function to the change in the constraint. We saw this in an earlier example. This is the sensitivity coefficient. In operations research, it is called the shadow price.

Let us look at a company that makes furniture. It makes only two types of products, tables and chairs. So a certain amount of wood is required for making one chair and a certain amount of wood is required for making one table. A certain amount of labor is required for making a chair and a certain amount of labor is required for making a table. The profit from a chair is C_1, while that from the table is C_2. Therefore, the total profit will be $C_1 x_1 + C_2 x_2$. We want to maximize the total profit subject to the condition that there is a finite amount of labor and material available. So in this problem, the two constraints are labor and material. So if more wood or labor is made available to the furniture company, how will the objective function y, which is the profit, change? So this is known as the shadow price. If we pump in more resources, what will the profit be? It is called a shadow price because it is not realized yet!

5.6 Tests for Maxima/Minima

For a one-variable problem:

$$\frac{dy}{dx} = 0 \tag{5.114}$$

$$\frac{d^2 y}{dx^2} > 0; \ \ y \text{ is a minimum} \tag{5.115}$$

$$\frac{d^2 y}{dx^2} < 0; \ \ y \text{ is a maximum} \tag{5.116}$$

$$\frac{d^2 y}{dx^2} = 0; \ \ \text{We have saddle/inflection point} \tag{5.117}$$

When $\frac{d^2 y}{dx^2} = 0$, it means the second-order test is insufficient or the function is moving very gently over there, which may be very good for us from an engineering viewpoint. Mathematically, we would like to know precisely the value of x at which y would become maximum or minimum. But this is sometimes dreaded by engineers. Since the measurements of all variables are subject to errors. We would love to have an optimum of y which is not very sensitive to the values of $x_1, x_2 \ldots x_n$ at the optimum, a robust optimum so to speak. We need to have some objective function that gives us some breathing space, where if the value of y is 100, it should be possible for us to get between 95 and 100 for a reasonable range of the independent variables. So we will say that any of the solutions that give y as 95 and above is fine. Hence, the sensitivity or rather its lack is very important for engineers.

When more than one variable is encountered, which is invariably the case in optimization problems, we need to go in for detailed tests. Let us consider a two-variable problem where $y = f(x_1, x_2)$. We seek the minimum of this function y. Let the point (a_1, a_2) be somewhere near the optimum or the optimum itself.

We expand $y(x_1, x_2)$ around (a_1, a_2) using Taylor series

$$
\begin{aligned}
y(x_1, x_2) = y(a_1, a_2) &+ \frac{\partial y}{\partial x_1}(x_1 - a_1) + \frac{\partial y}{\partial x_2}(x_2 - a_2) \\
&+ \frac{1}{2!}\frac{\partial^2 y}{\partial x_1^2}(x_1 - a_1)^2 + \frac{1}{2!}\frac{\partial^2 y}{\partial x_2^2}(x_2 - a_2)^2 \\
&+ \frac{\partial^2 y}{\partial x_1 \partial x_2}(x_1 - a_1)(x_2 - a_2) \\
&+ higher\ order\ terms
\end{aligned}
\tag{5.118}
$$

We ignore the higher order terms assuming that they do not contribute significantly to $y(x_1, x_2)$.

If $\partial y/\partial x_1$ or $\partial y/\partial x_2$ is a large value, we can simply move the point to a nearby value from (a_1, a_2) and increase the function or decrease the function depending on whether we are seeking a maximum or a minimum. In that case (a_1, a_2) will no longer be a solution to the problem. Therefore, the first-order derivative becoming zero is a mandatory condition.

For a minimum, if we move from (a_1, a_2), any perturbation from (a_1, a_2) should result in a value of y which is more than that at (a_1, a_2). Therefore, it is enough for us to prove that the second-order terms on the RHS of Eq. 5.118 result in a positive quantity for a minimum or alternatively we find out conditions such that the second-order terms are positive.

The second-order terms, thus, will decide whether $y(a_1, a_2)$ is a minimum

$$\text{Let } a_{11} = \frac{\partial^2 y}{\partial x_1^2} \tag{5.119}$$

$$a_{22} = \frac{\partial^2 y}{\partial x_2^2} \tag{5.120}$$

$$a_{12} = \frac{\partial^2 y}{\partial x_1 x_2} \tag{5.121}$$

We now start looking at the second-order terms

$$a_{11}\left[\Delta x_1^2 + 2\frac{a_{12}}{a_{11}}\Delta x_1 \Delta x_2 + \frac{a_{22}}{a_{11}}\Delta x_2^2\right] > 0 \tag{5.122}$$

If this condition is satisfied, we get a minimum regardless of the values of Δx_1 and Δx_2. So, how do we get the conditions? A very crude way is to keep changing Δx_1 and Δx_2 and find this out. This is very unimaginative. We take 20 values of Δx_1 and 20 values of Δx_2 and write a program to see if the condition given in Eq. 5.122 is violated. Apart from being an imperfect procedure it also becomes impossible if we have say 50 or 100 variables. There should be something that is mathematically more precise and which will tell us this!

Regardless of the values of Δx_1 and Δx_2, in fact we know that it is the coefficients a_{11}, a_{12} and a_{22} which ultimately decide the fate of this expression. We now use the concept of Hessian matrix of second derivatives

The above expression can be written as

$$a_{11}[\Delta x_1 + \frac{a_{12}}{a_{11}}\Delta x_2]^2 + [(\frac{a_{22}}{a_{11}} - (\frac{a_{12}}{a_{11}})^2)\Delta x_2^2] > 0 \qquad (5.123)$$

$$a_{11}[\Delta x_1^2 + 2\Delta x_1 \Delta x_2 \frac{a_{12}}{a_{11}} + (\frac{a_{12}}{a_{11}})^2 \Delta x_2^2 + a_{22}\Delta x_2^2 - (\frac{a_{12}}{a_{11}})^2 \Delta x_2^2] > 0 \qquad (5.124)$$

When we are liberally multiplying and dividing by a_{11}, the intrinsic assumption is that a_{11} is not equal to 0.

$$\text{Let } z_1 = \Delta x_1 + \frac{a_{12}}{a_{11}}\Delta x_2 \qquad (5.125)$$

and $z_2 = \Delta x_2$

$$a_{11}.z_1^2 + (a_{22} - \frac{a_{12}^2}{a_{11}})z_2^2 > 0 \qquad (5.126)$$

If for all values of z_1 and z_2, inequality in Eq. 5.126 has to be true, individually the terms have to be positive. For this to be true for any value of z_1 and z_2, a_{11} and the term within the brackets have to be greater than 0.

$$\therefore a_{11} > 0 \text{ and } (a_{22} - \frac{a_{12}^2}{a_{11}}) > 0 \qquad (5.127)$$

$$\text{If D} = \begin{bmatrix} a_{11} & a_{12} \\ a_{12} & a_{22} \end{bmatrix} \qquad (5.128)$$

$$\text{Then D} = \begin{bmatrix} \frac{\partial^2 y}{\partial x_1^2} & \frac{\partial^2 y}{\partial x_1 \partial x_2} \\ \frac{\partial^2 y}{\partial x_1 \partial x_2} & \frac{\partial^2 y}{\partial x_2^2} \end{bmatrix} > 0 \text{ and } a_{11} > 0, \text{ then y is a minimum} \qquad (5.129)$$

D happens to be the determinant of the matrix containing the partial derivatives of the second order. Therefore, if we have the Hessian matrix defined by

$$H = \begin{bmatrix} \frac{\partial^2 y}{\partial x_1^2} & \frac{\partial^2 y}{\partial x_1 \partial x_2} \\ \frac{\partial^2 y}{\partial x_1 \partial x_2} & \frac{\partial^2 y}{\partial x_2^2} \end{bmatrix} > 0 \qquad (5.130)$$

If the determinant is greater than 0 and $a_{11} > 0$, H is called a positive-definite matrix, then y is a minimum. If H is negative definite, where $D > 0$ and $a_{11} < 0$, y is a maximum. If H is indefinite, then y is a saddle point or an inflection point?

What will happen if $D = 0$? The solution becomes a critical point where the Hessian test is inconclusive. It very rarely happens. We can possibly set up mathematical equations such that $D = 0$, but in most engineering problems, this will not happen. In fact, in most engineering problems, without doing the test for the Hessian matrix, we will be in a position to decide whether the resulting optimum is a maximum or a minimum. However, we can use this test and be sure that the final extremum we have obtained is really a maximum or a minimum.

When the Hessian matrix of y is positive definite or positive semi-definite for all values of $x_1 \ldots x_n$, then the function f is called a *convex function*. If the Hessian is positive definite, then y is said to be strictly convex and has a unique minimum. By the same token, a function y is a *concave function* if and only if -y is a convex function (needless to say over the same range of each of the variables $x_1 \ldots x_n$).

Mathematically, a function of n variables y(X), where $X = (x_1, x_2 \ldots x_n)$ on a convex set R is said to be convex if and only if for any two points X^1 and $X^2 \in R$, and $0 \le \gamma \le 1$

$$y[\gamma X^1 + (1 - \gamma)X^2] \le \gamma y(X^1) + (1 - \gamma)y(X^2) \qquad (5.131)$$

Equation 5.131 tells us that the weighted average of the function at points X^1 and X^2 (RHS of Eq. 5.131) will always be equal to or more than the value of the function evaluated at the weighted average of X^1 and X^2 themselves (LHS of Eq. 5.131).

However, this is often cumbersome to test in a multivariable problem. The Hessian test may be more useful in such cases.

There is one more way of looking at it. We have the Hessian matrix of the partial derivatives of the second order. If all the eigenvalues are positive, we have a positive-definite matrix. If all the eigenvalues are negative, we have a negative definite matrix. If some values are positive and some negative, the matrix is indefinite.

Example 5.7 Minimize $y = f(x_1, x_2) = (x_1 - 8)^2 + (x_2 - 6)^2$ using Lagrange multipliers. Check for minimum using the Hessian matrix.

Solution

This is an unconstrained optimization problem, which we have already solved. It is a straightforward problem, where y is equal to square of the radius of the circle.

$$\frac{\partial y}{\partial x_1} = 0 = 2(x_1 - 8) = 0; \, x_1 = 8 \qquad (5.132)$$

$$\frac{\partial y}{\partial x_2} = 0 = 2(x_1 - 6) = 0; \, x_2 = 6 \qquad (5.133)$$

The second order derivatives are:

$$\frac{\partial^2 y}{\partial x_1^2} = 2 \tag{5.134}$$

$$\frac{\partial^2 y}{\partial x_2^2} = 2 \tag{5.135}$$

$$\frac{\partial^2 y}{\partial x_1 \partial x_2} = 0 \tag{5.136}$$

$$D = \begin{bmatrix} \frac{\partial^2 y}{\partial x_1^2} & \frac{\partial^2 y}{\partial x_1 \partial x_2} \\ \frac{\partial^2 y}{\partial x_1 \partial x_2} & \frac{\partial^2 y}{\partial x_2^2} \end{bmatrix} \tag{5.137}$$

$$D = \begin{bmatrix} 2 & 0 \\ 0 & 2 \end{bmatrix} \tag{5.138}$$

$$\frac{\partial^2 y}{\partial x_1^2} = a_{11} > 0 \tag{5.139}$$

$$D > 0 \tag{5.140}$$

Therefore y is a minimum. For many problems, the minimum is not so obvious as there are far too many variables and we do not know how the objective function is behaving. Therefore, it is good for us to have a condition like this, which is very rigorous, mathematical and systematic. The only thing is if we have more number of variables, getting the derivatives and the Hessian will be very cumbersome.

Example 5.8 Revisit the cylindrical solar water heater storage problem (see Example 5.3). Minimize $A = 2\pi r^2 + 2\pi rh$ subject to $\phi = \pi r^2 h - 4 = 0$. Establish that the solution is a minimum by evaluating the Hessian matrix.

Solution

The Lagrange multiplier equations are

$$\frac{\partial A}{\partial r} - \lambda \frac{\partial \phi}{\partial r} = 0 \tag{5.141}$$

$$\frac{\partial A}{\partial h} - \lambda \frac{\partial \phi}{\partial h} = 0 \tag{5.142}$$

$$\phi = 0 \tag{5.143}$$

$$A = 2\pi r^2 h + 2\pi rh \tag{5.144}$$

$$\frac{\partial A}{\partial r} = 4\pi r + 2\pi, \quad \frac{\partial^2 A}{\partial r^2} = 4\pi \tag{5.145}$$

$$\frac{\partial A}{\partial h} = 2\pi r, \quad \frac{\partial^2 A}{\partial h^2} = 0, \quad \frac{\partial^2 A}{\partial h \partial r} = 0 \tag{5.146}$$

$$H = \begin{bmatrix} \dfrac{\partial^2 A}{\partial r^2} & \dfrac{\partial^2 A}{\partial r \partial h} \\ \dfrac{\partial^2 A}{\partial r \partial h} & \dfrac{\partial^2 A}{\partial h^2} \end{bmatrix} \tag{5.147}$$

$$H = \begin{bmatrix} 4\pi & 2\pi \\ 2\pi & 0 \end{bmatrix} \tag{5.148}$$

So we find that D is negative and a_{11} is positive. Hence, the Hessian test is inconclusive as we said D has to be necessarily positive for us to conclude if it is a minimum or a maximum. $a_{22} = 0$ because it does not vary with the variables. When we said $a_{11} \neq 0$, it also means that $a_{22} \neq 0$. So if any of these become 0, then already the second derivative is becoming 0.

But when we considered this as a single-variable problem, we established the point to be a minimum. Or we can write a program and for all combinations of r and h, with the precision of 10^{-5} m, we can prove that we cannot get a solution that has an area smaller than what was obtained earlier. So, we should not get carried away by the Hessian test. Sometimes, what common sense tells us may not be revealed by the Hessian!.

It now becomes apparent that the Hessian needs to be tested on a new quantity L, the Lagrangian defined as $L = Y - \lambda\phi$, if we want to use it for a constrained optimization problem. We will see if the Hessian test is conclusive for L.

In this case $L = A - \lambda\phi$, $L = 2\pi r^2 + 2\pi rh + \lambda(\pi r^2 h - 4)$. On evaluating the Lagrangian, we have

$$L = \begin{bmatrix} 4\pi - 2\pi\lambda h & 2\pi - 2\pi\lambda r \\ 2\pi - 2\pi\lambda r & 0 \end{bmatrix}$$

$$determinant(L) = -(2\pi - 2\pi\lambda r)^2 < 0$$

As the determinant is negative, the test is inconclusive in this case too.

Example 5.9 This is a problem from fluid mechanics. Flow in a pipe network is being optimized using Lagrange multipliers. We have a circular duct, whose diameter varies as 3 levels d_1, d_2, and d_3. Air is flowing in.

Determine the diameters d_1, d_2, and d_3 of the circular duct shown below such that the static pressure drop between the inlet and the outlet is a minimum. The total quantity of sheet metal available is 120m^2 and the Darcy friction factors for pipes 1, 2, and 3 are to be calculated from the relation $f = 0.184 Re_D^{-0.2}$. The density of air is constant at $1.18\,kg/m^3$. Use the method of Lagrange multipliers. The kinematic viscosity of air may be assumed to be $15 \times 10^{-6} m^2/s$.

Solution

Formulation of the optimization problem

$$Minimize \ \Delta P = \Delta P_1 + \Delta P_2 + \Delta P_3 \tag{5.149}$$
$$subject \ to \ \pi d_1 L_1 + \pi d_2 L_2 + \pi d_3 L_3 = 120 \tag{5.150}$$

$$\Delta P_1 = \frac{f_1 L_1 v_1^2}{2g D_1} \rho g \tag{5.151}$$

$$\Delta P_2 = \frac{f_2 L_2 v_2^2}{2g D_2} \rho g \tag{5.152}$$

$$\Delta P_3 = \frac{f_3 L_3 v_3^2}{2g D_3} \rho g \tag{5.153}$$

$$m_1 = 4 = \rho A_1 v_1 = \frac{\rho \pi d_1^2}{4} v_1 \tag{5.154}$$

$$m_2 = 1.5 = \rho A_2 v_2 = \frac{\rho \pi d_2^2}{4} v_2 \tag{5.155}$$

$$m_3 = 1.5 = \rho A_3 v_3 = \frac{\rho \pi d_3^2}{4} v_3 \tag{5.156}$$

$$d_1^2 v_1 = 4.32 \tag{5.157}$$
$$d_2^2 v_2 = 2.7 \tag{5.158}$$
$$d_3^2 v_3 = 1.08 \tag{5.159}$$

Now we are able to see that there is considerable effort in formulating the optimization problem. This is what is normally encountered in engineering.

$$\Delta P_1 = \frac{f_1 L_1 v_1^2}{2g D_1} \rho g \tag{5.160}$$

$$\Delta P_1 = \frac{0.184(\frac{4m}{\pi \mu d_1})^{-0.2} 4(\frac{4.32}{d_1})^2 \rho}{2d_1} \tag{5.161}$$

$$\Delta P_1 = 0.971 \ d_1^{-4.8} \tag{5.162}$$

$$\Delta P_2 = 0.564 \ d_2^{-4.8} \tag{5.163}$$

$$\Delta P_3 = 0.135 \ d_3^{-4.8} \tag{5.164}$$

$$Min \ y = \Delta P_1 + \Delta P_2 + \Delta P_3 \tag{5.165}$$

Formulation of the optimization problem in d_1, d_2, and d_3

$$y = 0.971 \ d_1^{-4.8} + 0.564 \ d_2^{-4.8} + 0.135 \ d_3^{-4.8} \tag{5.166}$$

$$\text{subject to: } 6d_1 + 8d_2 + 10d_3 = (120/\pi) \tag{5.167}$$

$$3d_1 + 4d_2 + 5d_3 = (60/\pi) = 19.1 \tag{5.168}$$

$$\phi = 3d_1 + 4d_2 + 5d_3 - 19.1 = 0 \tag{5.169}$$

So we have not yet solved the problem, but just formulated it!! Now we have to use the Lagrange multiplier method.

$$\frac{\partial y}{\partial d_1} - \lambda \frac{\partial \phi}{\partial d_1} = 0 \tag{5.170}$$

$$\frac{\partial y}{\partial d_2} - \lambda \frac{\partial \phi}{\partial d_2} = 0 \tag{5.171}$$

$$\frac{\partial y}{\partial d_3} - \lambda \frac{\partial \phi}{\partial d_3} = 0 \tag{5.172}$$

$$\phi = 0 \tag{5.173}$$

There are 4 unknowns here. The resulting Lagrange equations are very simple and solvable. λ is unrestricted in sign. The only condition is ∇y and $\nabla \phi$ must be collinear. If we get d_1, d_2 and d_3 in terms of λ and substitute in the equation $\phi = 0$, we are done.

$$0.971(-4.8)d_1^{-5.8} - 3\lambda = 0 \tag{5.174}$$

$$0.564(-4.8)d_2^{-5.8} - 4\lambda = 0 \tag{5.175}$$

$$0.135(-4.8)d_3^{-5.8} - 5\lambda = 0 \tag{5.176}$$

$$d_1 = -1.07(\lambda)^{-0.172} \tag{5.177}$$

$$d_2 = -0.93(\lambda)^{-0.172} \tag{5.178}$$

$$d_3 = -0.68(\lambda)^{-0.172} \tag{5.179}$$

$$\phi = 0 \tag{5.180}$$

$$\phi = 3d_1 + 4d_2 + 5d_3 - 19.1 = 0 \tag{5.181}$$

$$(3)(1.07)(\lambda)^{-0.172} + (4)(0.93)(\lambda)^{-0.172}$$
$$+(0.68)(5)(\lambda)^{-0.172} = -19.1 \tag{5.182}$$

$$\lambda = -0.03097 \tag{5.183}$$

substituting for λ in the expressions of d_1, d_2 and d_3.

$$d_1^+ = 1.9641\text{m} \tag{5.184}$$

$$d_2^+ = 1.702\text{m} \tag{5.185}$$

$$d_3^+ = 1.2799\text{m} \tag{5.186}$$

$$y^+ = 0.1232 \tag{5.187}$$

For a constrained optimization problem, the Hessian must be evaluated on the Lagrangian L given by $L = Y - \lambda\phi$, as already discussed. Evaluating the Hessian matrix to check if the optimum is minimum.

$$L = \begin{bmatrix} 27.032d_1^{-6.8} & 0 & 0 \\ 0 & 15.701d_2^{-6.8} & 0 \\ 0 & 0 & 3.7583d_3^{-6.8} \end{bmatrix} \qquad (5.188)$$

$$L = \begin{bmatrix} 0.2743 & 0 & 0 \\ 0 & 0.4221 & 0 \\ 0 & 0 & 0.7017 \end{bmatrix} \qquad (5.189)$$

$$determinant\,(L) = 0.08127 > 0 \qquad (5.190)$$

L is positive definite and so the resulting solution is a minimum. In hindsight, it is to be noted that in Fig. 5.6, $d_3 > d_2 > d_1$. Eventually our solution was otherwise!

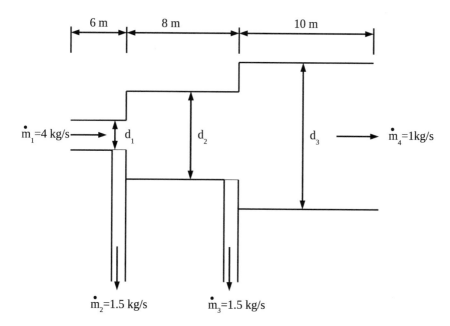

Fig. 5.6 Figure for Example 5.9

MATLAB code for Example 5.9

```
 1   clear;
 2   clc;
 3
 4   syms x1 x2 x3 x4
 5
 6   % Objective function
 7   y=0.971*(x1^(-4.8))+0.564*(x2^(-4.8))+0.135*(x3^(-4.8));
 8
 9   % Constraint
10   f=3*x1+4*x2+5*x3-19.1;
11
12   % First derivative of objective function with respect to x1
13   dydx1=diff(y,x1);
14
15   % Second derivative of objective function with respect to x1
16   dydx12=diff(y,x1,2);
17
18   % First derivative of objective function with respect to x2
19   dydx2=diff(y,x2);
20
21   % Second derivative of objective function with respect to x2
22   dydx22=diff(y,x2,2);
23
24   % First derivative of objective function with respect to x3
25   dydx3=diff(y,x3);
26
27   % Second derivative of objective function with respect to x3
28   dydx32=diff(y,x3,2);
29
30   % derivative of constraint with respect to x1
31   dfdx1=diff(f,x1);
32
33   % derivative of constraint with respect to x2
34   dfdx2=diff(f,x2);
35
36   % derivative of constraint with respect to x3
37   dfdx3=diff(f,x3);
38
39   eqns (x1, x2, x3, x4) = [dydx1-(x4*(dfdx1)),...
40       dydx2-(x4*(dfdx2)),dydx3-(x4*(dfdx3)),3*x1+4*x2+5*x3-19.1];
41
42   % Initial guess
43   x0 = [10, 10, 10, 10];
44
45   % Coverting symbols in symbolic equation to x(1), x(2)....
46   F = matlabFunction(eqns,'vars',{[x1,x2,x3,x4]});
47
48   % Will display the optimality tolerance values
49   options = optimoptions('fsolve','Display','iter');
```

```
50
51   % Function fsolve used to solve 4 nonlinear equations
52   [x,fval] = fsolve(F,x0,options);
53
54   % Optimised value of objective function
55   y_value = 0.971*(x(1)^(-4.8))+0.564*(x(2)^(-4.8))+0.135*(x(3)^(-4.8));
56
57   % Hessian matrix
58   H = [double(subs(dydx12,{x1,x2,x3},{x(1),x(2),x(3)})) 0 0;
59        0 double(subs(dydx22,{x1,x2,x3},{x(1),x(2),x(3)})) 0;
60        0 0 double(subs(dydx32,{x1,x2,x3},{x(1),x(2),x(3)}))];
61
62   detH = det(H);
63
64   % Printing optimised values
65   prt = ['y = ',num2str(y_value),...
66        ', d1 = ',num2str(x(1)),...
67        ', d2 = ',num2str(x(2)),...
68        ', d3 = ',num2str(x(3)),...
69        ', lambda = ',num2str(x(4)),...
70        ', Determinant of Hessian = ',num2str(detH)];
71   disp(prt)
```

The output of the program is:

```
y = 0.12323, lambda = –0.030969,
d1 = 1.9641, d2 = 1.702, d3 = 1.2799,
Determinant of Hessian = 0.081265
```

5.7 Handling Inequality Constraints

When it comes to inequality constraints, the simplest approach is to ignore the inequality constraint and solve the problem with only the equality constraints considered. We examine if the resulting solution violates any inequality constraint or not. If it does not violate, the constraint that was left out is no longer an active or binding constraint. If it is violated, on the other hand, the inequality constraint can be converted into an equality constraint and can be considered as an additional constraint in the Lagrange multiplier formulation. The resulting system can be solved and any way that constraint will be obeyed eventually.

Had we known this information in advance, (i.e., which constraints are active and binding), there is no need for us to go through this exercise. In advance, we would have made some inequality constraints into equality constraints and then completed the solution. But there is no way upfront or a priori to know whether a constraint is active or binding. This problem was thought about by Kuhn and Tucker and they finally came out with the Kuhn and Tucker Conditions (KTCs).[1]

Let us consider a nonlinear optimization problem involving n variables, m equality, and r inequality constraints.

$$\text{Minimize } y \quad = y(x_1, x_2.....x_n) \tag{5.191}$$
$$\text{subject to } \phi_1 = \phi_1(x_1, x_2.....x_n) = 0 \tag{5.192}$$
$$\phi_2 = \phi_1(x_1, x_2.....x_n) = 0 \tag{5.193}$$
$$\phi_m = \phi_m(x_1, x_2.....x_n) = 0 \tag{5.194}$$
$$\text{and } \psi_1 = \psi_1(x_1, x_2.....x_n) \geq 0 \tag{5.195}$$
$$\psi_2 = \psi_2(x_1, x_2.....x_n) \geq 0 \tag{5.196}$$
$$\psi_r = \psi_r(x_1, x_2.....x_n) \geq 0 \tag{5.197}$$

Now we write the KTCs applicable to this problem. These are given by

$$\nabla y - \sum_{i=1}^{m} \lambda_i \nabla \phi_i - \sum_{r=1}^{k} u_r \nabla \psi_r = 0 \tag{5.198}$$
$$\phi_i = 0 \text{ for all i} \leq m \tag{5.199}$$
$$\psi_r \geq 0 \text{ for all r} \tag{5.200}$$

Please note that the first two terms of Eq. 5.198 are the same as the Lagrange multipliers method. We have handled the objective function y and all the equality constraints where $\phi = 0$, applicable for $i = 1$ to m.

The last term accounts for the inequality constraints ψ, while u is similar to the Lagrange multiplier. It is a sensitivity coefficient, whose nature is not yet known to us.

Now for the new conditions.

$$u_r \psi_r = 0 \text{ for all r} \tag{5.201}$$
$$u_r \geq 0 \text{ for all r} \tag{5.202}$$

For a maximization problem, $-y$ has to be minimized. Equations 5.201 and 5.202 are important conditions introduced by Kuhn and Tucker. Equation 5.201 tells that

---

[1] A little history: Professor Tucker is no longer alive. He was the Ph.D. advisor of Prof. John Nash, who won the Nobel prize in 1994 for mathematics and also the subject of the movie "A beautiful mind ". Prof. Kuhn is a contemporary of Prof. Nash and continues to work at Princeton University and was the mathematics consultant for the movie. He was largely responsible for nominating John Nash for the Nobel prize and for getting the movie made. Prof. Tucker and Prof. Nash have done a lot of pioneering work in operations research, game theory and are particularly known for their work on the problem called Prisoners dilemma. They also developed the Hungarian method for the assignment problem and the traveling salesman problem in operations research.

either $\psi = 0$ or $u = 0$ or both are 0. The condition $u = 0$ means that the constraint is no longer active or binding and hence no longer affects the solution. We are worried about those cases where $u \neq 0$. If $\psi = 0$, it is a problem as it is an equality constraint and is a strict condition. So there is a complementarity involved in this. Either u or ψ is 0. Therefore, Eq. 5.201 is called the complementary slackness condition. Since we do not know a priori if the constraint is binding or not, only after one iteration, we come to know this. The u is very similar to λ except that it is restricted in sign while λ is unrestricted in sign. u has to be necessarily positive. More on this in Sect. 5.8.

Now let us consider a three-variable, one-equality constraint, one-inequality constraint optimization problem. Here, if ψ is active, Eq. 5.201 becomes an equality and u may take on a positive value. This is basically KTC. Reiterating this further, let us say we want to use the Lagrange multiplier method for a problem with m equality constraints and one inequality constraint that is nonlinear but differentiable. It is possible for us to completely ignore the inequality constraints, proceed with the solution and see if the solution violates the inequality constraint or not. If it is violated, we have to reinsert the inequality as an equality and proceed wherein, this time it will be obeyed.

Alternatively, we can use the KTC conditions, assume that the particular inequality constraint is active, solve the resulting set of equations. Apart from λ, we will also get u. We look at the nature of u. If u is negative, our original assumption that ψ is active is wrong. Therefore, it can be omitted from the calculations from the next iteration.

Example 5.10 Establish KTCs for the problem given below.

$$Min\ y = (x_1 - 8)^2 + (x_2 - 6)^2 \tag{5.203}$$

Subject to

$$\psi = x_1 + x_2 > 9 \tag{5.204}$$

Solution

The above problem is similar to Example 5.1. However, we now have given an additional constraint that $x_1 + x_2 > 9$. This is an inequality constraint and hence we want to use the KTC. The first step is to assume it as an active constraint and see whether u is positive or negative.

$$\frac{\partial y}{\partial x_1} - u\frac{\partial \psi}{\partial x_1} = 0 \tag{5.205}$$

$$\frac{\partial y}{\partial x_2} - u\frac{\partial \psi}{\partial x_2} = 0 \tag{5.206}$$

$$x_1 + x_2 = 9 \tag{5.207}$$

$$\psi = 0 \tag{5.208}$$

$$u \geq 0 \tag{5.209}$$

These are the five KTCs. Now, on solving we have

$$2(x_1 - 8) - u = 0 \tag{5.210}$$
$$2(x_2 - 6) - u = 0 \tag{5.211}$$
$$x_1 = 5.5, x_2 = 3.5 \tag{5.212}$$

From this, we get u = −5 and y^+ = 12.5. Since u is negative, therefore the original assumption that ψ is an active constraint is incorrect and ψ does not affect the solution to this problem. The same can be seen when depicted graphically in Fig. 5.7. We have a solution $y = 0$ corresponding to $x_1 = 8$ and $x_2 = 6$, which does not violate $x_1 + x_2 > 9$. The KTCs helped us to identify that having $x_1 + x_2 > 9$ is not an active constraint.

Now we rework the problem for $x_1 + x_2 > 18$.

$$Min \ y = (x_1 - 8)^2 + (x_2 - 6)^2 \tag{5.213}$$
$$subject \ to \ \psi = x_1 + x_2 > 18 \tag{5.214}$$

$$\frac{\partial y}{\partial x_1} - u\frac{\partial \psi}{\partial x_1} = 0 \tag{5.215}$$
$$\frac{\partial y}{\partial x_2} - u\frac{\partial \psi}{\partial x_2} = 0 \tag{5.216}$$
$$x_1 + x_2 = 18 \tag{5.217}$$
$$\psi = 0 \ , u \geq 0 \tag{5.218}$$

Fig. 5.7 Depiction of the solution to Example 5.10

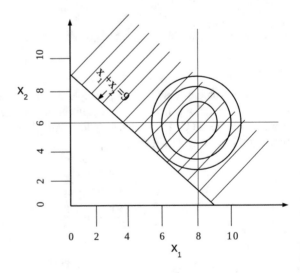

$$2(x_1 - 8) - u = 0 \tag{5.219}$$
$$2(x_2 - 6) - u = 0 \tag{5.220}$$
$$x_1 = 10, \ x_2 = 8 \tag{5.221}$$
$$\psi = 0, \ u = +4 \tag{5.222}$$

Since u has now become positive, ψ is a binding or active constraint. ψ has got to be treated as an equality constraint. $y^+ = 8$.

If we disregard the constraint, the solution is $x_1^+ = 8$ and $x_2^+ = 6$. If we substitute 8 and 6 in the constraint $x_1 + x_2 > 18$, we see that the constraint is violated and ignoring the constraint will lead to an erroneous solution!.

MATLAB code for Example 5.10

```
1   clear;
2   clc;
3
4
5   syms x1 x2 x3
6
7   % Objective function
8   y = ((x1–8)^2)+((x2–6)^2);
9
10  % Constraint Ï^ = x1 + x2 > 9
11  f = x1+x2–9;
12
13  % First derivative of objective function with respect to x1
14  dydx1=diff(y,x1);
15
16  % First derivative of constraint with respect to x1
17  dfdx1=diff(f,x1);
18
19  % First derivative of objective function with respect to x2
20  dydx2=diff(y,x2);
21
22  % First derivative of constraint with respect to x2
23  dfdx2=diff(f,x2);
24
25  eqns(x1,x2,x3)=[dydx1–(x3*(dfdx1)),dydx2–(x3*(dfdx2)),f];
26
27  % Initial guess
28  x0 = [10, 10, 10];
29
30  %Coverting symbols in symbolic equation to x(1),x(2)..
31  F = matlabFunction(eqns,'vars',{[x1,x2,x3]});
32
33  options = optimoptions('fsolve','Display','iter');
34
```

```
35   % Function fsolve used to solve 3 equations
36   [x,fval] = fsolve(F,x0,options);
37
38
39   % If u value is negative original assumption of
40   % Ϊˆ is an active constraint is incorrect
41   if (x(3) < 0)
42
43       prt1 = ['u = ',num2str(x(3))];
44       disp(prt1);
45
46       fprintf('Since u is negative, constraint is not binding\n')
47
48       eqns_new (x1, x2) = [dydx1, dydx2];
49
50       % Initial guess
51       x0_new = [10, 10];
52
53       F_new = matlabFunction(eqns_new,'vars',{[x1,x2]});
54
55       options_new = optimoptions('fsolve','Display','iter');
56
57       [x,fval] = fsolve(F_new,x0_new,options_new);
58
59       % Optimised value of objective function
60       y_value = ((round(x(1))−8)^2)+((round(x(2))−6)^2);
61
62       % Print
63       prt = [' x1 = ',num2str(x(1)),...
64              ', x2 = ',num2str(x(2)),...
65              ', y = ',num2str(y_value)];
66       disp(prt)
67
68   else
69
70       prt1 = ['u = ',num2str(x(3))];
71       disp(prt1);
72
73       fprintf('Since u is positive, constraint is binding\n');
74
75       % Optimised value of objective function
76       y_value = ((round(x(1))−8)^2)+((round(x(2))−6)^2);
77
78       % Print
79       prt = ['x1 = ',num2str(x(1)),...
80              ', x2 = ',num2str(x(2)),...
81              ', y = ',num2str(y_value)];
82       disp(prt)
83
84   end
```

The output of the program is

For the case of the constraint $x_1 + x_2 > 9$,
u = -5. Since u is negative, constraint is not binding.
x1 = 8, x2 = 6, y = 0

For the case of the constraint $x_1 + x_2 > 18$,
u = 4. Since u is positive, constraint is binding.
x1 = 10, x2 = 8, y = 8

5.8 Why Should U Be Positive?

Let us revisit Example 5.10

$$Minimize: \ y = (x_1 - 8)^2 + (x_2 - 6)^2 \tag{5.223}$$
$$subject \ to \ x_1 + x_2 - 9 > 0 \tag{5.224}$$

Let us look at why u should be positive in conjunction with this example.

$$u \approx \frac{\Delta y}{\Delta \psi} \tag{5.225}$$
$$\psi = x_1 + x_2 - 9, \ \psi > 0 \tag{5.226}$$

The solution to this problem was determined as (8, 6).

Suppose we change the condition as $\psi > 1$ or $\psi = x_1 + x_2 - 10$, the solution, in this case, is also (8, 6) and this line can be plotted as shown in Fig. 5.8.

When $\Delta \psi$ was positive or ψ was increased from 0 to 1, y did not change, Δy was 0. Therefore u = 0 and hence that constraint is inactive.

We should not take the increment as 20 or 200 and determine the change in y when ψ changes by 20 or 200. $\Delta \psi$ means a small change in ψ. For incremental changes in ψ, there is no change in the optimal solution and hence the constraint is not binding.

What is happening is here that compared to $\psi > 0$, when a constraint has changed to $\psi > 1$, we lose a certain portion of the feasible region and certain points too. Therefore any solution that satisfies $\psi > 0$ will anyway satisfy the condition $\psi > 1$. Therefore we cannot hope to find a new optimum having a smaller value of y than what would be found when $\psi > 0$. In this case, it so happened that for $\psi > 1$, we are still having the same optimum.

Now if we change it to $x_1 + x_2 - 18$, the situation totally changes. Since the feasible region is reducing, it is but logical that we can have the y minimum in the new feasible region but we cannot have a new y that is lower than what could

Fig. 5.8 Rationale behind
why u has to be positive

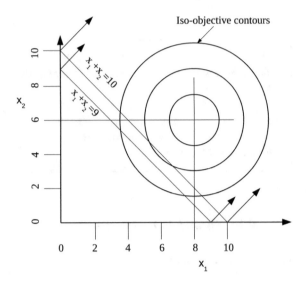

be captured by a less restrictive feasible region. When $\psi = x_1 + x_2 - 18 > 0$ is
considered, the graphical representation of this will look like what is given in Fig. 5.9
where the iso-objective contours are seen and the shaded portion is the feasible region.

The original solution to the problem (8, 6) is now seen to lie below the feasible
region. The constraint has cut the iso-objective curve in such a way that the original
solution (8, 6) is lost. (8, 6) is no longer a valid solution for the problem because
it violates the inequality constraint $x_1 + x_2 - 18 > 0$. The new solution was earlier
worked out to be $x_1^+ = 10$, $x_2^+ = 8$ and $y^+ = 8$. Now let us say that $\psi > 1$ or $x_1 +
x_2 - 19 \geq 0$. Now we are chopping off a portion of the feasible region. Reworking

Fig. 5.9 Reducing feasible
region with ψ moving away
from the origin

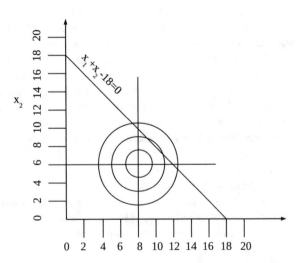

the solution for this, we get the optimum as (10.5, 8.5). When $\psi > 0$, $y_{old} = 8$;
When $\psi > 1$, $y_{new} = 12.5$

$$u \approx \frac{\Delta y}{\Delta \psi} = \frac{y_{new} - y_{old}}{1} \qquad (5.227)$$

$$\therefore u = (12.5 - 8)/1 = 4.5 \qquad (5.228)$$

When ψ was an inactive or a nonbinding constraint, the value of u was 0, while when
it is an active constraint, u is positive. There is no other possibility for u. **Therefore,
u(you!) should be positive, always!**

Problems

5.1 Determine the shortest distance from the point (5, 1) to the ellipse $\frac{x^2}{9} + \frac{y^2}{4} = 1$
by employing the method of Lagrange multipliers. Establish that the solution is
a minimum.

5.2 Consider the following minimization problem.

$$\text{Minimize } y = x_1^2 + x_2^2 + x_3^2$$
$$\text{Subject to: } 3x_1 + 2x_2 + x_3 = 10$$
$$x_1 + 2x_2 + 2x_3 = 6$$

Solve this using the Lagrange multipliers method. Obtain the values of the two
Lagrange multipliers. Confirm (mathematically) that the solution obtained is
indeed a minimum.

5.3 Rubber O-rings as shown in Fig. 5.10 are to be made for sealing a giant pressure
vessel used for an industrial application. The total quantity of molten rubber
available is $10\,m^3$ from which two O-rings need to be molded. The mean diameter
for the rings is 1 and 2 m, respectively. Find the optimum values of d_1 and d_2 for
maximum total surface area.

5.4 A circular orifice in a tank has a radius $r_1 = 0.25\,m$. Flow from the orifice is
to be accelerated by means of a convergent nozzle as shown in Fig. 5.11. The
nozzle, shaped like the frustum of a cone is to be made out of a stainless steel
sheet.

Fig. 5.10 Figure for
Problem 5.3

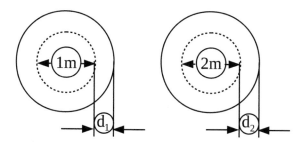

Fig. 5.11 Figure for
Problem 5.4

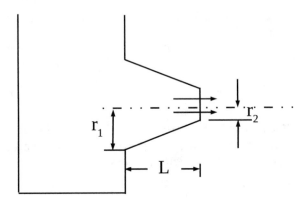

A total of $0.5\,\text{m}^2$ sheet is available. Find the maximum volume of the nozzle that
is achievable.

5.5 There are two electrical generators G_1 and G_2, whose power outputs are p_1 and
p_2 MW, respectively. The generators are connected to a load line such that

$$p_1 + p_2 = 800\text{MW}$$

The cost of producing power from the generator is given by following equations.

$$C_1 = a_1 p_1^2 + b_1 p_1 + c_1$$
$$C_2 = a_2 p_2^2 + b_2 p_2 + c_2$$

Determine the optimum value of p_1 and p_2 at which the total cost is minimum.

5.6 If in the previous problem, the cost functions are given by

$$C_1 = 300 + 7.3p_1 + 0.0034p_1^2$$
$$C_2 = 200 + 8p_2 + 0.0019p_2^2$$

determine p_1 and p_2 using the solution obtained to the previous problem. Determine the value of λ and comment on its significance.

5.7 A shell and a tube heat exchanger (shown in Fig. 5.12) is to be designed for the
minimum total cost. The shell diameter D, the length of the tubes L (which is
also the length of the shell, approximately), and the number of tubes "n" have
to be designed to minimize the total cost. The tubes are all 1 inch ($d = 0.025\text{m}$)
in diameter and have a single pass. The cost of the shell (in lakhs of rupees) is
given by $50D^{1.5}L^{1.25}$. The cost of each tube is $0.4L^{0.5}$, again in lakhs.
A few constraints in the problem are as follows:

(a) The total tube surface area needs to be $47\,\text{m}^2$.
(b) The packing density of the tubes (total tube volume/shell volume) shall not
exceed 50% to allow for shell-side fluid movement.
(c) The length of any tube (all tubes are of the same length) shall not exceed
$10\,\text{m}$ for ease of maintenance and replacement (*Note: D and L are in m*)

Fig. 5.12 Figure for the
shell and tube heat
exchanger Problem 5.7

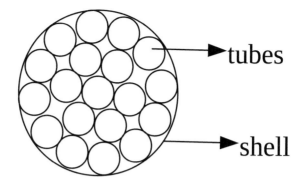

Set up the optimization problem for minimizing the total cost of the exchanger
and solve it as a constrained optimization problem using the Lagrange multiplier
method to determine the optimal solution. You may use constraint (b) as an
equality to reduce the number of variables (n, D, and L) to 2. This way you can
solve this as a two-variable, one-constraint optimization problem by substituting
for one of the variables from either constraint (a) or (b) into the objective function.
Furthermore, you may ignore constraint (c) and check if it is violated after
obtaining the optimum.

5.8 Establish the Kuhn–Tucker Conditions (KTC) for the following minimization
problem and solve it.

$$y = (x_1 - 1)^2 + (x_2 - 1)^2$$
$$\text{subject to } x_1 + x_2 \le 4$$

5.9 (a) Write down the Kuhn–Tucker conditions for the following nonlinear optimization problem (NLP).

$$\text{Minimize: } y = x_1^2 + 2x_2^2$$
$$\text{subject to: } 2x_1 - 2x_2 = 1$$
$$x_1^2 + x_2^2 \le 4$$

(b) Solve the above problem and check whether the inequality constraint is
binding.

Reference

Ravindran, A., And, Ragsdell K. M., & Reklaitis, G. V. (2006). *Engineering Optimization- Methods
and Applications*. New York, USA: Wiley.

Chapter 6
Search Methods

6.1 Introduction

Regardless of whether we want to use a calculus-based method or a search method, once we have an objective function, a set of constraints and bounds for variables, the goal is to invariably solve the optimization problem and obtain the maximum or minimum of the function. However, for some reason, if we do not want to use the Lagrange multiplier either because there are too many variables or the function and/or constraints are not differentiable, we may want to seek an alternative route to solving the optimization problem with increased computational resources. A search method is one key "alternative route" for doing this job.

So how do we go about a search method first? For example, if the problem in hand involves two variables we may first choose a few pairs of x_1 and x_2. We calculate the value of y at each of these points and among these, we look at which way y is increasing or decreasing and then choose the best point among this. We then take a few pairs of (x_1, x_2) around this point and proceed further. So we say that a point is better than the other, based on the value of the objective function alone! Derivatives do not come into the picture at all.

Recall that y becomes stationary even in the Lagrange multiplier method, we first solve to get $x_1...x_n$ at which y becomes stationary. Saying $x_1 = 4$ and $x_2 = 6$ is not the end of the problem. The main story is what is the value of y corresponding to this point. Calculation of the objective function is relegated to the very end in calculus-based methods usually. In fact, we are kind of "obsessed" with where the function becomes stationary in the Lagrange multiplier method!

But in real life, the function often times will not become stationary at all because the function itself may not be differentiable. For example, we want to get the minimum total cost of operating machines in a shop with many machines. We cannot possibly have an equation for us to get partial derivatives. Mathematically, it may be great to have a "function" for every optimization problem, but unfortunately it is not possible in all real-life scenarios!

© The Author(s) 2021
C. Balaji, *Thermal System Design and Optimization*,
https://doi.org/10.1007/978-3-030-59046-8_6

The key point is we look at certain combinations of independent variables, evaluate y at these points, come to some conclusions and then proceed further. This strategy is in stark contrast to the Lagrange multiplier method, where the calculation of y^+ is pushed to the end.

A search method is more physical as with every iteration we are able to see how y is changing. Of course, there are some issues like handling local minima. A search method is more an engineering approach to solving the problem, where mostly we are happy with a "nearly optimal" solutions. But there is a conceptual difficulty with this search method—the true optimum can never be determined. For example, for the solar water heater storage problem, we will be able to say that the optimum lies between r = 0.8 and 0.9 m, if we decide on an accuracy of 0.1 m. If we want an accuracy of 0.01 m, we will say r lies between 0.85 m and 0.86 m. But we are always giving bounds. *Hence, the true optimum can never be reached, only be approached, while employing a search technique.*

In search methods, we specify the final interval of uncertainty. We say that y^+ or y optimum lies between two values of x for a single variable problem. So the search method will essentially be an iterative technique, wherein we originally start with some interval of uncertainty, which may come from our experience or some background knowledge. For the cylindrical storage problem, we do not have to say that we need to start from 1 mm to 300 m. After all it is a storage tank that should be kept outside, and we can say it may be as small as 40–50 cm and cannot exceed say 5 m. We bracket the solution and start using the search method. The original interval of uncertainty is specified before starting the problem and comes from our engineering knowledge. Then we use a search method and hopefully, with every iteration, the interval of uncertainty will go down. Let us now see an example.

Example 6.1 Revisit the solar water heater storage problem (see Example 5.3) of minimizing the heat losses from a cylindrical water heat storage tank. Treating it as a single variable, the unconstrained optimization problem in radius r, with an initial uncertainty of $0.5 \leq r \leq 3.5$ m, solve the problem by searching with a uniform step size of 0.5 m by evaluating A(r) at 7 intermediate points.

Solution:

Minimize $A = 2\pi r^2 + 2\pi r h$; $V = \pi r^2 h = 4$; subject to $\pi r^2 h = 4$

$$A = 2\pi r^2 + 2\pi r (4/\pi r^2) \tag{6.1}$$
$$A = 2\pi r^2 + (8/r) \tag{6.2}$$

We now tabulate A(r) for 7 values of r, as indicated in Table 6.1. The graphical variation is depicted in Fig. 6.1.

Table 6.1 A(r) for various values of r for Example 6.1

S. No	r, m	A(r), m^2
1	0.5	17.57
2	1	14.28
3	1.5	19.47
4	2	29.13
5	2.5	42.47
6	3	59.21
7	3.5	79.25

Fig. 6.1 Variation of A(r) with r for Example 6.1

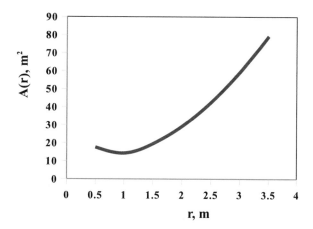

From both Table 6.1 and Fig. 6.1, we are able to see that the solution lies between r = 0.5 and r = 1.5 m. We do not know whether the minimum is reached between 0.5 and 1 or 1 and 1.5. However, we know that it takes a turn between 0.5 and 1.5 m. What did we achieve? We did 7 function evaluations and reduced the interval of uncertainty from (3.5–0.5) m to (1.5–0.5) m. It is not great but not very bad either. What we have done now is called the equal interval exhaustive search method. The original interval of uncertainty = b–a = 3.5–0.5 = 3 m. The original interval of uncertainty can be written as $a \leq r \leq b$.

Number of intermediate points = m (for this problem, m = 5).

Interval spacing = $(b - a)/(m + 1)$ (in this problem: 0.5 m).

The final interval of uncertainty = $2(b - a)/(m + 1)$ (in this problem this is 1 m).

The ratio of the original interval of uncertainty to the final interval of uncertainty RR is known as the **reduction ratio**. The RR in the current example is (3/1) = 3.

The RR tells by how much the original interval of uncertainty has reduced. It is a performance metric. It is like mileage for the car or CGPA for a student's academic performance. The RR has to be read in conjunction with the number of observations.

If m is the number of observations, then the reduction ratio = (m+1)/2. If we perform the equal interval exhaustive search method as shown above. Therefore, it is also reasonable to assume, scientists would have developed more advanced methods, where, for a given m observation, we get a reduction ratio RR far superior to what could be achieved by the exhaustive search method. The exhaustive search method is highly unimaginative to say the least. But if we do not know anything about the nature of the function, for starters we can use this. We can use the exhaustive search method to bracket between the interval between $r = 0.5$ m and $r = 1.5$ m as we have done now. After we get here, we can switch over to very sophisticated methods to quickly reach the optimum value.

What is the relationship between RR and the number of observations or the number of functional evaluations? We say observations, because they could also be experimental. Here, the objective function, A is just a formula. But in reality, A could be the output of a CFD software program or some experiments or could be the data from elsewhere. Therefore we are interested in the number of observations. If n is the number of observations (n = 7 here): $n = m + 2$ (where m is the number of intermediate points).

6.1.1   A Smarter Way of Solving Example 6.1

Can we solve the same problem in a little smarter fashion without going for a sophisticated algorithm? We can take 3 points at a time starting from the left. We have $r = 0.5, 1, 1.5, 2, 2.5, 3$, and 3.5. Let us take the first 3 points 0.5, 1, and 1.5. If $f(x_1) > f(x_2)$ and $f(x_2) < f(x_3)$, then we are home. But it is fortuitous that at one side of the interval we got the solution for this problem. That is the solution lies between x_1 and x_3.

If the above condition is not satisfied, we make $x_1 = x_2$, $x_2 = x_3$, and $x_3 = x_4$. We keep doing this till we get to the right end of the interval. If we reach the other end point and still do not get an optimum, either the function does not have an optimum or the optimum lies at one of the two boundaries.

On an average, if the solution is likely to be around the middle, it may take only half the number of observations. The RR of $(m + 1)/2$, which we saw a little while ago, is the worst-case scenario for the exhaustive search.

Algorithm for the alternative approach for Example 6.1 with 3 points taken at a time will be as follows:

1 $\Delta r = \frac{(b-a)}{(m+1)}$ where m is the number of intermediate points
2 $r_2 = r_1 + \Delta r$, $r_3 = r_2 + \Delta r$
3 If $A(r_1) \geq A(r_2) \leq A(r_3)$, optimum lies in the range $r_1 \leq r \leq r_3$ - Stop
4 Else $r_1 = r_2$; $r_2 = r_3$; $r_3 = r_2 + \Delta r$, proceed to step 3
5 If $r_3 = b$, and if stopping criterion is not satisfied, optimum does not lie between (a,b) or may lie at the boundary.

6.2 Monotonic and Unimodal Functions

First, we need to look at the nature of the objective function. There are some important definitions like a monotonic function, a unimodal function, the concept of global minimum and local minimum that need to be fleshed out. These are best understood in relation to a single variable problem . Once we understand them with respect to a single variable problem, we can extrapolate these definitions for a multivariable problems without much difficulty.

6.2.1 Monotonic Function

For any x_1 and x_2, with $x_1 \leq x_2$, a function $f(x)$ is said to be

- monotonically increasing, if $f(x_1) \leq f(x_2)$
- monotonically decreasing if $f(x_1) \geq f(x_2)$

Even in the case of (Fig. 6.2) where the function is discontinuous, it is still monotonic.

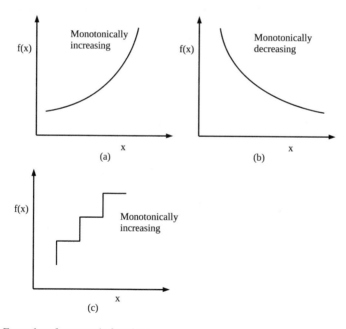

Fig. 6.2 Examples of monotonic functions

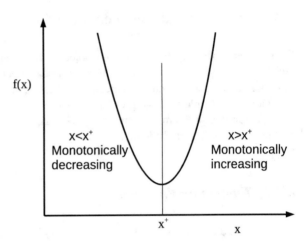

Fig. 6.3 Example of a unimodal function f(x)

6.2.2   Unimodal Function

A function $f(x)$ is said to be unimodal in $a \leq x \leq b$, if and only if it is monotonic on either side of x^+, where x^+ is the optimal point. Consider a parabolic function f(x) depicted graphically in Fig. 6.3.

For this function, for all $x > x^+$, f(x) is monotonically increasing. For all $x < x^+$, f(x) is monotonically decreasing.

Example 6.2 Examine if $y = |x|$ is unimodal.

Solution:

The function is shown in Fig. 6.4a, and its derivative is shown in Fig. 6.4b. This function ($y = |x|$) exhibits no discontinuity. However, dy/dx is discontinuous at x = 0 and so y = |x| is not differentiable at x = 0. Yet, y = |x| is a unimodal function.

Suppose a function is multi-modal as given in Fig. 6.5, we have to divide it into intervals and seek the unimodal optimum in a particular interval.

6.3   Concept of Global and Local Minimum

6.3.1   Global and Local Minimum

A function $f(x)$ on a domain R is said to attain its **global minimum** at a point $x^{++} \in R$ if and only if $f(x^{++}) \leq f(x)$ for all $x \in R$. There is no point x at which

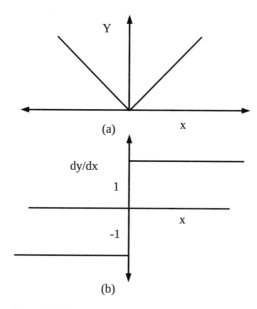

Fig. 6.4 Plot of **a** y=|x|, and **b** d(|x|)/dx

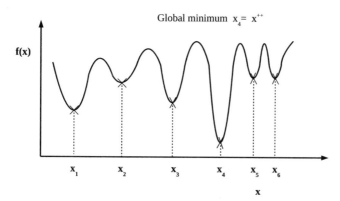

Fig. 6.5 Graph indicating the local and global minimum

y can take a value that is lower than the value taken by y at the point x^{++}. A typical multi-modal function with the global optimum at x_4 is given Fig. 6.5.

A function $f(x)$ on a domain R is said to attain a **local minimum** at a point $x^+ \in R$ if and only if $f(x^+) \le f(x)$ for all x which lie within a reasonable distance from x^+.

If we locate a local optimum and perturb the independent variables around that, we will not get a value of "f" that is better than this. However, it is not global because if we go far away from this x^+, there could be other values of x where $f(x)$ will be significantly lower than this. This is the basic difference between a global minimum and a local minimum.

Among the 6 points marked in Fig. 6.5, 5 ($x_1 \ldots x_6$, sans x_4) are local minima. x_4 is the global minimum here. There are 3 important points one needs to remember.

1. For a unimodal function, the local minimum and the global minimum coincide.
2. For multi-modal objective functions, several local minima exist. We need to evaluate y at all these local optima and select the lowest value of y, which is the global minima. Hence, it is much harder to solve a multi-modal function.
3. The definition for local and global minimum can be modified for a maximization problem.

Algorithms have been specifically developed to handle multi-modal functions that can finally give us the global minima. The problem with Lagrange multipliers or any calculus-based technique is that if we start from the left of the function that is shown in Fig. 6.5, the first minimum is determined as the solution. Even if we go on either side of this, the algorithm will say that this is indeed the solution.

6.3.2 Concept of Elimination and Hill Climbing

In Example 6.1, we employed an exhaustive, equal interval search to remove a portion of the original interval of uncertainty. After bracketing the solution we can again subdivide the new interval of uncertainty and keep removing portions of the interval. Such a method is frequently referred to as a **"region elimination method"** or simply as an **"elimination method"**.

There is another possibility to accomplish a search. For example, consider a 2 variable minimization problem. We draw the contours of the objective function as depicted in Fig. 6.6.

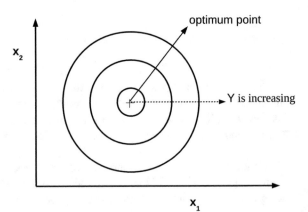

Fig. 6.6 Typical iso-objective contours for a 2 variable problem

Fig. 6.7 Concept of Lattice method

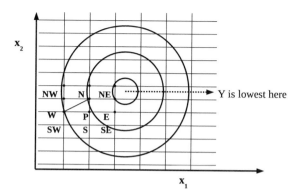

Now, it is possible for us to come out with a rectangular uniform grid like what is shown in Fig. 6.7, where $\Delta x_1 = \Delta x_2$. We start from a point, say P.

For a two-dimensional problem, where $y = f(x_1, x_2)$, we evaluate y at P and at its 8 neighboring nodes. So in total, we do 9 function evaluations. Depending on where y is lowest, we can assign a new P. So if this is NE, say, then the NE point becomes new P and we take 8 points around it and proceed further.

Conceptually, it is quite a different story compared to the elimination of the interval by using the exhaustive search method. If the optimum can be considered as a summit or a hilltop by adopting the strategy discussed above, we are systematically climbing toward it. So this is called a **hill climbing technique**. We are not cutting away portions of the original interval, We just start with one guess value and keep moving around the guess value till the minimum of y is reached. What we saw above is known as the **Lattice method**. If we do not have time and money to invest in optimization and want a quick solution, instead of an exhaustive search, one can go for this. Without using calculus, we just keep getting y. If getting y is difficult, we just set up a neural network with 10 values of x_1, 10 values of x_2 and corresponding 100 values of y such that when the network is supplied with any value of (x_1, x_2), it will automatically give us the values of y. This may be called a neural-network-based lattice search method for solving a two variable optimization problem. In sum, conceptually, we have 2 approaches: regional elimination methods, and the hill climbing methods. We can now come out with a broad framework for solving optimization problems using search methods (refer to Fig. 6.8). We can superpose the nature of an optimization problem (single or multivariable, constrained or unconstrained) with the type of search method and several possibilities emerge, some of which are shown in Fig. 6.8.

However, it is instructive to mention that some of these techniques will use calculus. We may use calculus to find out the best direction to proceed. Yet, there is a difference in using the search method for finding the optimum solution. We start with a guess point. What is the best movement from position 1 to position 2? For that can we take the help of calculus. We try to answer this question. However, we are not taking the help of calculus to make the whole function stationary like we did in the Lagrange multiplier method.

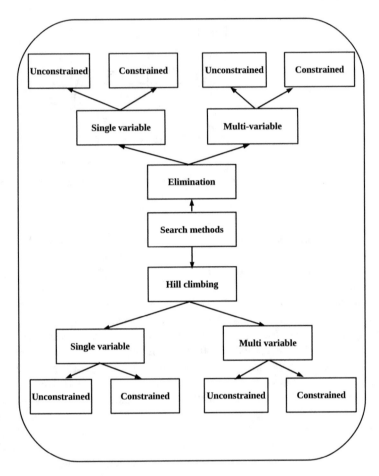

Fig. 6.8 Broad overview of search methods

6.4 Elimination Method

6.4.1 The Two Point Test

Consider a unimodal function f(x). We are seeking a minimum in the interval $a \leq x \leq b$. The challenge is to determine the optimum without taking recourse to a calculus technique. The simple two point test involves the evaluation of y at 2 points and looking at the nature of y, whether $y(x_1)$ is greater than or less than or equal to $y(x_2)$.

1. Based on the simple two point test, as seen from Fig. 6.9a, when we are seeking a minimum; if $f(x_1) > f(x_2)$, the region to the left of x_1 can be eliminated. We can only say that since the function is decreasing, the minimum cannot lie to the left of x_1.

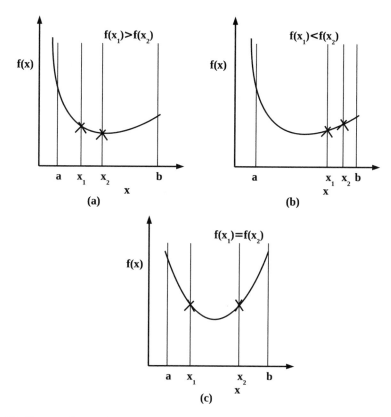

Fig. 6.9 Concept of two point test

2. For the case shown in Fig. 6.9b, $f(x_1) < f(x_2)$, the region to the right of x_2 can be eliminated.
3. For Case 3 shown in Fig. 6.9c, $f(x_1) = f(x_2)$, the regions to the left of x_1, and right of x_2 can be eliminated.

Hence for a unimodal function $f(x)$ in the closed interval $a \le x \le b$, for $x_1 < x_2$:

1. If $f(x_1) > f(x_2)$, minimum does not lie in (a, x_1).
2. $f(x_1) < f(x_2)$, minimum does not lie in (x_2, b).
3. $f(x_1) = f(x_2)$, minimum lies in between (x_1, x_2)—but this is very rare in practice. Numerically, it is not very difficult to achieve this except for simple functions. Even so, it can be a mathematical possibility.

We can use these rules to develop algorithms for solving the optimization problem at hand.

6.4.2   The Dichotomous Search Method

Consider a unimodal function f(x), whose extremum we are seeking.

Consider an interval $a \leq x \leq b$, whose length is I_0. Let us take two points on either side of its center at a distance of $\epsilon/2$ from the center, as shown in Fig. 6.10. The point to the left is x_1 and the point on the right is x_2. Now the ϵ can be made very small.

The distance between x_1 and x_2 and the two ends $(x = a, x = b)$ is then $(I_0+\epsilon)/2$.

We can apply the two point test on x_1 and x_2 to eliminate a portion of I_0. For the moment, we forget about the case when $f(x_1) = f(x_2)$. So if we are seeking a maximum and $f(x_1) \geq f(x_2)$, the interval to the right of x_2 can be eliminated. The eliminated portion is $(I_0-\epsilon)/2$. If ϵ is much smaller compared to I_0, then the interval reduction is almost $I_0/2$. Therefore, every 2 function evaluations will reduce the interval by 50% or every function evaluation will reduce the interval by 25%. Since we have a two point test, we cannot have an odd number of evaluations. In just 8 evaluations, the interval of uncertainly reduces to 1/16 of the initial interval of uncertainty. In exhaustive search, in 8 evaluations the interval reduces only to 1/3.5 of the initial value. The dichotomous search thus is indeed an extremely powerful algorithm.

The dichotomous search—the formal algorithm

1. Divide I_0 (original interval of uncertainty) given by $a \leq x \leq b$ into 2 parts.
2. Around the center, choose two points x_1 and x_2 at $\epsilon/2$ distance from the center, where ϵ is very small.

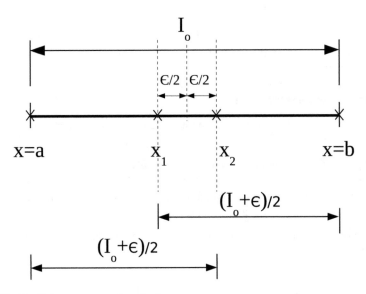

Fig. 6.10 The dichotomous search method

3. Evaluate f(x) at x_1 and x_2.
4. Use the 2 point rule to eliminate (a, x_1) or (x_2, b). Numerically, the third case of $f(x_1) = f(x_2)$ is very very rare, unless the function is so symmetric.
5. Proceed in the same way till desired accuracy is achieved.

In the dichotomous search method, whenever we employ the two point test, the points considered will always be different. They will never be the same across iterations.

After two evaluations:

The new interval of uncertainty = $(I_0 + \epsilon)/2$.

Reduction ratio after 2 evaluations or 1 iteration = $2I_0/(I_0 + \epsilon)$.

If $\epsilon \ll I_0$, (e.g. $I_0 = 1$ m, $\epsilon = 5$ mm), RR = 2^1.

If we go through r iterations, RR = 2^r

What is the relation between the number of iterations and the number of evaluations? If n is the number of evaluations and r is the step number of the iteration, $n = 2r$.

Therefore $RR \approx 2^{n/2}$

Whether the "$\approx$"sign can be replaced by an equality sign depends on how small ϵ is. Since ϵ is finite and nonzero, the actual reduction ratio achieved is always smaller than $2^{n/2}$.

If we compare the RR for the exhaustive search method and the dichotomous search methods for various values of n, we get the following:

n	Dichotomous	Exhaustive
8	16	3.5
16	256	7.5

If n is very small, we do not see much difference between the RR of the two methods. But as n increases, either because we want an increased accuracy or because the function is so complicated that it is not very easy for us to figure out the optimum, it can be seen that the reduction ratio of the dichotomous search method is far superior compared to the single exhaustive search.

Invariably, we want to solve an optimization problem to a desired accuracy, rather than say that we want to do 30 evaluations, 64 evaluations, and so on. So we start off with a search method from the last step.

$$RR = \frac{I_0}{I_n} \tag{6.3}$$

RR is also given by

$$RR = f(n) \tag{6.4}$$

where n is the number of evaluations. From RR we can evaluate n. So, upfront when we use a search technique we know how many function evaluations are required.

Example 6.3 Consider the cylindrical solar water heater storage problem (see Example 5.3). Minimize A. Use dichotomous search and perform 8 evaluations.

Solution:

$A = 2\pi r^2 + 2\pi rh$; $V = \pi r^2 h = 4$ m^3; $0.5 \le r \le 3.5$ m ; We take $\epsilon = 0.02$ m

$$A = 2\pi r^2 + 2\pi rh \tag{6.5}$$

$$\pi r^2 h = 4 \tag{6.6}$$

$$h = \frac{4}{\pi r^2} \tag{6.7}$$

$$A = 2\pi r^2 + 2\pi r \cdot \frac{4}{\pi r^2} \tag{6.8}$$

$$= 2\pi r^2 + \frac{8}{r} \tag{6.9}$$

First iteration:

The first two points have to be close to the center ($r = 2$ m), we choose

$$r_1 = 1.99 \text{ m}, r_2 = 2.01 \text{ m} \tag{6.10}$$

$$A(r_1) = 28.9 \text{ m}^2, A(r_2) = 29.36 \text{ m}^2 \tag{6.11}$$

$$A(r_1) < A(r_2) \tag{6.12}$$

The region to the right of r_2 can be eliminated.

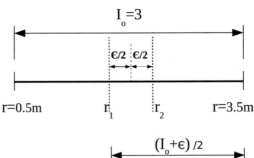

Second iteration:

$$\text{Next two points:} \quad r_3 = 1.25 \text{ m}, r_4 = 1.26 \text{ m} \tag{6.13}$$

$$A(r_3) = 16.2 \text{ m}^2, A(r_4) = 16.32 \text{ m}^2 \tag{6.14}$$

$$A(r_3) < A(r_4) \tag{6.15}$$

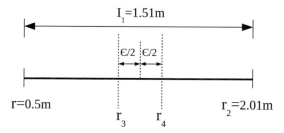

The region to the right of r_4 can be eliminated.

Third iteration:

$$r_5 = 0.87 \text{ m}, r_6 = 0.89 \text{ m} \tag{6.16}$$

$$A(r_5) < A(r_6) \tag{6.17}$$

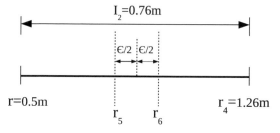

The region to the right of r_6 can be eliminated.

Fourth iteration:

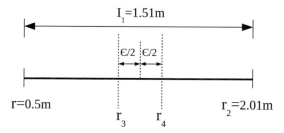

$$r_7 = 0.69 \text{ m}, r_8 = 0.70 \text{ m} \tag{6.18}$$

$$A(r_7) > A(r_8) \tag{6.19}$$

Region to the left of r_7 can be eliminated.

The RR(at the end of four iterations) = $3/0.2 = 15$ [close to $2^{\frac{8}{2}} = 16$].

Therefore, the final interval of uncertainty is $0.69 \leq r \leq 0.89$ m ($I_4 = 0.2$ m).

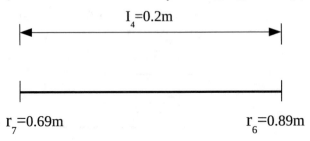

$$I_4 = 0.2\text{m}$$

$$r_7 = 0.69\text{m} \qquad r_6 = 0.89\text{m}$$

The RR is 15 and not 16 because ϵ is not 0. If we make ϵ say like 0.00001, the RR would have been 15.8 or so.

MATLAB code for Example 6.3

```
1   clear;
2   clc;
3
4   a=0.5;                      % lower limit of radius
5   b=3.5;                      % upper limit of radius
6
7   countmax=8;                 % number of evaluations (even number)
8   e=0.02;
9
10
11  for i=1:countmax/2
12
13      I=b-a;
14
15      % 1st point selection according to Dichotomus method
16      r1=b-(I+e)/2;
17
18      % 2nd point selection according to Dichotomus method
19      r2=a+(I+e)/2;
20
21      % surface area according to 1st point
22      A1=6.28*r1^2+8/r1;
23
24      % surface area according to 2nd point
25      A2=6.28*r2^2+8/r2;
26
27      if A1>A2
28          a=r1;
29      else
30          b=r2;
31      end
```

```
32
33     % Print
34        prt = ['Itr = ',num2str(i),...,
35           ', a = ',num2str(a),...
36           ', b = ',num2str(b)];
37        disp(prt)
38   end
```

The output of the program is

```
Itr = 1, a = 0.5, b = 2.01
Itr = 2, a = 0.5, b = 1.265
Itr = 3, a = 0.5, b = 0.8925
Itr = 4, a = 0.68625, b = 0.8925
```

A natural question that arises is why so much of fuss about having a sophisticated technique for a single variable problem, while most optimization problems are multivariable ones. The answer to this question is that each of these problems can be broken down to single variable problems. We can keep all variables except one variable at some constant value in an iteration and then solve the resulting optimization problem using the most powerful single variable search algorithm for that variable. We can then change the variable under consideration and continue to use the best optimization technique available to us. After we are finished with one round of iterations for all the variables, we go to the next round. Hence, it makes eminent sense to research on more efficient single variable searches.

6.4.3 The Fibonacci Search Method

Can one increase RR beyond this? Intuitively the answer is no as it appears that the best one can do is cut the interval by 50%. Anything less than that is suboptimal. So if at all there is a technique which claims that it has a reduction ratio which is superior to the dichotomous search, then what should support that claim? What is the logic that makes such a proposition possible? We do not say that we can use a three point test. Each function evaluation comes with a cost, cost of computational time. So if somebody claims that he/she has come up with an algorithm that gives a better reduction ratio than the dichotomous search, will it be similar to violating the Kelvin–Planck statement of the second law of thermodynamics? We now try to get answers to these questions.

First, there are indeed algorithms that are superior to the dichotomous search method. One such is the **Fibonacci search method**. Needless to say this method works on the Fibonacci series. Any number in the Fibonacci series can be written as

$$F_n = F_{n-1} + F_{n-2} \qquad n \geq 2 \tag{6.20}$$
$$F_0 = 1, \; F_1 = 1 \tag{6.21}$$

The first few numbers of the Fibonacci series are

$$1, 1, 2, 3, 5, 8, 13, 21, 34, 55, 89 \ldots \tag{6.22}$$

In the Fibonacci search method, we use the Fibonacci series to divide the interval into two and choose two points. In other words, if the original interval of uncertainty is $I_0 = (b-a)$, the first two points are given by

$$I_1 = I_0 \frac{F_{n-1}}{F_n} \tag{6.23}$$

or

$$I_1 = (b - a) \frac{F_{n-1}}{F_n} \tag{6.24}$$

from both the ends of the interval.

Consider a unimodal function. As usual, we start with the original interval of uncertainty I_0. We choose two points x_1 and x_2 which are symmetric about the end points at a distance of $I_1 = I_0(F_{n-1}/F_n)$, as shown in Fig. 6.11. We then use a simple two point test about the two points, which are symmetric from the two ends of the interval.

In the dichotomous search: $I_1 \sim I_0/2$ when $\varepsilon \sim 0$.

For the Fibonacci search, the algorithm is

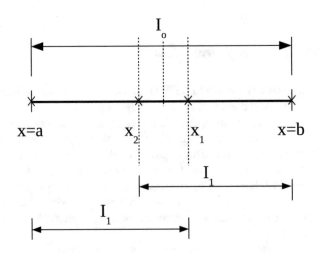

Fig. 6.11 The Fibonacci search method

$$I_1 = I_0 \frac{F_{n-1}}{F_n} = (b-a)\frac{F_{n-1}}{F_n} \tag{6.25}$$

So we choose a particular n. How do we do that? In order to figure out n, we need to first work out the reduction ratio (RR) of the algorithm.

The reduction ratio (RR)

$$I_0 = (b-a) \tag{6.26}$$

$$I_1 = I_0 \times \frac{F_{n-1}}{F_n} \tag{6.27}$$

$$I_2 = I_1 \times \frac{F_{n-2}}{F_{n-1}} = I_0 \frac{F_{n-2}}{F_n} \tag{6.28}$$

$$\vdots$$

$$I_n = \frac{I_0}{F_n} \tag{6.29}$$

The RR of the algorithm should then be

$$RR = \frac{I_0}{I_n} = \frac{I_0}{I_0}F_n \tag{6.30}$$

$$RR = F_n \tag{6.31}$$

Therefore the Fibonacci number itself becomes the reduction ratio of the algorithm.

Let us now look at the algorithm a little more closely for a unimodal function $f(x)$ in the interval $a \leq x \leq b$. Consider the first two points in the Fibonacci search

$$x_1 = a + (b-a)\frac{F_{n-1}}{F_n} \tag{6.32}$$

$$x_2 = b - (b-a)\frac{F_{n-1}}{F_n} \tag{6.33}$$

$$x_1 = \frac{a F_n + (b-a)F_{n-1}}{F_n}$$

$$= \frac{[a(F_{n-1} + F_{n-2}) + (b-a)F_{n-1}]}{F_n} \tag{6.34}$$

$$x_1 = \frac{(a F_{n-2} + b F_{n-1})}{F_n} \tag{6.35}$$

similarly

$$x_2 = \frac{[b(F_{n-1} + F_{n-2}) - (b-a)F_{n-1}]}{F_n} \tag{6.36}$$

$$x_2 = \frac{(b F_{n-2} + a F_{n-1})}{F_n} \tag{6.37}$$

We then perform the two point test and depending on whether $y(x_1) < y(x_2)$, we eliminate the region to the right of x_1 or the region to the left of x_2, depending on whether the problem under consideration is a maximization or a minimization problem. There is nothing great so far!!.

Now for the purpose of demonstration, we have to assume how the curve looks and decide on the region to be eliminated. So let us assume that the region to the left of x_2 (as shown in the figure below) is eliminated and proceed to the next iteration.

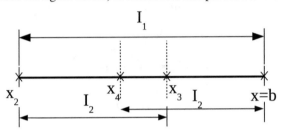

So the end points of the new interval are x_2 and $x=b$ and the new interval is I_1. Now let us choose 2 points x_3 and x_4 such that they are at a distance I_2 from the 2 new ends. I_2 then becomes

$$I_2 = I_1 \frac{F_{n-2}}{F_{n-1}} \tag{6.38}$$

$$I_1 = I_0 \frac{F_{n-1}}{F_n} \tag{6.39}$$

$$I_2 = I_0 \frac{F_{n-1}}{F_n} \frac{F_{n-2}}{F_{n-1}} = I_0 \frac{F_{n-2}}{F_n} \tag{6.40}$$

$$I_2 = (b - a) \frac{F_{n-2}}{F_n} \tag{6.41}$$

$$x_3 = x_2 + I_2 \tag{6.42}$$

$$x_4 = b - I_2 \tag{6.43}$$

Substituting for x_2 and I_2 in Eqs. 6.42 and 6.43, we have

$$x_3 = \frac{aF_{n-1} + bF_{n-2}}{F_n} + (b - a) \frac{F_{n-2}}{F_n} \tag{6.44}$$

$$x_4 = b - (b - a) \frac{F_{n-2}}{F_n} \tag{6.45}$$

$$x_4 = \frac{b(F_{n-1} + F_{n-2}) - bF_{n-2} + aF_{n-2}}{F_n} = \frac{bF_{n-1} + aF_{n-2}}{F_n} = x_1 \tag{6.46}$$

Therefore, $x_1 = x_4$

Out of two points (x_3, x_4), only one is really new!! This happens iteration after iteration and this contributes to the magical efficiency of the method.

Example 6.4 Execute the Fibonacci search algorithm for the cylindrical solar water heater storage problem (see Example 5.3). Perform 8 function evaluations. Start with $F_n = 34$.

Solution:

$$A = 2\pi r^2 + 2\pi rh \tag{6.47}$$
$$V = \pi r^2 h = 4 \text{ m}^3 \tag{6.48}$$
$$0.5 \leq r \leq 3.5 \text{ m} \tag{6.49}$$

Taking the original interval 3 m, $F_{n-1}/F_n = 21/34$. Therefore $I_1 = 3 \times 21/34$.

Taking I_1 from both the ends we get 2 points. We calculate the value of A at both the points and see which portion of the interval can be eliminated.

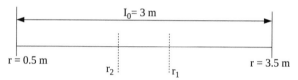

$$I_1 = I_0 \frac{F_{n-1}}{F_n} = 3 \times \frac{21}{34} = 1.85 \text{ m} \tag{6.50}$$
$$r_1 = 2.35 \text{ m} \tag{6.51}$$
$$r_2 = 1.65 \text{ m} \tag{6.52}$$

We do not worry if r_1 is greater than r_2 right now. Our notation is that whatever is chosen from the left is r_1 and whatever is chosen from the right is r_2.

$$A(r_1) = 38.1 \text{ m}^2; A(r_2) = 21.95 \text{ m}^2 \tag{6.53}$$
$$A(r_1) > A(r_2) \tag{6.54}$$

therefore region to the right of $r_1 = 2.65$ m can be eliminated

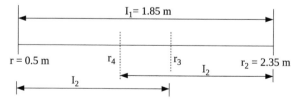

First, we used $21/34$ for getting two points from the ends, now we use $13/21$. We move from the right to the left of the Fibonacci series to get the ratios, till we hit 1, when x_n and x_{n+1} will coincide. That is when we stop the Fibonacci search.

There may be a small problem with the decimals, but we adjust them so that we are able to use one value of x at every iteration from the previous iteration.

$$I_2 = I_1 \frac{F_{n-2}}{F_{n-1}} = 1.85 \times \frac{13}{21} = 1.15 \text{ m} \tag{6.55}$$

$$r_3 = a + I_2 = 1.65 \text{ m} = r_2 \tag{6.56}$$

$$r_4 = r_1 - I_2 = 1.20 \text{ m} \tag{6.57}$$

$$A(r_3) = 21.95 \text{ m}^2 \tag{6.58}$$

$$A(r_4) = 15.2 \text{ m}^2 \tag{6.59}$$

$\therefore$ Region to the right of r_3 can be eliminated.

Going on to the next iteration, the end points of interval are $r = 0.5$, $r_3 = 1.65$ and $I_2 = 1.15$ m.

Now we consider 2 points r_5 and r_6 such that

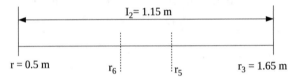

$$r_5 = 0.5 + 1.15 \times \frac{8}{13} = 1.2 \text{ m} \tag{6.60}$$

$$r_6 = 1.65 - 1.15 \times \frac{8}{13} = 0.95 \text{ m} \tag{6.61}$$

$$A(r_5) = 15.7 \text{ m}^2 \tag{6.62}$$

$$A(r_6) = 14.1 \text{ m}^2 \tag{6.63}$$

The region to the right of r_5 can be eliminated. Please note that we now use a ratio of (8/13) for getting two points in the next iteration.

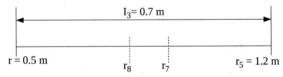

Though we seem to be working out a lot, it is just one evaluation per step, because the other point is already in. In this case, it so happens that the function evaluation is easy and the other steps seem to be as laborious or as time consuming as the function evaluation. However, many times in thermal sciences, the function evaluation may be the result of a solution to the Navier–Stokes equations and the equation of energy. In such a case, the other steps will be much less time consuming compared to the function evaluation.

$$r_7 = 0.5 + 0.7 \times \frac{5}{8} = 0.95 \text{ m} \tag{6.64}$$

$$r_8 = 1.2 - 0.7 \times \frac{5}{8} = 0.76 \text{ m} \tag{6.65}$$

$$A(r_7) = 14.1 \text{ m}^2 \tag{6.66}$$

$$A(r_8) = 14.2 \text{ m}^2 \tag{6.67}$$

The region to the left of r_8 can be eliminated

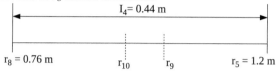

$$r_9 = 0.76 + 0.44 \times \frac{3}{5} = 1.02 \text{ m} \tag{6.68}$$

$$r_{10} = 1.2 - 0.44 \times \frac{3}{5} = 0.94 \text{ m} \tag{6.69}$$

$$A(r_9) = 14.38 \text{ m}^2 \tag{6.70}$$

$$A(r_{10}) = 14.1 \text{ m}^2 \tag{6.71}$$

The region to the right of r_9 can be eliminated

$$r_{11} = 0.76 + 0.26 \times \frac{2}{3} = 0.94 \text{ m} \tag{6.72}$$

$$r_{12} = 1.02 - 0.26 \times \frac{2}{3} = 0.84 \text{ m} \tag{6.73}$$

$$A(r_{11}) = 14.1 \text{ m}^2 \tag{6.74}$$

$$A(r_{12}) = 13.95 \text{ m}^2 \tag{6.75}$$

$$A(r_{11}) > A(r_{12}) \tag{6.76}$$

The region to the right of r_{11} can be eliminated

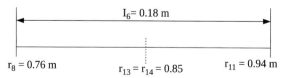

The next point is right in the center and the two points coincide. The final interval of uncertainty is 0.18m. The solution (r^+) now lies between 0.76 m and 0.94 m. With just 7 function evaluations, we are able to reduce the final interval of uncertainty to 18 cm from 300 cm! So we got a remarkable reduction ratio.

The original interval of uncertainty is 3 m.

$\therefore$ RR $= 3/0.18 = 16.67$

The final interval of uncertainty is 0.94–$0.76 = 0.18$ m.

Now, what is the theoretical RR? The theoretical RR is supposed to be $f_n = 34$. But we do not seem to get this performance.

Why are we getting this difference? Because we actually stopped with 7 evaluations. We can go for one more and do a three point test in the last step. So we can halve the interval further which will make our RR $= 32$. We will take this issue up in Sect. 6.4.4. In an exhaustive search, after 7 evaluations, the final interval of uncertainty would have been 75 cm.

MATLAB code for Example 6.4

```matlab
1   clear;
2   clc;
3   a=0.5;                    % lower limit of radius
4   b=3.5;                    % upper limit of radius
5   countmax=12;             % number of evaluations (even number)
6
7   ratio = zeros(countmax/2,1);
8
9   %f = fibonacci(3:3+countmax/2);
10
11  f(1) = 1;
12  f(2) = 1;
13
14  for i = 3 : 3+countmax/2
15
16      f(i) = f(i-1) + f(i-2);
17
18  end
19
20  f(1:2)=[];
21
22  for j = 2:length(f)
23      ratio(j-1)=f(j-1)/f(j);
24  end
25
26  for i=1:countmax/2
27
28      I=b-a;
29
30      % 1st point selection according to Fibonacci method
31      r1=b-I*ratio(countmax/2 - i + 1);
```

```
32
33        % 2nd point selection according to Fibonacci method
34        r2=a+I*ratio(countmax/2 - i + 1);
35
36        % surface area according to 1st point
37        A1=6.28*r1^2+8/r1;
38
39        % surface area according to 2nd point
40        A2=6.28*r2^2+8/r2;
41
42        if A1>A2
43            a=r1;
44        else
45            b=r2;
46        end
47
48        % Print
49        prt = ['Itr = ',num2str(i),...
50            ', a = ',num2str(a),...
51            ', b = ',num2str(b)];
52        disp(prt)
53
54    end
```

The output of the program is

```
Itr = 1, a = 0.5, b = 2.3529
Itr = 2, a = 0.5, b = 1.6471
Itr = 3, a = 0.5, b = 1.2059
Itr = 4, a = 0.76471, b = 1.2059
Itr = 5, a = 0.76471, b = 1.0294
Itr = 6, a = 0.76471, b = 0.94118
```

6.4.4 A Critical Assessment of the Fibonacci Search From the Previous Example

There are some higher order searches which are possible. For example, after we have gotten the center point, $r_{13} = r_{14}$, the Fibonacci search ends there. Now if we are pretty confident that we have bracketed the minimum and it is indeed lying between $r = 0.76$ m and $r = 0.94$ m, it is possible for us to use the Lagrange interpolation formula and have a local polynomial connecting these points and then make the function stationary. That will take us close to the true solution. While it may be an overkill for a simple problem like this, it is possible. If $F_n = 34$, we have to do the last step which is the three point test. So if both the points coincide, we do the evaluation and take the three point test. Else if we have a desirable accuracy required, we go on to the next nearest Fibonacci number so that this problem of "getting stuck" in the last iteration does not arise.

Though theoretically the RR is supposed to be F_n, we were not able to reach it because in the last iteration, both the points were at the center and coincided and hence we were not able to proceed any further. So the time has come critically to revisit our reduction ratio.

Actual Reduction ratio of Fibonacci search method

The theoretical RR of F_n which we derived in Eq. 6.31 can never be reached!.

If we take $F_8 = 34$ and we start dividing the interval as 21/34, 15/21, and so on, we are not able to do the 8th evaluation as both the points were at the center and it reduces to half. Hence, if $F_n = 34$, it corresponds to n = 8 in the Fibonacci series. When n = 8, we are able to do only (n–1) function evaluations. So, we can say that RR of the Fibonacci search with n function evaluations is given by

$$RR = F_{n+1}/2 = F_{7+1}/2 = F_8/2 = 34/2 = 17$$

which was what we got!

There is nothing wrong theoretically with the formula $1/F_n$, except that it cannot be reached unless we do some small tricks, like slightly moving one point away so that we get two points, but then it is no longer the Fibonacci search. So, the actual RR of the Fibonacci search method is

$$RR = F_{n+1}/2 = I_0/I_n$$

We could have started out this problem as follows. Minimize the surface area of the cylindrical water storage heater with an initial interval of uncertainty as 0.5–3.5 m our and get a final interval of uncertainty of 0.17 or 0.18 m.

$$\therefore RR = 3/0.18 = 17 = F_{n+1}/2; \text{ and}$$
$$F_{n+1} = 34, n = 8$$

34 corresponds to the 8th number in the Fibonacci series. Therefore n = 7 and we will do 7 evaluations. We do not have to count the number of evaluations being done as anyway after 7, it will automatically come to a halt. Please note that we have to work backwards from the desired accuracy we need on our solution.

We will now go on to yet another powerful method in our quest for developing efficient single variable searches. Thus far from whatever we have seen, there are some basic cardinal principles. The points we choose for the two point test must be close to the center of the interval. The 2 points must be equally spread from both sides of the interval. Common sense tells us that the 2 points must be at the same distance from the ends of the interval, because a priori we do not know if the left-hand or the right-hand side will be eliminated based on the two point test. Any disturbance to this, that is, the distance from the left side not being equal to the distance from the right side, may sometimes be advantageous while sometimes could be terrible. Therefore, in all these methods, including the Fibonacci method, the points are at the same distance from both the ends.

The Fibonacci method actually obeys this. In the two point test, the two points are always symmetrically placed about the center and about the 2 ends. We get a good reduction every time.

Disadvantages of the Fibonacci search method

There are a few disadvantages with the Fibonacci search method. The value n is small, as for example, 8 in this problem, Even so, it is small only for trivial problems. When we solve very complex problems, may be a 100 variable optimization problem, and we desire a high level of accuracy, n may be of the order of 300 or 400. So we have to calculate and store the Fibonacci series numbers first. Then we have to recall these numbers in every iteration. Furthermore, the reduction in the interval of uncertainty is not the same in all the iterations.

Now the question is, can we think of some method that enjoys all the advantages of the Fibonacci search method, but having the same interval reduction for all iterations, that is, when we go from I_0 to I_1, whatever be the reduction in the interval of uncertainty, will be exactly the same as we would get from I_1 to I_2 and I_2 to I_3 and so on.

6.4.5 The Golden Section Search

Let us consider an interval whose end points are $x = 0$ and $x = 1$ in appropriate units which may be meters, centimeters, kilometers, rupees, kilograms, and so on. We take 2 points x_1 and x_2, and we say that these are at a distance τ from both the ends.

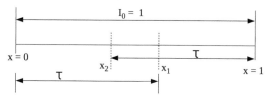

$$x_1 = \tau, x_2 = 1 - \tau \tag{6.77}$$

We are still talking about a unimodal function. Based on the two point test, one portion of the interval can be eliminated. Let us say the region to the left of x_2 can be eliminated. What is I_1 now ? $I_1 = \tau$

Our objective is to now come out with an algorithm such that if $I_1 = \tau \times 1$, $I_2 = \tau \times I_1$ or $\tau^2 \times 1$. Now consider 2 points x_3 and x_4 such that they are at a distance τ^2 from the ends.

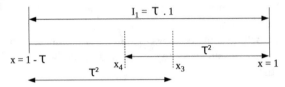

$$x_3 = 1 - \tau + \tau^2 \tag{6.78}$$
$$x_4 = 1 - \tau^2 \tag{6.79}$$

We started out with 2 points in the interval 0 and 1 and we are saying that in the first iteration, the percentage of reduction is τ and that this percentage of reduction has to be maintained. So in the second iteration, we take 2 new points x_3 and x_4, but we want to retain all the advantages of the Fibonacci search method. Therefore we want x_1 to be x_4.

If x_1 has to be equal to x_4

$$\tau = 1 - \tau^2 \tag{6.80}$$
$$\tau^2 + \tau - 1 = 0 \tag{6.81}$$
$$\tau = \frac{-1 \pm \sqrt{1+4}}{2} = \frac{\sqrt{5}-1}{2} \tag{6.82}$$
$$\tau = 0.618 \tag{6.83}$$

What is so great about 0.618? The reciprocal of 0.618, that is $1/0.618$, is 1.618.

That is why it is satisfying the property $x_1 = x_4$. The number 1.618 is called the **Golden ratio or the Golden number**.

If ϕ is the Golden ratio, it satisfies the following property:

$$\frac{1}{1 + \Phi} = \Phi \tag{6.84}$$
$$\Phi^2 + \Phi - 1 = 0, \ \Phi = 0.618 \tag{6.85}$$

So instead of struggling with the Fibonacci search method, we multiply I_0 every time with 0.618 and 38.2% of the interval will be gone. This is called the **Golden section search**. We take 2 points at distance $0.618\,I_0$ from the ends. Next time, it will be $0.618\,I_1$ from both the ends. We get a very good reduction ratio with this algorithm.

The letter Φ was introduced by the Greek sculptor Phidias.

The ratio of the height to the base of the Egyptian pyramids is 0.618. There are also claims by Leonardo da Vinci and others that the height to width ratio of a beautiful face must be 1.618.

What is the connection between the golden section and Fibonacci search methods?

$$\lim_{n \to \infty} \frac{F_{n-1}}{F_n} = 0.618 \qquad (6.86)$$

As we approach larger and larger numbers in the Fibonacci series, F_{n-1}/F_n reduces asymptotically to the value 0.618. So the two search methods are interconnected. So, if we want to enjoy all the advantages of the Fibonacci search, but do not want to calculate and store the numbers and also do not want unequal interval reduction across iterations, we can use the Golden section search.

Example 6.5 Consider the cylindrical solar water heater storage problem (Minimize surface area A for a given volume V = 4000 litres). Perform 7 function evaluations with the Golden section search method for one variable. Initial interval of uncertainty is $0.5 \le r \le 3.5$ m.

Solution:

$$A = 2\pi r^2 + 2\pi r h; \ V = \pi r^2 h = 4; \ 0.5 \le r \le 3.5 \ \text{m} \qquad (6.87)$$

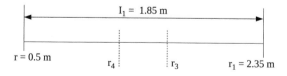

$$r_1 = 0.5 + 0.618 \times 3 = 2.35 \ \text{m} \qquad (6.88)$$
$$r_2 = 3.5 - 0.618 \times 3 = 1.65 \ \text{m} \qquad (6.89)$$
$$A(r_1) = 38.1 \ \text{m}^2, \ A(r_2) = 21.91 \ \text{m}^2 \qquad (6.90)$$
$$A(r_1) > A(r_2) \qquad (6.91)$$

Hence region to the right of r_1 can be eliminated.

$$r_3 = 0.5 + 0.618 \times 1.85 = 1.65 \text{ m } (r_3 \text{ is the same as } r_1) \qquad (6.92)$$
$$r_4 = 2.35 - 0.618 \times 1.85 = 1.21 \text{ m} \qquad (6.93)$$
$$A(r_3) = 21.91 \text{ m}^2, A(r_4) = 15.7 \text{ m}^2 \qquad (6.94)$$
$$A(r_3) > A(r_4) \qquad (6.95)$$

The region to the right of r_3 can be eliminated.

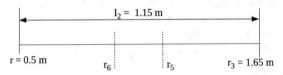

$$r_5 = 1.21 \text{ m}, A(r_5) = 15.7 \text{ m}^2 \ (r_5 \text{ is the same as } r_4) \qquad (6.96)$$
$$r_6 = 0.94 \text{ m}, A(r_6) = 14.1 \text{ m}^2 \qquad (6.97)$$
$$A(r_5) > A(r_6) \qquad (6.98)$$

The region to the right of r_5 can be eliminated.

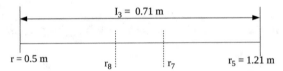

$$r_7 = 0.94 \text{ m}, A(r_7) = 14.06 \text{ m}^2 \ (r_7 \text{ is the same as } r_6) \qquad (6.99)$$
$$r_8 = 0.77 \text{ m}, A(r_8) = 14.12 \text{ m}^2 \qquad (6.100)$$
$$A(r_7) < A(r_8) \qquad (6.101)$$

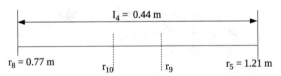

The region to the left of r_8 can be eliminated.

$$r_9 = 1.04 \text{ m}, A(r_9) = 14.48 \text{ m}^2 \ (r_{10} \text{ is the same as } r_7) \qquad (6.102)$$
$$r_{10} = 0.94 \text{ m}, A(r_{10}) = 14.06 \text{ m}^2 \qquad (6.103)$$
$$A(r_{10}) < A(r_9) \qquad (6.104)$$

The region to the right of r_9 can be eliminated.

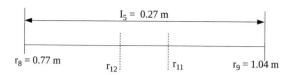

$$r_{11} = 0.94 \text{ m}, A(r_{11}) = 14.06 \text{ m}^2 \ (r_{11} \text{ is the same as } r_7) \tag{6.105}$$
$$r_{12} = 0.87 \text{ m}, A(r_{12}) = 13.94 \text{ m}^2 \tag{6.106}$$
$$A(r_{12}) < A(r_{11}) \tag{6.107}$$

The region to the right of r_{11} can be eliminated. We have done 7 function evaluations now.

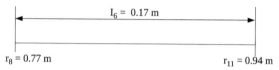

The final interval of uncertainty is $0.77 \leq r \leq 0.94$ m and is the same as what we got using the Fibonacci search method. Now let us calculate the reduction ratio RR.

$$RR = 3/0.17 = 17.65 \tag{6.108}$$
$$RR = (1/0.618)^{n-1} = 17.9, \text{ where } n = 7 \tag{6.109}$$

Where n is the number of function evaluations. In this formula, we get $(n-1)$ and not n because in the first iteration, we did 2 evaluations!

MATLAB code for Example 6.5

```
1   % Example 6.5: Golden Section search method
2   clear;
3   clc;
4
5   a=0.5;                  % lower limit of radius
6   b=3.5;                  % upper limit of radius
7
8   countmax=15;            % number of evaluations
9
10  for i = 1:countmax
11
12      I = b–a;
13
14      % 1st point selection according to Golden method
15      r1=a+I*0.618;
16
17      % 2nd point selection according to Golden method
```

```
18        r2=b-l*0.618;
19
20        % surface area according to 1st point
21        A1=6.28*r1^2+8/r1 ;
22
23        % surface area according to 2nd point
24        A2=6.28*r2^2+8/r2 ;
25
26        if A1>A2
27            b=r1 ;
28        else
29            a=r2 ;
30        end
31
32        % Print
33        prt = ['Itr = ',num2str(i) ,...
34             ', a = ',num2str(a) ,...
35             ', b = ',num2str(b) ];
36        disp(prt)
37    end
```

The output of the program is

```
Itr = 1, a = 0.5, b = 2.354
Itr = 2, a = 0.5, b = 1.6458
Itr = 3, a = 0.5, b = 1.2081
Itr = 4, a = 0.77049, b = 1.2081
Itr = 5, a = 0.77049, b = 1.0409
Itr = 6, a = 0.77049, b = 0.93762
Itr = 7, a = 0.83433, b = 0.93762
```

6.4.6 Improvements in Single Variable Searches

By now, we have learnt a few powerful techniques for performing single variable searches. They do not exploit any information on the derivatives or the higher order derivatives or we do not even look at the nature of the variation of the y between 2 points. We just see if $y_1 > y_2$ or $y_1 < y_2$. Obviously, optimization specialists and mathematicians would not have left it at this!

There are other powerful methods which will exploit all this information. Let us take a sneak peek of how powerful it can get when we employ some superior or more advanced techniques to this problem.

We expand the first derivative around x_{i+1} using the Taylor series expansion. Now we want to make this $f'(x_{i+1})$ stationary, because we are looking at an optimization problem.

$$f'(x_{i+1}) = f'(x_i) + f''(x_i)(x_{i+1} - x_i)$$
$$+ \text{Higher order terms} \tag{6.110}$$

$$x_{i+1} - x_i = -\frac{f'(x_i)}{f''(x_i} \tag{6.111}$$

$$x_{i+1} = x_i - \frac{f'(x_i)}{f''(x_i)} \tag{6.112}$$

This is the **Newton–Raphson method for an optimization problem**. This is different from the Newton–Raphson method for getting the roots of an equation for which the algorithm is

$$x_{i+1} = x_i - \frac{f(x_i)}{f'(x_i)} \tag{6.113}$$

But in an optimization problem we have $f'(x)/f''(x)$ in the algorithm because we are trying to make $f'(x)$ stationary and not $f(x)$ stationary. Since the Newton–Raphson method for optimization is using information on the first and second derivatives, it is demonstrably superior compared to the other search techniques. But where is the catch? It should be possible for us to evaluate $f'(x)$ and $f''(x)$. Suppose we are solving a CFD problem or we are trying to determine stress in a system, getting the value of f(x) itself is difficult. Outside of that we want to get f' and f''. So calculus is not the only route to solving optimization problems ! Calculus may look very attractive. But it is simply not possible to get the higher order derivatives, when we are working with complicated optimization problems, where each function evaluation itself is so difficult. Even getting the derivatives numerically does not help in many situations. All these have been responsible for the development of calculus-free techniques like Genetic Algorithms and Simulated Annealing, which can be applied to a wide class of engineering optimization. More on these in Chap. 8.

Let us consider the last step of Example 6.5 problem and apply the Newton–Raphson method. Refer to Fig. 6.12 that displays the final interval of uncertainty for Example 6.5.

Fig. 6.12 Final interval of uncertainty for Example 6.5

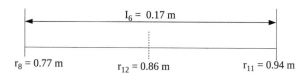

$I_6 = 0.17$ m

$r_8 = 0.77$ m $\qquad r_{12} = 0.86$ m $\qquad r_{11} = 0.94$ m

$$\text{Let } r_i = 0.86 \text{ m } (r_{12}) \tag{6.114}$$

$$A(r_i) = 2\pi r^2 + \frac{8}{r} \tag{6.115}$$

$$A'(r_i) = 4\pi r - \frac{8}{r^2} \tag{6.116}$$

$$A''(r_i) = 4\pi + \frac{16}{r^3} \tag{6.117}$$

Substituting the value for r_i as 0.86 m, we calculate the values of the above 3 expressions.

$$r_{(i+1)} = r_i - \frac{A'(r_i)}{A''(r_i)} \tag{6.118}$$

$$r_{(i+1)} = 0.86 - \frac{(-9.57 \times 10^{-3})}{37.7} \tag{6.119}$$

$$r_{(i+1)} = 0.861 \text{ m} \tag{6.120}$$

$A'(r)$ has become a very small value which shows that $A(r)$ has reached the optimum.

What we have done now is essentially a **hybrid optimization technique**. We start with a non-calculus search technique and toward the end we use a calculus-based optimization technique which exploits the power of the first and second derivatives in order to narrow down the solution.

There is yet another route to solving this problem. We have the end points r = 0.77 m and r = 0.94 m at the end of the Fibonacci/Golden section search method and the center of this interval is 0.86 m. We can fit a Lagrange interpolation formula between the 3 points and Newton's divided difference formula and get the derivative of this function and make it stationary and determine the solution by finding out where it becomes 0. All these are required if we want to get a very sophisticated or a very narrow interval.

The assumption behind fitting a function to the 3 points is that if the 3 points are sufficiently close and if we approximate it by a second degree or polynomial, not much error is incurred. This cannot be done on the original interval of uncertainty which varies from say, r = 1–200 m. However, when we have sufficiently narrowed down the interval and we are pretty sure that the interval is already very tight, it is possible to do the polynomial interpolation. This puts an end to our tour of search techniques for single variable optimization problems.

6.5 Search Methods for Multivariable Unconstrained Optimization Problem

We now look at solutions to multivariable optimization problems using search techniques. Though multivariable optimization problems are eventually broken down into single variable problems and for solving each of these, we eventually choose a very efficient single variable technique, there are some protocols involved in handling multivariable problems. A multivariable problem can either be a constrained and unconstrained one. Earlier, we did solve a set of multivariable constrained problems, where the number of constraints, which are equality constraints, is less than the number of variables and so on. The Lagrange multiplier is very elegant when the derivatives are not very cumbersome to evaluate. The major problem with the Lagrange multiplier technique is, after getting the derivatives, solving the set of simultaneous equations is tough, especially when there are 20, 30, or more variables.

So one possibility is to use the information on the derivatives and see if there is some other way of getting the solution instead of solving the resultant simultaneous equations. This means that instead of solving them simultaneously, we search for the solution.

Multivariable search techniques are very important in thermal sciences because we rarely encounter single variable problems. The problem of optimization of a shell and tube heat exchanger is a typical two variable problem, where the objective function is to minimize the surface area (and so the cost) of the exchanger subject to the constraints of heat duty and allowable pressure drop. The length and diameter of the tubes are important variables here.

As foresaid, multivariable optimization problems are of two types—unconstrained and constrained. Needless to say, constrained multivariable optimization problems are a lot harder to solve compared to unconstrained multivariable optimization problems.

Let us look at a typical two variable problem in x_1 and x_2. One possibility is to start from an initial, draw the grid around it, as seen before. We have a node P and mark the eight neighboring points in the 8 directions E, SE, S, SW, W, NW, N, and NE. We evaluate the function at the 8 neighboring points and find out in which direction the rate of change of y is maximum (see Fig. 6.13). For example, if we are seeking a maximum here and this is at the center of the iso-objective lines, then from node P,

Fig. 6.13 Lattice search method

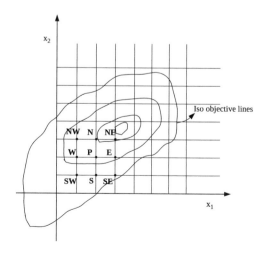

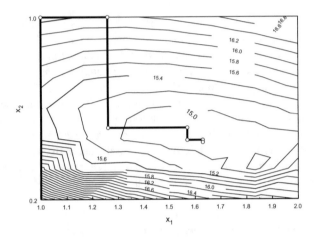

Fig. 6.14 Depiction of the unidirectional search method

hopefully we go to NE and take that as the next node. We then take 8 points around the new point and continue with our search. This is known as the lattice method , which we discussed briefly earlier.

As we go closer and y keeps increasing, the rate of change of y decreases, we can make the grid finer and finer. We can have adaptive meshing, wherein, we can start with coarse grids and then refine the grid. The number of function evaluations in each iteration for a 2 variable problem is 9.

The number of function evaluations at every iteration can then be generalized as 3^n, where n is called the dimensionality of the problem or the number of variables.

So computationally, the above approach is a very costly affair. If each of these solutions represents a CFD solution or a solution arising from a set of nonlinear partial differential equations from any branch of science and engineering, the algorithm is terribly expensive. We just keep evaluating though there clearly is some "method in the madness". Please recall that a technique like this falls under the category of "hill climbing".

Let us say that we want a reasonable solution to a two-dimensional CFD problem with two design variables using this method and let each solution take 2 hours on a powerful workstation. Every iteration will require 9 function evaluations. This will help us get an idea of the total time required to get to the optimum.

A logical question that arises now is whether one can solve for one variable at a time? Such an approach is called a **unidirectional search**. Let us start with a point in the domain as shown in Fig. 6.14. We keep x_1 fixed and try to get x_2 at which the function becomes minimum or maximum. In the next iteration, we keep x_2 fixed and determine x_1 at which the function again is a minimum or a maximum.

We then determine the value of x_2 at which y becomes minimum. This may not be the final solution because this is the final solution corresponding to the value of x_1 obtained after iteration 1 which itself is not the solution. Now we have got the new value of y and x_2. We keep this x_2 fixed and get the new value of x_1. We now keep this x_1 fixed and get back x_2 and keep going this way. So instead of doing 9 evaluations, we are going one at a time. So if y is a function of several variables, we assume values for all but one variable and write the objective function in terms of that variable, say x. We then take dy/dx$_1$, equate it to 0 and solve the equation, if it is quadratic or cubic and so on. If the function is too complex to be solved this way, we apply the Golden section search or the Fibonacci search or the dichotomous search and determine that value of x_1 which is the optimum solution when x_2 is fixed. In the next step, we solve for, say, x_2 and eventually we will reach the final solution.

Example 6.6 Optimize y, where:

$$y = 8 + \frac{x_1^2}{2} + \frac{2}{x_1 x_2} + 6x_2 \tag{6.121}$$

subject to $(x_1, x_2) \geq 0$
Solve it using a unidirectional search method with an initial guess of $x_1 = 1$ and $x_2 = 1$. Decide on an appropriate stopping criteria.

Solution:

$$\frac{\partial y}{\partial x_1} = x_1 - \frac{2}{x_1^2 x_2} \tag{6.122}$$

$$\frac{\partial y}{\partial x_2} = 6 - \frac{2}{x_1 x_2^2} \tag{6.123}$$

$$x_1^3 x_2 = 2 \tag{6.124}$$

$$x_1^+ = \left(\frac{2}{x_2}\right)^{\frac{1}{3}} \tag{6.125}$$

$$x_1 x_2^2 = \frac{1}{3} \tag{6.126}$$

$$x_2^+ = \left(\frac{1}{3x_1}\right)^{\frac{1}{2}} \tag{6.127}$$

We made the 2 derivatives stationary because of which, we got the expression for x_1 in terms of x_2 and for x_2 in terms of x_1. In any iteration, if we have x_2, using $\frac{\partial y}{\partial x_1} = 0$ and making y stationary with respect to x_1, we have an opportunity to calculate x_1 in terms of x_2. Once we calculate x_2, we can substitute in the other equation which helps us calculate x_2 in terms of x_1.

So we start with $x_1 = 1$ and determine x_2. For these values of x_1 and x_2, we calculate the value of y.

Table 6.2 Unidirectional search method for Example 6.6

Iter.no	x_1	x_2	y	$(x_{1,i+1} - x_{1,i})^2 + (x_{2,i+1} - x_{2,i})^2$
1	1	0.85	15.43	–
2	1.51	0.47	14.78	0.405
3	1.62	0.45	14.76	0.013
4	1.64	0.45	14.75	4×10^{-4}

Using this value of x_2, we try to get the value of x_1 by substituting for x_2 in Eq. 6.125. We continue working on this till convergence is achieved. The first four iterations are tabulated in the accompanying Table 6.2 and it can be seen that at the end of the fourth iteration.

$$R = (x_{1,i+1} - x_{1,i})^2 + (x_{2,i+1} - x_{2,i})^2 \text{ has reduced to } 4 \times 10^{-4}$$

and so iterations can be stopped here. Hence the optimal solution to this problem is

$$x_1^+ = 1.64, x_2^+ = 0.45, y^+ = 14.75.$$

We can check for the positive definiteness of the Hessian matrix of second derivatives of y to confirm that the solution obtained is indeed a minimum.

If the expression for x_1^+ in terms of x_2^+ or vice versa were to involve a difficult to evaluate function, it would be a chore to just evaluate x_1 and x_2 at each step. We then use the Golden section search to find out x_1 from x_2. Again we will use the Golden section search to find out x_2 from x_1. The advantage here is we do not have to solve the equations simultaneously. If we have many variables and it becomes messy to solve simultaneous equations, we can use this method instead of the Lagrange multiplier method. In this method, if the differential cannot be obtained as a closed-form expression, we can get it numerically too. Though this method also employs derivatives, it is better than the Lagrange multiplier method because we do not have to simultaneously solve the equations. They need to be solved only sequentially.

6.5.1   Cauchy's Method: The Method of Steepest Ascent/Descent

Cauchy's method is a very powerful and popular search technique for multivariable unconstrained problems. If we start a little away from the optimum, the method converges rapidly but as we get closer to the optimum, it becomes very very sluggish. Needless to say, because of the name steepest ascent or descent, we know that we are going to use the information on the derivatives in order to develop the algorithm.

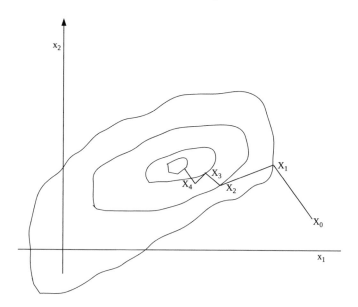

Fig. 6.15 Illustration of Cauchy's method

The algorithm:

$$y = y(x_1, x_2 \ldots x_n) \tag{6.128}$$

$$\nabla y = \frac{\partial y}{\partial x_1}.i_1 + \frac{\partial y}{\partial x_2}.i_2 + \cdots + \frac{\partial y}{\partial x_n}.i_n \tag{6.129}$$

where $i_1, i_2, \ldots i_n$ are all unit vectors.

The maximum rate of change of a function occurs in a direction orthogonal to the function at the point under consideration. For example, if we have a function like the one given in Fig. 6.15, the gradient vector is orthogonal to the function at every point. From vector calculus, it follows that the direction in which we move from the initial point will obey the following:

$$\frac{\Delta x_1}{\left(\frac{\partial y}{\partial x_1}\right)} = \frac{\Delta x_2}{\left(\frac{\partial y}{\partial x_2}\right)} = \cdots = \frac{\Delta x_n}{\left(\frac{\partial y}{\partial x_n}\right)} \tag{6.130}$$

In the above equation, if Δx_1 is fixed, all other Δx's can be evaluated at every iteration. We start with $X^0 = (x_1, x_2 \cdots x_n)^0$. We assume the values of $x_1, x_2, \ldots$ up to x_n. We also assume that y is continuous and differentiable. So, all the derivative values can be calculated at X^0 and the values of all the denominators in Eq. 6.130 are known. Therefore

$$x_{1,\text{new}} = x_{1,\text{old}} + \Delta x_1 \tag{6.131}$$

Likewise, all the other values of x can be got and hence X^1 can be obtained. Since from vector calculus, we know that when we are orthogonal to the iso-objective line, there is a maximum rate of change of the function. If we follow this, we are moving along this direction. But there is a problem here and, that is, how far should we go? The direction may be alright.

The vector X represents x_1 to x_n. This represents the direction of movement. But how far will we go in this direction?

A very simple way of making this algorithm will be to make $\frac{\Delta x_1}{\left(\frac{\partial y}{\partial x_1}\right)} = \frac{\Delta x_2}{\left(\frac{\partial y}{\partial x_2}\right)} = \frac{\Delta x_n}{\left(\frac{\partial y}{\partial x_n}\right)} = 1$. If after sometime, there is some problem with the function, the Δx_1 can be appropriately changed. This is a very simple interpretation of the Cauchy's method. A slightly more advanced version comes next. Let us call $\frac{\Delta x_n}{\left(\frac{\partial y}{\partial x_n}\right)} = \alpha$. Though we can decide the direction on this basis, in every iteration we calculate the value of α. What is that value of α which will minimize y at the current point? It is possible to answer this question. But now the problem is getting more difficult as in every iteration, we want to calculate the value of α. But if we do this and proceed with the algorithm, it will be exceedingly fast. Now if we are making $\alpha = 1$, we are pre-setting the value of Δx_1 and calculating all the other Δx, as they have to obey Eq. 6.130.

Example 6.7 Revisit Example 6.6 and solve it using the method of steepest ascent or descent (steepest ascent is for maximization problem and steepest descent is for minimization problem)

Solution:

$$y = 8 + \frac{x_1^2}{2} + \frac{2}{x_1 x_2} + 6x_2 \tag{6.132}$$

$$\frac{\partial y}{\partial x_1} = x_1 - \frac{2}{x_1^2 x_2} \tag{6.133}$$

$$\frac{\partial y}{\partial x_2} = 6 - \frac{2}{x_1 x_2^2} \tag{6.134}$$

(a) Method where $\alpha=1$

The starting values are $x_1 = x_2 = 1$ and let us assign $\Delta x_1 = 0.3$. Here we assume $\alpha = 1$. Now this $\Delta x_1 = 0.3$ appears fine, But whether we should correct x_1 as $(1+0.3)$ or $(1–0.3)$ should be decided. If y is increasing and we want a maximum, then 0.3 will be added to x_1. If y is increasing and we are seeking a minimum, then $x_1 = 1–0.3$.

The step size $\Delta x_1 = 0.3$ is not universal, and after some iterations we can change it to 0.2 or 0.1. If we choose to have a small step size of 0.1 now itself, may be it takes a long time to complete and hence we have made it 0.3. When we take

Table 6.3 Cauchy's method for Example 6.7 (method (a))

S.No	x_1	x_2	$\frac{\partial y}{\partial x_1}$	$\frac{\partial y}{\partial x_2}$	Δx_1	Δx_2
1	1	1	-1	4	$+0.1$	-0.4
2	1.1	0.6	-1.65	0.95	$+0.1$	-0.06
3	1.2	0.54	-1.37	0.28	$+0.1$	-0.02
4	1.3	0.52	-0.98	0.31	$+0.1$	-0.03
5	1.4	0.49	-0.68	0.05	$+0.1$	-7.4×10^{-3}
6	1.5	0.48	-0.35	0.21	$+0.05$	-0.03
7	1.55	0.45	-0.31	-0.37	$+0.02$	$+0.025$
8	1.57	0.475	-0.14	0.35	$+0.01$	-0.025
$\vdots$	$\vdots$	$\vdots$	$\vdots$	$\vdots$	$\vdots$	$\vdots$
15	1.631	0.45	-0.04	-0.05	$+0.001$	$+0.001$

$\Delta x_1 = 0.3$, we determine the new value of x_2 to be -0.2. But the constraint is x_1 and x_2 have to be greater than 0. So let us rework the problem assuming $\Delta x_1 = 0.1$.

Now it may start behaving funny. Now y will become very close to 0 because we are very close to the solution. Even after 8 or 9 iterations, we do not have convergence. If we are far away from the optimum, suppose we started out with $x_1 = 5$, it will quickly come down to 1. But it will struggle to come from $x_1 = 1$ to 1.6, which is the solution in this case. After the second iteration, x_2 is tantalizingly close to the correct answer but x_1 is far off. When x_1 starts inching forward, x_2 will move away!!

Let us now make $\Delta x_1 = 0.05$ as we are close to convergence. In step 7, $\partial y / \partial x_2$ is changing sign which tells us that we have overshot the answer. At the end of 8 iterations, $x_1 = 1.57$ and $x_2 = 0.475$. At the end of 15 iterations, it is seen that $x_1^+ = 1.631$, $x_2^+ = 0.45$, and $y^+ = 14.76$.

So the method seems to be going in the right direction but is now very slowly converging. The solution is given in Table 6.3.

(b) Method where α is calculated at every iteration.

$$y = 8 + \frac{x_1^2}{2} + \frac{2}{x_1 x_2} + 6x_2 \tag{6.135}$$

Initial guess $(1, 1)$

$$\frac{\partial y}{\partial x_1} = x_1 - \frac{2}{x_1^2 x_2} \tag{6.136}$$

$$\frac{\partial y}{\partial x_2} = 6 - \frac{2}{x_1 x_2^2} \tag{6.137}$$

Table 6.4 Cauchy's method for Example 6.7 (method (b))

S.No	x_1	x_2	$\frac{\partial y}{\partial x_1}$	$\frac{\partial y}{\partial x_2}$	α	Δx_1	Δx_2
1	1	1	−1	4	−0.12	+0.12	−0.48
2	1.12	0.52	−1.95	−0.6	−0.12	+0.234	0.072
3	1.354	0.592	−0.49	1.79	−0.06	+0.03	−0.107
4	1.384	0.485	−0.77	−0.14	−0.36	+0.28	0.05
5	1.664	0.54	0.33	1.88	−0.05	−0.0165	−0.094
6	1.6475	0.446	−0.004	−0.10			

$$\text{At } (1,1) \quad \frac{\partial y}{\partial x_1} = -1 \tag{6.138}$$

$$\frac{\partial y}{\partial x_2} = 4 \tag{6.139}$$

Now we choose α such that $y(1 - \alpha, 1 + 4\alpha)$ is minimized.

$$\Delta x_1 = \alpha dy/dx_1; \tag{6.140}$$
$$\Delta x_2 = \alpha dy/dx_2; \tag{6.141}$$

So $x_{\text{new}} = x_{\text{old}} + \Delta x_1$.

We solve it for α and once we do that, we have the new values of x_1 and x_2. The results are presented in Table 6.4. It can be seen that at the end of 5 iterations itself, the solution has converged. Hence, if we determine α at every iteration, the computational cost per iteration goes up but the number of iterations required goes down dramatically. Let us look at yet another example

Example 6.8 Minimize $y = (x_1 - 8)^2 + (x_2 - 6)^2$

Solution:

This is the equation of a circle and the optimum is $(8, 6)$ if we are seeking a minimum to y which we know (y^+ at the optimum is $y^+ = 0$).

We now solve it using Cauchy's method.

Let us start with $(2, 2)$. As the initial guess value,

$$\partial y/\partial x_1 = 2(x_1 - 8) = -12 \tag{6.142}$$
$$\partial y/\partial x_2 = 2(x_2 - 6) = -8 \tag{6.143}$$

Now we have to calculate y at the new points and optimize.

$$y(2 + \alpha \partial y/\partial x_1, 2 + \alpha \partial y/\partial x_2) = y(2 - 12\alpha, 2 - 8\alpha) \tag{6.144}$$

$$y = [(2 - 12\alpha - 8)^2 + (2 - 8\alpha - 6)^2] \tag{6.145}$$

$$y = [(-6 - 12\alpha)^2 + (-4 - 8\alpha)^2] \tag{6.146}$$

$$= [36(1 + 2\alpha)^2 + 16(1 + 2\alpha)^2] \tag{6.147}$$

$$dy/d\alpha = 0; \tag{6.148}$$

$$\alpha = -1/2 \tag{6.149}$$

New values of (x_1, x_2) are

$x_1 = 2 + (12/2) = 8; \ x_2 = 2 + (8/2) = 6; \ y^+$ at $(8, 6) = 0$

We straightaway get $(8,6)$ as the answer. We can start anywhere, but will get the answer because the function is just a simple quadratic.

Example 6.9 Minimize $y = 2x_1^2 - 2x_1x_2 + x_2^2$ with an initial guess of $(3,5)$ using Cauchy's Steepest Descent Method and perform at least 4 iterations

Solution:

We now solve it using the Cauchy's method.

$$\partial y/\partial x_1 = 4x_1 - 2x_2 \tag{6.150}$$

$$\partial y/\partial x_2 = -2x_1 + 2x_2 \tag{6.151}$$

Now we have to calculate y at the new points and optimize.

Iteration 1:

$$\partial y/\partial x_1 = 4x_1 - 2x_2 = 4 \times 3 - 2 \times 5 = 2 \tag{6.152}$$

$$\partial y/\partial x_2 = -2x_1 + 2x_2 = -2 \times 3 + 2 \times 5 = 4 \tag{6.153}$$

$$y(3 + \alpha \partial y/\partial x_1, 5 + \alpha \partial y/\partial x_2) = y(3 + 2\alpha, 5 + 4\alpha) \tag{6.154}$$

$$y = [2 \times (3 + 2\alpha)^2 - 2 \times (3 + 2\alpha) \times (5 + 4\alpha) + (5 + 4\alpha)^2] \tag{6.155}$$

$$dy/d\alpha = 0; \tag{6.156}$$

$$\alpha = -1.25 \tag{6.157}$$

New values of (x_1, x_2) are

$x_1 = 3 + 2 \times (-1.25) = 0.5; \ x_2 = 5 + 4 \times (-1.25) = 0; \ y^+$ at $(0.5, 0)$

Iteration 2:

$$\partial y/\partial x_1 = 4 \times 0.5 - 2 \times 0 = 2 \tag{6.158}$$

$$\partial y/\partial x_2 = -2 \times 0.5 + 2 \times 0 = -1 \tag{6.159}$$

$$y(0.5 + \alpha \partial y/\partial x_1, 0 + \alpha \partial y/\partial x_2) = y(0.5 + 2\alpha, -\alpha) \qquad (6.160)$$

$$y = [2 \times (0.5 + 2\alpha)^2 - 2 \times (0.5 + 2\alpha) \times (-\alpha) + (-\alpha)^2] \qquad (6.161)$$

$$dy/d\alpha = 0; \qquad (6.162)$$

$$\alpha = -0.1923 \qquad (6.163)$$

New values of (x_1, x_2) are

$x_1 = 0.5 + 2 \times (-0.1923) = 0.1154$; $x_2 = - \times (-0.1923) = 0.1923$; y^+ at $(0.1154, 0.1923)$

Iteration 3:

$$\partial y/\partial x_1 = 4x_1 - 2x_2 = 0.077 \qquad (6.164)$$

$$\partial y/\partial x_2 = -2x_1 + 2x_2 = 0.1538 \qquad (6.165)$$

$$y(0.1154 + \alpha \partial y/\partial x_1, 0.1923 + \alpha \partial y/\partial x_2)$$
$$= y(0.1154 + 0.077\alpha, 0.1923 + 0.1538\alpha)$$
$$y = [2 \times (0.1154 + 0.077\alpha)^2 - 2 \times (0.1154 + 0.077\alpha) \times$$
$$(0.1923 + 0.1538\alpha) + (0.1923 + 0.1538\alpha)^2] \qquad (6.166)$$

$$dy/d\alpha = 0; \qquad (6.167)$$

$$\alpha = -1.25 \qquad (6.168)$$

New values of (x_1, x_2) are

$x_1 = 0.1154 + 0.077 \times (-1.25) = 0.0192$; $x_2 = 0.1923 + 0.1538 \times (-1.25) = 0$; y^+ at $(0.0192, 0)$

Iteration 4:

$$\partial y/\partial x_1 = 4x_1 - 2x_2 = 0.0768 \qquad (6.169)$$

$$\partial y/\partial x_2 = -2x_1 + 2x_2 = -0.0384 \qquad (6.170)$$

$$y(0.0192 + \alpha \partial y/\partial x_1, 0 + \alpha \partial y/\partial x_2)$$
$$= y(0.0192 + 0.0768\alpha, -0.0384\alpha)$$
$$y = [2 \times (0.0192 + 0.0768\alpha)^2 - 2 \times (0.0192 + 0.0768\alpha) \qquad (6.171)$$
$$\times (-0.0384\alpha) + (-0.0384\alpha)^2]$$

$$dy/d\alpha = 0; \qquad (6.172)$$

$$\alpha = -0.1923 \qquad (6.173)$$

New values of (x_1, x_2) are

Table 6.5 Cauchy method for Example 6.9

S.No	x_1	x_2	$\frac{\partial y}{\partial x_1}$	$\frac{\partial y}{\partial x_2}$	α	Δx_1	Δx_2
1	3	5	2	4	−1.25	−2.5	−5
2	0.50	0	2	−1	−0.1923	−0.3846	0.1923
3	0.1154	0.1923	0.077	0.1538	−1.25	−0.09625	−0.19225
4	0.192	0	0.768	−0.384	−0.1923	−0.1477	0.071

$$x_1 = 0.0192 + 0.0768 \times (-0.1923) = 0.00443; \; x_2 = -0.0384 \times (-0.1923)$$
$$= 0.00739; \; y^+ \text{ at } (0.00443, 0.071)$$

The results are tabulated in Table 6.5.

MATLAB code for Example 6.9

```
1   clear;
2   clc;
3
4   X1=3;    % guess value of x1
5   X2=5;    % guess value of x2
6
7   syms x1 x2 a
8   y=2*(x1^2)-2*x1*x2+x2^2; % objective function
9
10  % derivative of objective function with respect to x1
11  dydx1=diff(y,x1);
12
13  % derivative of objective function with respect to x2
14  dydx2=diff(y,x2);
15
16  % value of derivative
17  DYDX1=single(subs(dydx1,{x1,x2},{X1,X2}));
18
19  % value of derivative
20  DYDX2=single(subs(dydx2,{x1,x2},{X1,X2}));
21
22  countmax=15; % maximum count for iteration
23  errTol=10^-10; % specified tolerance for convergence
24  label=0;
25  count=0;
26
27  while label==0
28
29      count=count+1;
30
31      DYDX1=single(subs(dydx1,{x1,x2},{X1,X2}));
32      DYDX2=single(subs(dydx2,{x1,x2},{X1,X2}));
33
34      % objective function for minimisation
35      y_new=4*((X1+DYDX1*a)^2)-4*(X1+DYDX1*a)*(X2+DYDX2*a)+2*(X2+DYDX2*a)^2;
36
37      dy_newda=diff(y_new,a);
```

```
38
39
40      % Equating the derivative of the objective function
41      %is equal to zero for minimisation
42      r=solve(dy_newda,a);
43
44      % value of alpha at which the objective function is minimum
45      alpha=single(r);
46
47      X1new=X1+DYDX1*alpha;    % new value of x1
48      X2new=X2+DYDX2*alpha;    % new value of x2
49
50      err=(X1new-X1)^2+(X2new-X2)^2; % residual
51
52      X1=X1new;        % update value of x1
53      X2=X2new;            % update value of x2
54
55      if count==countmax || err < errTol
56
57          label=1;
58      end
59
60      % Print
61      prt = ['Itr = ',num2str(count),...
62          ', x1 = ',num2str(X1),...
63          ', x2 = ',num2str(X2),...
64          ', err = ',num2str(err)];
65      disp(prt)
66
67  end
```

The output of the program is

- *Itr = 1, x1 = 0.5, x2 = 0, Y = 26, err = 31.25*
- *Itr = 2, x1 = 0.11538, x2 = 0.19231, Y = 1, err = 0.18491*
- *Itr = 3, x1 = 0.019231, x2 = 1.4901e-08, Y = 0.038462, err = 0.046228*
- *Itr = 4, x1 = 0.0044379, x2 = 0.0073965, Y = 0.0014793, err = 0.00027354*
- *Itr = 5, x1 = 0.00073965, x2 = 4.6566e-10, Y = 5.6896e-05, err = 6.8385e-05*
- *Itr = 6, x1 = 0.00017069, x2 = 0.00028448, Y = 2.1883e-06, err = 4.0464e-07*
- *Itr = 7, x1 = 2.8448e-05, x2 = 5.8208e-11, Y = 8.4166e-08, err = 1.0116e-07*
- *Itr = 8, x1 = 6.565e-06, x2 = 1.0942e-05, Y = 3.2372e-09, err = 5.9859e-10*
- *Itr = 9, x1 = 1.0942e-06, x2 = 5.457e-12, Y = 1.2451e-10, err = 1.4965e-10*
- *Itr = 10, x1 = 2.525e-07, x2 = 4.2084e-07, Y = 4.7889e-12, err = 8.8552e-13*

6.5.2 How Does the Cauchy's Method (Steepest Descent) Work?

Let

$$y = y(x_1, \cdots, x_n) \tag{6.174}$$

$$c = \Delta y = \frac{\partial y}{\partial x_1} i_1 + \frac{\partial y}{\partial x_2} i_2 + \cdots + \frac{\partial y}{\partial x_n} i_n \tag{6.175}$$

$$d = -c \tag{6.176}$$

d is the negative of the gradient vector. The modulus of the C vector is given by

$$|c| = \sqrt{\left(\frac{\partial y}{\partial x_1}\right)^2 + \left(\frac{\partial y}{\partial x_2}\right)^2 . + \cdots + \left(\frac{\partial y}{\partial x_n}\right)^2} \tag{6.177}$$

$$X^0 = (x_1, \cdots, x_n)^0 \Rightarrow \text{Initial guess} \tag{6.178}$$

$$X^1 = X^0 + \alpha d^0 \tag{6.179}$$

We solve for α, get X^1, then d^1 and so this is how the algorithm works. What we have done so far is to write the algorithm in compact mathematical notation.

Let us now look at the product of c^i and c^{i+1}. We are trying to see if this product is 0 so that the new direction is always orthogonal to the old direction. The idea is to see if $c^i.c^{i+1} = 0$.

How did we get α?

We minimize $y(X^i + \alpha d^i)$

$$\frac{\partial y}{\partial \alpha}|_{i+1} = 0 = \left(\frac{\partial y}{\partial x}\Big|_{i+1}\right)\left(\frac{\partial x}{\partial \alpha}\Big|_{i+1}\right) \tag{6.180}$$

$$\frac{\partial y}{\partial x}|_{i+1} = c^{i+1} \tag{6.181}$$

$$\frac{\partial x}{\partial \alpha}|_{i+1} = \frac{\partial}{\partial \alpha}(x^i + \alpha d^i) = d^i = -c^i \tag{6.182}$$

So if $dy/d\alpha$ has to be 0, this itself is a product of 2 partial derivatives $\partial y/\partial x$ and $\partial x/\partial \alpha$.

$\partial y/\partial x$ turns out to be c^{i+1} whereas $\partial x/\partial \alpha$ turns out to be c^i. From this we get, $c^i.c^{i+1} = 0$. Therefore at the $(i + 1)$th iteration, the direction of movement is orthogonal to the previous direction. This is the proof of the steepest descent method.

6.5.3   Conjugate Gradient Method

We saw a very crude form of the conjugate gradient method when we discussed the Levenberg–Marquardt algorithm in Chap. 3, when we tried to do nonlinear regression, where we introduced the damping factor, λ.

Consider a two variable problem. Let us start with $x_0 = (x_{1,0}, x_{2,0})$ and hit $x_1 = (x_{1,1}, x_{2,1})$. We keep proceeding until we reach the optimum. This is Cauchy's method. The next logical question to ask is: can we directly go to x_3 bypassing x_2? It is possible to jump some steps. One possibility is to deflect the direction in which we are moving instead of always going orthogonal to the previous direction and come up with a deflected steepest descent method. This **deflected steepest descent method is called the conjugate gradient method**.

So the first 2 steps are the same as the Cauchy's method. We start with x_0 and go to x_1 using the steepest descent. Then from x_1, we directly go to x_3. This is also called the **Fletcher–Powell method**. It is one of the most powerful algorithms ever used.

Conjugate gradient method—the Algorithm

- The first 2 steps are the same as in Cauchy's method
- $d^i = -c^i + \beta_i d^{i-1}$

  where $\beta_i = \left[\dfrac{|c_i|}{|c_{i-1}|} \right]^2$

  Calculate the length of Δy at the current and previous iteration, take the ratio of this, square it and this becomes β.
- Evaluate α_i to minimize $y(x^i + \alpha d^i)$
- Update x^{i+1}; If $\beta = 0$, it becomes the Cauchy's method.

To calculate β, we require the value of c at 2 steps. That is why we will start with Cauchy's method, get 2 values of C and then move on to the conjugate gradient method.

Let us revisit Example 6.7 The first two steps from Cauchy's method are shown in Table 6.6.

$$|c_1| = \sqrt{(-1)^2 + (4)^2} = 4.12 \tag{6.183}$$

$$|c_2| = \sqrt{(-1.95)^2 + (-0.6)^2} = 2.04 \tag{6.184}$$

$$\beta_2 = (\frac{2.04}{4.12})^2 = 0.25 \tag{6.185}$$

$$d_2 = -c_2 + \beta_2 d_1 \tag{6.186}$$

$$d_2 = -[-1.95i_1 - 0.6i_2] + 0.25[-\{(-1)i_1 + 4i_2\}] \tag{6.187}$$

$$d_2 = (2.2i_1 - 0.4i_2) \tag{6.188}$$

We use the exhaustive search technique or Golden section search method to determine the value of α that minimizes $y(1.12 + 2.2\alpha, 0.52 - 0.4\alpha)$.

Table 6.6 First two iterations from Table 6.4

S.No	x_1	x_2	$\frac{\partial y}{\partial x_1}$	$\frac{\partial y}{\partial x_2}$	α	Δx_1	Δx_2
1	1	1	−1	4	−0.12	+0.12	−0.48
2	1.12	0.52	−1.95	−0.6			

Table 6.7 Conjugate gradient method for Example 6.7

S.No	x_1	x_2	$\frac{\partial y}{\partial x_1}$	$\frac{\partial y}{\partial x_2}$	α	Δx_1	Δx_2
1	1	1	−1	4	−0.12	+0.12	−0.48
2	1.12	0.52	−1.95	−0.6	0.23	+0.50	−0.09
3	1.626	0.43	−0.1	−0.6	0.03	+0.008	+0.02
4	1.634	0.45	−0.04	−0.04			

Fig. 6.16 Depiction of the conjugate gradient method

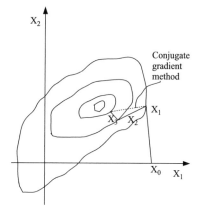

As can be seen from Table 6.7, the solution is reached at the third iteration itself compared to 14 iterations for method (a) of the steepest descent and 5 iterations for method (b) of the steepest descent method.

The progression of the solution using the conjugate gradient method is depicted in Fig. 6.16.

The flowchart for solving a typical two variable (x_1, x_2) optimization problem is shown in Fig. 6.17.

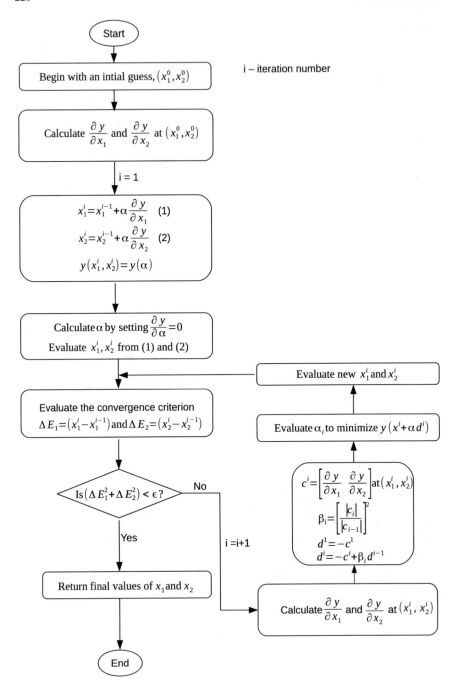

Fig. 6.17 Flowchart for the conjugate gradient method for a two variable unconstrained optimization problem

Example 6.10 Minimize $y = 2x_1^2 - 2x_1x_2 + x_2^2$ with an initial guess of (3, 5) using the conjugate gradient Method

Solution:

We now solve it using the conjugate gradient method.
    Let us start with (3, 5). As the initial guess value,

$$\partial y/\partial x_1 = 4x_1 - 2x_2 \tag{6.189}$$
$$\partial y/\partial x_2 = -2x_1 + 2x_2 \tag{6.190}$$

Now we have to calculate y at the new points and optimize.

Iteration 1:

$$\partial y/\partial x_1 = 4x_1 - 2x_2 = 4 \times 3 - 2 \times 5 = 2 \tag{6.191}$$
$$\partial y/\partial x_2 = -2x_1 + 2x_2 = -2 \times 3 + 2 \times 5 = 4 \tag{6.192}$$

$$y(3 + \alpha\partial y/\partial x_1, 5 + \alpha\partial y/\partial x_2) = y(3 + 2\alpha, 5 + 4\alpha) \tag{6.193}$$
$$y = [2 \times (3 + 2\alpha)^2 - 2 \times (3 + 2\alpha) \times (5 + 4\alpha) + (5 + 4\alpha)^2] \tag{6.194}$$

$$dy/d\alpha = 0; \tag{6.195}$$
$$\alpha = -1.25 \tag{6.196}$$

New values of (x_1, x_2) are
    $x_1 = 3 + 2 \times (-1.25) = 0.5$; $x_2 = 5 + 4 \times (-1.25) = 0$; y^+ at $(0.5, 0)$
Iteration 2:

$$\partial y/\partial x_1 = 4x_1 - 2x_2 = 2 \tag{6.197}$$
$$\partial y/\partial x_2 = -2x_1 + 2x_2 = -1 \tag{6.198}$$

Up to this point procedure is same as that of Cauchy's method.

$$|c_k| = \sqrt{\left(\frac{\partial y}{\partial x_1}\right)^2 + \left(\frac{\partial y}{\partial x_2}\right)^2} \tag{6.199}$$

where, k is the iteration no.
$$|c_1| = \sqrt{(2)^2 + (4)^2} = 4.72 \tag{6.200}$$
$$|c_2| = \sqrt{(2)^2 + (-1)^2} = 2.236 \tag{6.201}$$
$$\beta_2 = \left(\frac{2.236}{4.72}\right)^2 = 0.25 \tag{6.202}$$
$$d_2 = -c_2 + \beta_2 d_1 \tag{6.203}$$

$$d_1 = -c_1 = -[2i_1 + 4i_2] = -2i_1 - 4i_2 \qquad (6.204)$$

$$c_2 = 2i_1 - i_2 \qquad (6.205)$$

$$d_2 = -[2i_1 - i_2] + 0.25[\{-2i_1 - 4i_2\}] \qquad (6.206)$$

$$d_2 = (-2.5i_1 + 0i_2) \qquad (6.207)$$

$$dy/d\alpha = 0; \qquad (6.208)$$

$$\alpha = 0.2 \qquad (6.209)$$

New values of (x_1, x_2) are:
   $x_1 = 0.5 - 2.5 \times (0.2) = 0; \;\; x_2 = 0 + 0 \times (0.2) = 0; \;\; y^+$ at $(0, 0)$

Iteration 3:

$$|c_2| = \sqrt{(2)^2 + (-1)^2} = 2.236 \qquad (6.210)$$

$$|c_3| = \sqrt{(0)^2 + (0)^2} = 0 \qquad (6.211)$$

$$\beta_3 = \left(\frac{0}{2.236} \right)^2 = 0 \qquad (6.212)$$

$$d_3 = -c_3 + \beta_3 d_2 = -c_3 \qquad (6.213)$$

$$c_3 = 0i_1 + 0i_2 \qquad (6.214)$$

$$d_3 = 0i_1 + 0i_2 \qquad (6.215)$$

New values of (x_1, x_2) are
   $x_1 = 0; \;\; x_2 = 0; \;\; y^+$ at $(0, 0)$
The results are presented in Table 6.8.

Table 6.8 Conjugate gradient method for Example 6.10

S.No	x_1	x_2	$\frac{\partial y}{\partial x_1}$	$\frac{\partial y}{\partial x_2}$	α	Δx_1	Δx_2
1	3	5	2	4	−1.25	−2.5	−5
2	0.50	0	2	−1	0.2	−0.50	0
3	0	0	0	0	−	0	0

MATLAB code for Example 6.10

```
1   clear;
2   clc;
3
4   X1=3;                    % guess value of x1
5   X2=5;                    % guess value of x2
6
7   syms x1 x2 a
8
9   % objective function
10  y=2*(x1^2)-2*x1*x2+x2^2;
11
12  % derivative of objective function wrt x1
13  dydx1=diff(y,x1);
14
15  % derivative of objective function wrt x2
16  dydx2=diff(y,x2);
17
18  % value of derivatives
19  DYDX1=single(subs(dydx1,{x1,x2},{X1,X2}));
20  DYDX2=single(subs(dydx2,{x1,x2},{X1,X2}));
21
22  countmax=10;  % maximum cout for iteration
23  errTol=10^-10; % specified tolerance for convergence
24  label=0;
25  count=0;
26
27  while label==0
28
29      count=count+1;
30
31      DYDX1=single(subs(dydx1,{x1,x2},{X1,X2}));
32      DYDX2=single(subs(dydx2,{x1,x2},{X1,X2}));
33
34
35      if count==1
36          d2 = [DYDX1; DYDX2]; % update gradient vector
37      else
38          c2 = [DYDX1; DYDX2]; % update gradient vector
39          b=((norm(c2))/(norm(c1)))^2;
40          d2 = -c2+b*d1;
41      end
42
43      % objective function for minimisation
44      y_new=4*((X1+d2(1)*a)^2)-4*(X1+d2(1)*a)*(X2+d2(2)*a)+2*(X2+d2(2)*a)^2;
45
46      dy_newda=diff(y_new,a);
47
48      % Equating the derivative of the objective function
49      %is equal to zero for minimisation
```

```
50        r=solve(dy_newda,a);
51
52        % value of alpha at which the objective function
53        %is minimum
54        alpha=single(r);
55
56        X1new=X1+d2(1)*alpha; % new value of x1
57        X2new=X2+d2(2)*alpha; % new value of x2
58
59        err=(X1new-X1)^2+(X2new-X2)^2; % residual
60        X1=X1new;                % update value of x1
61        X2=X2new;                % update value of x2
62        c1 = [DYDX1; DYDX2];
63        y=4*((X1+DYDX1*alpha)^2)...
64            -4*(X1+DYDX1*alpha)*(X2+DYDX2*alpha)...
65            +2*(X2+DYDX2*alpha)^2;
66
67        if count==1
68            d1 = -c1;
69        else
70            d1=d2;
71        end
72        if count==countmax || err<errTol
73            label=1;
74        end
75        % Print
76        prt = ['Itr = ',num2str(count),...
77            ', x1 = ',num2str(X1),...
78            ', x2 = ',num2str(X2),...
79            ', Y = ',num2str(y),...
80            ', err = ',num2str(err)];
81        disp(prt)
82  end
```

The output of the program is

```
Itr = 1, x1 = 0.5, x2 = 0, Y = 26, err = 31.25
Itr = 2, x1 = 0, x2 = 0, Y = 1.04, err = 0.25
Itr = 3, x1 = 0, x2 = 0, Y = 0, err = 0
```

6.6 Constrained Multivariable Optimization Problems

Consider the following minimization problem:

$$\text{Minimize } y = y(x_1, x_2 \ldots x_n) \tag{6.216}$$

$$\text{subject to the constraints } \phi_1 = \phi_1(x_1, x_2 \ldots x_n) = 0 \tag{6.217}$$

$$\phi_2 = \phi_2(x_1, x_2 \ldots x_n) = 0 \tag{6.218}$$

$$\vdots$$

$$\phi_m = \phi_m(x_1, x_2 \ldots x_n) = 0 \tag{6.219}$$

We assume that we have only equality constraints.

We now transform the objective function in a such a way that the constrained optimization problem is converted into an equivalent unconstrained optimization problem. We introduce a composite objective function V, which is defined as follows:

$$V = y + (P_1\phi_1^2 + P_2\phi_2^2 + \cdots P_m\phi_m^2) \tag{6.220}$$

where $P_1, P_2 \ldots P_m$ are all scalars and are positive.

Each of these (ϕ's) constraints is expected to be 0, because each one is an equality constraint. If each of these ϕ's take on a value that is different from 0, we square it and multiply by a positive quantity to get quantities $P_1\phi_1^2, P_2\phi_2^2, \ldots P_m\phi_m^2$. It is to be noted that all of these are positive quantities. These are all added to the original cost for a minimization problem. Therefore V will be substantially higher than function y, if the ϕ's are not equal to 0. Stated more explicitly, the objective function is penalized for violation of constraints. This is the penalty! If the violation is too much, then ϕ_1, ϕ_2, ϕ_m will all be much more than 0, and these are multiplied by P (P is a scalar decided by the analyst, it can be 10, 100, 200, or even more). If the constraints are satisfied, there will be no penalty. Equation 6.220 presents what is known as a parabolic penalty. We minimize V by assigning some values to each of $P_1, P_2, \ldots, P_m$. We slowly increase the values of these Ps until V becomes invariant with respect to the Ps. At this stage, the ϕ's will be more or less exactly satisfied. The function V in Eq. 6.220 is frequently referred to as the composite objective function.

For a maximization problem, V becomes thus

$$V = y - (P_1\phi_1^2 + P_2\phi_2^2 + \cdots P_m\phi_m^2) \tag{6.221}$$

Though the minus and plus signs appear to be incongruous for the maximization and minimization problems, it needs to be reiterated that we gave a positive penalty for a cost function and negative penalty for a profit function. Other penalty functions are also possible. For example, we can have

$$V = y + \sum_{j \in R} P_j|\psi_j| \tag{6.222}$$

Here j is an element of the set R which contains all the violated constraints. We can employ P_j or a universal P. We take the modulus of ψ and if it violates, take the total sum of the violations, multiply by P, which will be a very huge quantity like 10^{20} or 10^{25}. This constraint could be $\psi_j > 0$.

When $\psi_j > 0$, the penalty is 0, while if $\psi_j < 0$, then P takes on the value 10^{20} or 10^{25}. P always looks at the value of ψ, whenever it becomes negative, P is active and gets added to the cost. Such a penalty is known as an **infinite barrier penalty**.

Immediately, a host of new ideas may strike us. Instead of $|\psi|$, we can have $\log(\psi)$ or $\log|\psi|$ and so on. We can come up with our own concept of penalty or barrier. Basically, we add to the cost and ensure that ψ never becomes negative. If ψ becomes negative, the product becomes huge and we will automatically correct it such that ψ is satisfied.

In Eq. 6.222, R is the set of violated constraints and we eventually ensure that they are not violated. Instead of checking each time whether the constraints are violated or not, we make them a part of the objective function, set up an infinite barrier penalty so that they are not violated, eventually when convergence is reached.

To start with, we will assign a small value to the Ps so that convergence is fast. We will get some solution and then keep changing the value of Ps and increase them till the solution proceeds in such a way that the Φ/s are not violated too much. A stage will come when the Φ/s are satisfied more or less exactly and regardless of the value of Ps, i.e., we will get the same value of V. That means we have reached convergence. In view of the above, it is evident that it is a lot more hard work to solve an optimization problem using the penalty function method, as the solution proceeds with guess values of Ps. However, on a computer, the solution is elegant and the penalty function method is a powerful tool to solve practical optimization problems.

Example 6.11 Revisit the cylindrical solar water heater (Minimize A, $V = 4m^3$) storage problem (see Example 5.3). Solve it using the penalty function method.

$$A = 2\pi r^2 + 2\pi rh \tag{6.223}$$

Minimize A, subject to

$$V = \pi r^2 h = 4 \tag{6.224}$$

Solution:

The composite objective function is given by

$$y = A + P(V - 4)^2 \tag{6.225}$$
$$y = 2\pi r^2 + 2\pi rh + P(\pi r^2 h - 4)^2 \tag{6.226}$$

Table 6.9 The output of the program for Example 6.11

S.No	P	y^+	A^+	r^+	h^+
1	1	12.46	10.65	0.75	1.51
2	10	13.81	13.67	0.85	1.71
3	50	13.92	13.89	0.86	1.71
4	100	13.94	13.94	0.85	1.76
5	500	13.95	13.95	0.86	1.72

where P is the penalty function. The resulting unconstrained optimization problem shown in Eq. 6.226 can be solved by an efficient single variable search, like Golden section search or the Fibonacci search, by considering one variable at a time. The resulting solution for various values of P is given in Table 6.9. From this Table, it is seen that at $P = 10$ itself, the y and A are quite close to each other, and at $P = 500$, the two are indistinguishable, thereby signifying that the volume constraint is almost exactly satisfied.

Hence, the final solution to this problem using the penalty function method is $r^+ = 0.86$ m, $h^+ = 1.72$ m, and $A^+ = 13.95$ m^2, which is very close to the solutions obtained earlier with a host of other methods.

MATLAB code for Example 6.11

```
1    clear;
2    clc;
3
4    a=0.5;   % lower limit of radius
5    b=1;     % upper limit of radius
6    c=1;     % lower limit of height
7    d=2;     % upper limit of height
8    p=100;   % penalty value
9
10   countmax=100; % number of evaluations
11   label=0;
12   count=0;
13
14   while label==0
15
16       count=count+1;
17       % 1st point selection according to Golden method
18       I=b-a;
19       J=d-c;
20       r1=a+I*0.618;
21       h1=c+J*0.618;
22       % 2nd point selection according to Golden method
23       r2=b-I*0.618;
24       h2=d-J*0.618;
25
26       % For first iteration fixing height as mid point
```

```
27      %of its range.
28      if count==1
29          h=(c+d)/2;
30      end
31      % objective function accoridng to 1st point
32      %of r for fixed h
33      y1=2*pi*r1^2+2*pi*r1*h+p*(pi*r1^2*h-4)^2;
34
35      % objective function accoridng to 2nd point
36      %of r for fixed h
37      y2=2*pi*r2^2+2*pi*r2*h+p*(pi*r2^2*h-4)^2;
38
39      if y1>y2
40          b=r1;
41          r=r1;
42      else
43          a=r2;
44          r=r2;
45      end
46      % objective function accoridng to 1st point
47      %of h for fixed r
48      y3=2*pi*r^2+2*pi*r*h1+p*(pi*r^2*h1-4)^2;
49
50      % objective function accoridng to 2nd point
51      %of h for fixed r
52      y4=2*pi*r^2+2*pi*r*h2+p*(pi*r^2*h2-4)^2;
53      if y3>y4
54          d=h1;
55          h=h1;
56      else
57          c=h2;
58          h=h2;
59      end
60      A= 2*pi*r^2+2*pi*r*h;
61      if count==countmax
62
63          label=1;
64      end
65
66      % Print
67      prt = ['Itr = ',num2str(count) ,...
68          ', y1 = ',num2str(y1) ,...
69          ', y2 = ',num2str(y2) ,...
70          ', A = ',num2str(A) ,...
71          ', r1 = ',num2str(r1) ,...
72          ', r2 = ',num2str(r2) ,...
73          ', h1 = ',num2str(h1) ,...
74          ', h2 = ',num2str(h1)];
75      disp(prt)
76
77  end
```

6.7 Multi-objective Optimization

So far in this book, we have looked at optimization problems with a single objective. However, often in engineering as in life, multiple objectives are present and these are invariably in conflict with each other. Consider the problem of choosing a personal car. There are many objectives but comfort and cost often hold the key. Unfortunately, these are orthogonal as can be seen from the figure below.

From Fig. 6.18, it can be seen that multiple solutions exist and we cannot minimize cost and maximize comfort simultaneously. Some higher order information will be required to choose the most desirable car. This, for example, may be weightage for cost and weightage for comfort. Let us now consider a problem in thermal engineering involving heat exchangers. The two key objectives in a heat exchanger are

1. Heat transfer, Q
2. Pressure drop, ΔP.

The goal would often be to maximize Q and minimize ΔP. However, unfortunately, these two objectives are at conflict with each other. If one tries to increase the velocity so as to increase the heat transfer coefficient and thereby increase Q, the ΔP will simultaneously increase. Therefore, an optimization procedure would typically yield a lot of solutions called Pareto-optimal solutions instead of just a single solution. Pareto-optimal solutions are those in which no solution is better than the other solutions on the front, as an automatic degradation of one objective results if we try to improve the other objective. Stated more explicitly, a unique optimum does not exist. These typically look like what is shown in Fig. 6.19.

The abscissa of the plot is ΔP and the ordinate is 1/Q. It is apparent that we are seeking a minimum to both these objectives. However, from the Pareto plot, it is seen that it is impossible to get a solution where both the quantities are minimum simultaneously. Each point on the Pareto front actually represents the optimal solution with

Fig. 6.18 Plot showing comfort level versus cost required

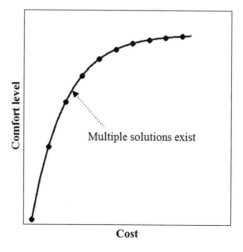

Fig. 6.19 Plot showing a
Pareto front, positive ideal,
and negative ideal solutions

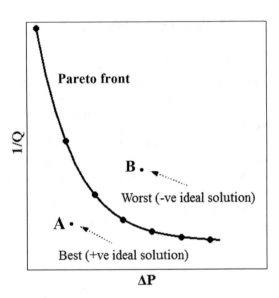

different weightages or preferences attached to two objectives. Solutions at the top
end of the curve imply more preference to low ΔP, while solutions at the lower end
of the curve represent a higher preference to maximize Q. On this plot, a hypothetical
solution which represents maximum Q and minimum ΔP that are obtained if optimization problem is solved separately for maximizing Q and minimizing ΔP is also
shown as the positive ideal solution. On the same plot, the lowest Q and the highest
ΔP are represented as the negative ideal solution. Therefore, one way of solving
this multi-objective problem will be to minimize the distance from the positive ideal
solution and maximize the distance from the negative ideal solution. Let us now look
at the typical formulation for a n-variable M-objective optimization problem with
I-equality constrains and J-inequality constraints.

$$\text{Min (or Max)} Y_m(x_1, x_2, \ldots x_n), \quad m = 1, 2, \ldots, M \quad (6.227)$$
$$\text{subject to } \phi_i(x_1, x_2, \ldots x_n) = 0, \quad i = 1, 2, \ldots, I \quad (6.228)$$
$$\psi_j(x_1, x_2, \ldots x_n) \geq 0, \quad j = 1, 2, \ldots, J \quad (6.229)$$
$$x_k^L \leq x_k \leq x_k^U \quad k = 1, 2, \ldots, n \quad (6.230)$$

A simpler way of solving a multi-objective problem is to convert into an equivalent
single objective problem by introducing a composite objective function that gives
different preferences to each of the objectives. For the heat exchanger problem under
consideration, the composite objective function may be proposed as follows:

$$\text{Min } V = \gamma \left(\frac{\left(\frac{1}{Q}\right)}{\left(\frac{1}{Q_{max}}\right)} \right) + (1 - \gamma) \left(\frac{\Delta P}{\Delta P_{min}} \right) \quad (6.231)$$

Here γ is preference. $0 \le \gamma \le 1$. In the equation, Q is the solution corresponding to the maximum heat transfer, and ΔP_{min} is the solution corresponding to the minimum pressure drop. The resulting single objective problem may then be solved using any of the known methods. It is to be noted that the final solution is critically dependent on the choice of γ which is a user-defined input.

Example 6.12 Consider the multi-objective optimization problem given below.

$$y_1(x_1, x_2) = x_1^2 + x_2^2$$
$$y_2(x_1, x_2) = (x_1 - 1)^2 + x_2^2$$

By using the preference method (i.e., by introducing a composite objective function), determine the optimal solution for each of the weight vectors of the form $(\gamma, 1-\gamma)$ that are given below.

(i)   (0, 1)
(ii) (0.25, 0.75)
(iii) (0.5, 0.5)
(iv) (0.75, 0.25)
(v)   (1, 0)

Solution:

Let us write a composite objective function in the following form.

$$\text{Minimize } y = \gamma y_1 + (1 - \gamma) y_2 \tag{6.232}$$

We now solve it using the Cauchy's method.

(i) $\boxed{\gamma = 0, (1 - \gamma) = 1}$

$y = (x_1 - 1)^2 + x_2^2$
Let us start with (2, 2). As the initial guess value,

$$\partial y / \partial x_1 = 2x_1 - 2 \tag{6.233}$$
$$\partial y / \partial x_2 = 2x_2 \tag{6.234}$$

Now we have to calculate y at the new points and optimize.

Iteration 1:

$$\partial y / \partial x_1 = 2 \tag{6.235}$$
$$\partial y / \partial x_2 = 4 \tag{6.236}$$

$$y(2 + \alpha \partial y/\partial x_1, 2 + \alpha \partial y/\partial x_2) = y(2 + 2\alpha, 2 + 4\alpha) \qquad (6.237)$$

$$y = [(2 + 2\alpha - 1)^2 + (2 + 4\alpha)^2] \qquad (6.238)$$

$$dy/d\alpha = 0; \qquad (6.239)$$

$$\alpha = -0.5 \qquad (6.240)$$

New values of (x_1, x_2) are:

$x_1 = 2 + 2 \times (-0.5) = 1$; $x_2 = 2 + 4 \times (-0.5) = 0$; y^+ at $(1, 0)$

Iteration 2:

$$\partial y/\partial x_1 = 2x_1 - 2 = 0 \qquad (6.241)$$

$$\partial y/\partial x_2 = 2x_2 = 0 \qquad (6.242)$$

$$\Delta x_1 = 0 \qquad (6.243)$$

$$\Delta x_2 = 0 \qquad (6.244)$$

New values of (x_1, x_2) are:

$x_1 = 1$; $x_2 = 0$;

y^+ at $(1, 0)$

(ii) $\boxed{\gamma = 0.25, (1 - \gamma) = 0.75}$

$y = 0.25 \times (x_1^2 + x_2^2) + 0.75 \times ((x_1 - 1)^2 + x_2^2)$

Let us start with (2,2). As the initial guess value,

$$\partial y/\partial x_1 = 2x_1 - 1.5 \qquad (6.245)$$

$$\partial y/\partial x_2 = 2x_2 \qquad (6.246)$$

Now we have to calculate y at the new points and optimize.

Iteration 1:

$$\partial y/\partial x_1 = 2.5 \qquad (6.247)$$

$$\partial y/\partial x_2 = 4 \qquad (6.248)$$

$$y(2 + \alpha \partial y/\partial x_1, 2 + \alpha \partial y/\partial x_2) = y(2 + 2.5\alpha, 2 + 4\alpha)$$
$$(6.249)$$

$$y = [0.25 \times ((2 + 2.5\alpha)^2 + (2 + 4\alpha)^2) + 0.75 \times ((2 + 2.5\alpha - 1)^2 + (2 + 4\alpha)^2)]$$
$$(6.250)$$

$$\mathrm{d}y/\mathrm{d}\alpha = 0; \tag{6.251}$$
$$\alpha = -0.5 \tag{6.252}$$

New values of (x_1, x_2) are
$x_1 = 2 + 2.5 \times (-0.5) = 0.75; \quad x_2 = 2 + 4 \times (-0.5) = 0; \quad y^+$ at $(0.75, 0)$
Iteration 2:

$$\partial y/\partial x_1 = 2x_1 - 1.5 = 0 \tag{6.253}$$
$$\partial y/\partial x_2 = 2x_2 = 0 \tag{6.254}$$
$$\Delta x_1 = 0 \tag{6.255}$$
$$\Delta x_2 = 0 \tag{6.256}$$

New values of (x_1, x_2) are:
$x_1 = 0.75; \quad x_2 = 0;$
y^+ at $(0.75, 0)$

(iii) $\boxed{\gamma = 0.5, (1 - \gamma) = 0.5}$

$y = 0.5 \times (x_1^2 + x_2^2) + 0.5 \times ((x_1 - 1)^2 + x_2^2)$
Let us start with $(2, 2)$. As the initial guess value,

$$\partial y/\partial x_1 = 2x_1 - 1 \tag{6.257}$$
$$\partial y/\partial x_2 = 2x_2 \tag{6.258}$$

Now we have to calculate y at the new points and optimize.

Iteration 1:

$$\partial y/\partial x_1 = 3 \tag{6.259}$$
$$\partial y/\partial x_2 = 4 \tag{6.260}$$

$$y(2 + \alpha \partial y/\partial x_1, 2 + \alpha \partial y/\partial x_2) = y(2 + 3\alpha, 2 + 4\alpha) \tag{6.261}$$
$$y = [0.5 \times ((2 + 3\alpha)^2 + (2 + 4\alpha)^2) + 0.5 \times ((2 + 3\alpha - 1)^2 + (2 + 4\alpha)^2)] \tag{6.262}$$

$$\mathrm{d}y/\mathrm{d}\alpha = 0; \tag{6.263}$$
$$\alpha = -0.5 \tag{6.264}$$

New values of (x_1, x_2) are
$x_1 = 2 + 3 \times (-0.5) = 0.5; \quad x_2 = 2 + 4 \times (-0.5) = 0; \quad y^+$ at $(0.5, 0)$

Iteration 2:

$$\partial y/\partial x_1 = 2x_1 - 1 = 0 \qquad (6.265)$$
$$\partial y/\partial x_2 = 2x_2 = 0 \qquad (6.266)$$
$$\Delta x_1 = 0 \qquad (6.267)$$
$$\Delta x_2 = 0 \qquad (6.268)$$

New values of (x_1, x_2) are:
$x_1 = 0.5$; $x_2 = 0$;
y^+ at $(0.5, 0)$

(iv) $\boxed{\gamma = 0.75, (1 - \gamma) = 0.25}$

$y = 0.75 \times (x_1^2 + x_2^2) + 0.25 \times ((x_1 - 1)^2 + x_2^2)$
Let us start with $(2, 2)$. As the initial guess value,

$$\partial y/\partial x_1 = 2x_1 - 0.5 \qquad (6.269)$$
$$\partial y/\partial x_2 = 2x_2 \qquad (6.270)$$

Now we have to calculate y at the new points and optimize.

Iteration 1:

$$\partial y/\partial x_1 = 3.5 \qquad (6.271)$$
$$\partial y/\partial x_2 = 4 \qquad (6.272)$$

$$y(2 + \alpha \partial y/\partial x_1, 2 + \alpha \partial y/\partial x_2) = y(2 + 3.5\alpha, 2 + 4\alpha) \qquad (6.273)$$
$$y = [0.75 \times ((2 + 3.5\alpha)^2 + (2 + 4\alpha)^2) + 0.25 \times ((2 + 3\alpha - 1)^2 + (2 + 4\alpha)^2)] \qquad (6.274)$$

$$dy/d\alpha = 0; \qquad (6.275)$$
$$\alpha = -0.5 \qquad (6.276)$$

New values of (x_1, x_2) are:
$x_1 = 2 + 3.5 \times (-0.5) = 0.25$; $x_2 = 2 + 4 \times (-0.5) = 0$; y^+ at $(0.25, 0)$
Iteration 2:

$$\partial y/\partial x_1 = 2x_1 - 0.5 = 0 \qquad (6.277)$$
$$\partial y/\partial x_2 = 2x_2 = 0 \qquad (6.278)$$
$$\Delta x_1 = 0 \qquad (6.279)$$
$$\Delta x_2 = 0 \qquad (6.280)$$

New values of (x_1, x_2) are
$x_1 = 0.25;\ x_2 = 0;$
y^+ at $(0.25, 0)$

(v) $\boxed{\gamma = 1,\ (1 - \gamma) = 0}$

$y = (x_1^2 + x_2^2)$
Let us start with $(2, 2)$. As the initial guess value,

$$\partial y/\partial x_1 = 2x_1 \tag{6.281}$$
$$\partial y/\partial x_2 = 2x_2 \tag{6.282}$$

Now we have to calculate y at the new points and optimize.

Iteration 1:

$$\partial y/\partial x_1 = 4 \tag{6.283}$$
$$\partial y/\partial x_2 = 4 \tag{6.284}$$

$$y(2 + \alpha \partial y/\partial x_1, 2 + \alpha \partial y/\partial x_2) = y(2 + 4\alpha, 2 + 4\alpha) \tag{6.285}$$
$$y = [(2 + 4\alpha)^2 + (2 + 4\alpha)^2)] \tag{6.286}$$

$$\mathrm{d}y/\mathrm{d}\alpha = 0; \tag{6.287}$$
$$\alpha = -0.5 \tag{6.288}$$

New values of (x_1, x_2) are:
$x_1 = 2 + 4 \times (-0.5) = 0;\ x_2 = 2 + 4 \times (-0.5) = 0;\ y^+$ at $(0, 0)$
Iteration 2:

$$\partial y/\partial x_1 = 2x_1 = 0 \tag{6.289}$$
$$\partial y/\partial x_2 = 2x_2 = 0 \tag{6.290}$$
$$\Delta x_1 = 0 \tag{6.291}$$
$$\Delta x_2 = 0 \tag{6.292}$$

New values of (x_1, x_2) are
$x_1 = 0;\ x_2 = 0;$
y^+ at $(0, 0)$

The optima for the five Cases (i)–(v) are shown in Table 6.10
The method seen above has actually not treated the multi-objective problem in its full strength. In fact, the multi-objective problem has been converted to an equivalent

Table 6.10 Optimum values for given $(\gamma, 1 - \gamma)$

S.No	γ	$1 - \gamma$	y^+	x_1^+	x_2^+
1	0	1	0	1	0
2	0.25	0.75	0.1875	0.75	0
3	0.5	0.5	0.25	0.5	0
4	0.75	0.25	0.1875	0.25	0
5	1	0	0	0	0

single objective problem with the introduction of weights. While the introduction of weights seems to be intellectually appealing, often one is confronted with conflicting multiple objectives that have different dimensions and units.

In view of this, if the equivalent weighted single objective function is to be meaningful, all the objectives have to be scaled with respect to their extrema. This is tantamount to solving n single objective problems where n is the number of objectives. Over and above this, the choice of weights is entirely subjective which can cut both ways if one is looking at an engineering solution to the multi-objective problem.

6.7.1 TOPSIS Method

In the light of the foregoing discussion, it is quite clear that the optimization community would have developed better methods to solve the multi-objective problems. One such powerful method is TOPSIS (The Technique for Order of Preference by Similarity to Ideal Solution) method which is a multi-criteria decision analysis method. This method was first developed by Ching-Lai Hwang and Yoon in 1981 and was further developed by Yoon in 1987 and Hwang et al. in 1993.

The algorithm for this method is as follows:

- Create a matrix containing m different alternatives and n objectives. Each element a_{ij} represents the value of ith alternative for jth objective, where $i = 1, 2, \ldots, m$ and $j = 1, 2, \ldots, n$. The matrix can be represented as $(a_{ij})_{m \times n}$.
- The elements of the matrix are normalized as follows:

$$an_{ij} = \frac{a_{ij}}{\sqrt{\Sigma_{k=1}^{m} a_{kj}^2}} \tag{6.293}$$

$i = 1, 2, \ldots, m$ and $j = 1, 2, \ldots, n$

- Use the pre-decided weights for the objective and create a weight normalized decision matrix. The weights must be decided such that $\Sigma_{j=1}^{n} w_j = 1$ The weight normalized decision matrix is calculated as

$$awn_{ij} = an_{ij}.w_j \tag{6.294}$$

$$i = 1, 2, \ldots, m \text{ and } j = 1, 2, \ldots, n$$

- Determine the best alternative, i.e., positive ideal solution (I_j^+) and worst alternative, i.e., negative ideal solution (I_j^-) with respect to the objectives.
- Calculate the Euclidean distances from the positive ideal solution (d_i^+) and the negative ideal solution (d_i^+).

$$d_i^+ = \sqrt{\Sigma_{j=1}^n (awn_{ij} - I_j^+)^2} \tag{6.295}$$

$$d_i^- = \sqrt{\Sigma_{j=1}^n (awn_{ij} - I_j^-)^2} \tag{6.296}$$

- The proximity of each alternative to positive ideal solution is calculated by

$$D_i^+ = \frac{d_i^-}{d_i^- + d_i^+} \tag{6.297}$$

- Rank the alternatives with respect to D_i^+ such that the alternative with the highest D_i^+ values is the best ranked.

The above algorithm will become clear after solving the Example 6.13.

Example 6.13 Consider a thermal system, whose heat dissipation rate is given by $Q = 2.5 + 6.2v^{0.8}$, where Q is in kW and v is the velocity in m/s of the fluid being used as the medium for accomplishing the heat transfer. The accompanying pumping power is given by $P = 1.3 + 0.04v^{1.8}$, again in kW with v in m/s (in both the expressions, the constants ensure that both Q and P are in kW). It is desired to maximize Q and minimize P. Solve this multi-objective optimization problem using the TOPSIS method.

Solution:

Given,

$$\text{Maximize } Q = 2.5 + 6.2v^{0.8} \tag{6.298}$$
$$\text{Minimize } P = 1.3 + 0.04v^{1.8} \tag{6.299}$$

$$\text{subject to, } 3 \leq v \leq 12 \tag{6.300}$$

This is a two objective, one variable problem. From Fig. 6.20, one can observe that the two objectives are conflicting with each other. The TOPSIS algorithm described above can be used to solve this multi-objective optimization problem.

Select some values for velocity in the given range. Evaluate the values of P and Q for all the selected values of velocity. The values of P and Q are normalized to

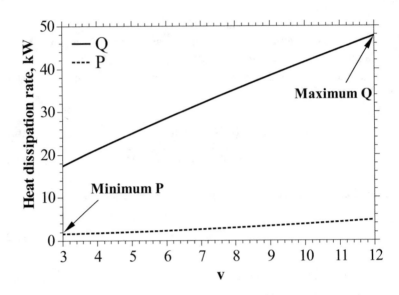

Fig. 6.20 Heat dissipation rate (Q) and Pumping power (P) versus velocity (v)

Table 6.11 Values obtained after applying TOPSIS method

v	P	Q	Pn	Qn	Pwn	Qwn	d^+	d^-	D
3	1.589	17.430	0.160	0.159	0.080	0.079	0.138	0.161	0.538
4	1.785	21.294	0.179	0.194	0.089	0.097	0.121	0.153	0.557
5	2.024	24.968	0.203	0.228	0.102	0.114	0.106	0.144	0.574
6	2.306	28.496	0.232	0.260	0.116	0.130	0.095	0.135	0.587
7	**2.628**	**31.908**	**0.264**	**0.292**	**0.132**	**0.146**	**0.089**	**0.128**	**0.588**
8	2.989	35.223	0.301	0.322	0.150	0.161	0.090	0.122	0.573
9	3.387	38.457	0.341	0.351	0.170	0.175	0.100	0.119	0.544
10	3.823	41.619	0.385	0.380	0.192	0.190	0.116	0.121	0.510
11	4.296	44.718	0.432	0.409	0.216	0.204	0.137	0.127	0.481
12	4.804	47.762	0.483	0.437	0.241	0.218	0.161	0.138	0.461

Pn and Qn based on Eq. 6.293. For the values given in Table 6.11, $\sqrt{\Sigma P_i^2} = 9.9301$ and $\sqrt{\Sigma Q_i^2} = 109.2875$. Therefore, we get the value of Pn and Qn by dividing P and Q with 9.9301 and 109.2875, respectively.

Let us give equal weightage to both the objectives, i.e., $w_p = 0.5$ for minimizing P and $w_q = 0.5$ for maximizing Q. We can now evaluate the weight normalized values Pwn and Qwn using Eq. 6.294. The values are shown in Table 6.11.

The positive ideal solution for this multi-objective problem is given by the maximum value of Qwn, i.e., 0.218 and minimum value of Pwn, i.e., 0.080. The negative ideal solution is given by the minimum value of Qwn, i.e., 0.079 and maximum value

Table 6.12 TOPSIS solutions for different weights

S. No	w_p	w_q	v^+	P^+	Q^+
1	0	1	12	4.8	47.76
2	0.25	0.75	11.2	4.39	45.33
3	0.5	0.5	7	2.63	31.91
4	0.75	0.25	4	1.78	21.29
5	1	0	3	1.59	17.43

of P_{wn}, i.e., 0.241. Therefore, the distances (d_i^+ and d_i^-) of a point (P, Q) from the positive ideal solution (0.080, 0.218) and negative ideal solution (0.241, 0.079) are calculated as shown in Table 6.11.

The proximity to the positive ideal solution (D^+) for all the points is calculated and it is observed that $D^+ = 0.588$ is the maximum and it is obtained for $v = 7$ m/s.

Therefore, the solution to this multi-objective problem using the TOPSIS method is

$$v^+ = 7 \text{ m/s},$$
$$P^+ = 2.628 \text{ kW}$$
$$Q^+ = 31.908 \text{ kW}$$

The above solution is obtained when both the objectives are given equal weightage, i.e., ($w_p = w_q = 0.5$). The solution changes if both the objectives are given a different weightage as shown in Table 6.12.

Additionally, the TOPSIS method requires specification of weights which is quite intuitive from an engineering perspective. If we look at Row 1 from Table 6.12, it corresponds to the highest velocity which results in the highest heat dissipation (Q) and the highest pumping power (P). Whereas, if we look at Row 5 from Table 6.12, it corresponds to the lowest velocity which results in the lowest heat dissipation (Q) and the lowest pumping power (P).

Hence TOPSIS method has helped us to arrive at a compromise where there is a penalty involved in deviating from the solution corresponding to maximum Q and minimum P. This is done by minimizing the distance from the positive ideal solution.

MATLAB code for Example 6.13 is given below.

<u>MATLAB code for Example 6.13</u>

```
1   clear;
2   clc;
3
4   v1 = 3;
5   v2 = 12;
6
7   dv = 1;
```

```
8   n = (v2–v1)/dv + 1;
9
10  v = linspace(v1,v2,n);
11
12  P = 1.3 + 0.04.*v.^1.8;
13  Q = 2.5+6.2.*v.^0.8;
14
15  wp = 0.5;
16  wq = 1–wp;
17
18  Pn = P./sqrt(sum(P.^2));
19  Qn = Q./sqrt(sum(Q.^2));
20
21  Pwn = Pn*wp;
22  Qwn = Qn*wq;
23
24  p_best = min(Pwn);
25  p_worst = max(Pwn);
26  q_best = max(Qwn);
27  q_worst = min(Qwn);
28
29  d_best = sqrt((Pwn – p_best).^2 + (Qwn – q_best).^2 );
30  d_worst = sqrt((Pwn – p_worst).^2 + (Qwn – q_worst).^2 );
31
32  D = d_worst./(d_worst + d_best);
33
34  [P_sort,i] = sort(D,'descend');
35
36  % Print
37  prt = ['velocity (v) = ',num2str(v(i(1))) ,...
38       ', Heat dissipation rate (Q) = ',num2str(Q(i(1))) ,...
39       ', Pumping power = ',num2str(P(i(1)))];
40  disp(prt)
```

The output of the program is

Solution from TOPSIS method is:
velocity (v) = 7 m/s,
Heat dissipation rate (Q) = 31.9083 kW,
Pumping power (P) = 2.6281 kW

Problems

6.1 Determine the minimum of the function $y = x^2 - [(40x^2 + 1)/x] + 6$ in the interval $5 \leq x \leq 45$ using the Fibonacci search method. The required final uncertainty in x should be less than 1.

6.2 Conduct a single variable search to minimize the following function $y = 2x^2 + \frac{7}{x^3} - 4$, in the interval $0.5 \le x \le 5$, using the equal interval exhaustive search and the dichotomous search by performing 12 function evaluations. Compare the reduction ratio (RR) of the two algorithms at the end of the 12 function evaluations.

6.3 In a constrained optimization problem involving two costs, the first term is $0.08x_1x_2$ while the second term is $1.04 \times 10^{-7}x_1^3$. There is a constraint that appears as $x_1^2 - x_2 = 1.23 \times 10^6$. Convert this into a single variable unconstrained optimization problem in total cost. If the initial interval of uncertainty is $100 \le x_1 \le 1000$, using the Golden section search carry out a single variable search, until a final uncertainty of less than 50 in x_1 is reached.

6.4 Solve problem 6.3 using the dichotomous search for the required final uncertainty of less than 50 on x_1.

6.5 Consider a thermal system, whose heat dissipation rate is given by $Q = 2.5 + 6.2v^{0.8}$, where Q is in kW and v is the velocity in m/s of the fluid being used as the medium for accomplishing the heat transfer. The accompanying pumping power is given by $P = 1.3 + 0.04v^{1.8}$, again in kW with v in m/s (in both the expressions, the constants ensure that both Q and P are in kW). It is desired to maximize the performance parameter Q/P. Conduct a one variable search in v to obtain the optimal velocity by using

(a) Dichotomous search,
(b) Golden section search,
(c) Fibonacci search with the initial interval of uncertainty being $3 \le v \le 12$ m/s. A final interval of uncertainty of 0.25 m/s or less on v is desired.

6.6 A fully closed rectangular box is to be made of sheet metal of area 3 m². It is desired to make a box with the maximum volume. The three dimensions of the box are length-x, breadth-y, and height z (all in m). (a) Convert this into a two variable unconstrained optimization problem in x and y (b) Using the steepest ascent method with an initial guess of $x = y = 0.8$ m, determine the optimal dimensions of the box. Employ a reasonable stopping criterion.

6.7 Revisit exercise Problem 6.3. Solve it using the penalty function method for $P = 1, 10, 100$. Use an appropriate search technique for solving the resultant unconstrained optimization problem. You are encouraged to solve this problem on a computer by writing your own code, or using an analysis tool like MATLAB.

6.8 Revisit exercise Problem 5.7. Convert the problem into a 2 variable, 1 constraint optimization problem as mentioned in Problem 5.7. Solve it using the penalty function method for $P = 1, 10, 100$. Use Cauchy's steepest descent technique for solving the resultant unconstrained optimization problem. You are encouraged to solve this problem on a computer by writing your own code, or using an analysis tool like MATLAB.

6.9 Revisit Problem 6.5. (i) Develop a composite objective function, which is to be maximized, for this multi-objective optimization problem using the weighted sum approach and with the help of dimensionless objective functions (i.e., Q/Q-max and $\Delta P_{min}/\Delta P$) along with a weighting factor of γ. (For convenience, you may want to use $1/\Delta P$ in the composite objective function). (ii) For $\gamma = 0, 0.5$, and 1, solve the multi-objective problem with the Golden section search (single variable in velocity, v), wherever the solution is not obvious from common sense. Take the initial interval of uncertainty as $3 \le v \le 12$ m/s. A level of uncertainty of 0.5 m/s on the velocity is required.

Chapter 7
Linear Programming and Dynamic Programming

7.1 Introduction

More or less, we, by now, have seen the techniques that are applicable to problems
frequently encountered in thermal sciences. We now look at two techniques that are
not so frequently used in thermal sciences namely (i) Linear programming and (ii)
Dynamic programming. Linear programming is a very important technique used in
areas like management sciences, operations research, and sociology. Yet there are
some problems in engineering that can be eminently handled using linear programming or dynamic programming and it would be instructive to know how they work.

7.2 Linear Programming or LP

**An LP problem is one in which both the objective function and all the con-
straints can be represented as linear combinations of the variables.** So we will
not encounter $sinh(x)$, $tanh(x)$, e^x terms either in the objective function or the constraints. Obviously, now we can see why it is not frequently used in heat transfer, fluid
mechanics, and thermal sciences, as there are only very few situations where both
the objective function and constraints can be written as linear combinations of the
variables. But if there is such a situation, then there is a body of knowledge which is
available and can be used, instead of trying to rely on either the Lagrange multiplier
method or the penalty function method or a conventional search technique.

LP was first used in World War II for the optimal allocation of men or ammunition,
aircraft, and artillery for maximizing the strategy. It was first tried by the allied forces
and origins can be traced to the UK in the 1930s and 1940s. Subsequently, the subject
of operations research, in which LP occupies a pride of place, has now become a
very mature field in its own right.

LP is also used in sociology and industrial engineering but has limited applications
in engineering. Some applications in mechanical engineering where this technique

© The Author(s) 2021
C. Balaji, *Thermal System Design and Optimization*,
https://doi.org/10.1007/978-3-030-59046-8_7

can be applied are the optimal allocation of jobs to a special purpose machine like a CNC lathe, optimal use of labor, optimal use of several machines on a shop floor, and so on. There is a total machine availability and there are cost and time allocations that have to be done.

We also constantly do this linear programming in our minds. Students often try to maximize the pleasure such that their CGPA does not fall below say 7.5, when CGPA is written as a constraint. Alternatively, students may try to maximize the CGPA subject to minimum effort or maximize the CGPA subject to minimum pleasure. By nature, everybody wants to optimize! Optimization lies not only at the root of engineering but at the root of life itself. Consider the example of an optimal product mix in a petroleum refinery. Here, we have raw material costs, refining costs, and selling price. With the goal being maximizing profit, the challenge before an analyst is to decide whether everything we want to sell is only petrol or diesel or kerosene or a mix of these. This is a classic LP problem in which the objective function and constraints arising due to mass balance, refining capacities, uneven demand for different products, transportation costs can all be written as linear combinations of the variables under consideration.

7.2.1 Formulation of an LP Problem

An objective function, y, whose extremum is sought is first defined as follows:

$$y = C_1 x_1 + C_2 x_2 + C_3 x_3 + \cdots + C_n x_n \tag{7.1}$$

where the function y is written as a linear combination of the variables x_1 to x_n, which could probably represent the products of a company or manufactures. For example, for a furniture company, x_1 and x_2 can be tables and chairs, respectively. For a petroleum refinery, x_1 can be petrol, x_2 can be diesel, and so on. The constants C_1 to C_n may be the profit associated with each product or the cost of each, depending on whether the objective function y has to be maximized, if y is the profit or has to be minimized, if y is the total cost. The maximization or minimization is subject to some constraints. These are

$$\psi_1 = a_{11} x_1 + a_{12} x_2 + \cdots + a_{1n} x_n \geq r_1 \tag{7.2}$$
$$\psi_2 = a_{21} x_1 + a_{22} x_2 + \cdots + a_{2n} x_n \geq r_1 \tag{7.3}$$
$$\vdots$$
$$\psi_j = a_{j1} x_1 + a_{j2} x_2 + \cdots + a_{jn} x_n \geq r_j \tag{7.4}$$
$$\text{where } x_1, x_2 \ldots x_n \geq 0 \tag{7.5}$$

It is possible to replace all the above inequalities by equalities, by introducing slack variables, but for starters, we can retain the inequalities here. In the above equations,

$x_1, x_2, x_3 \ldots x_n$ are all positive. Therefore, we have to write out an extra constraint that none of these variables can become negative. These are called the non-negativity constraints. If we solve the resultant system of equations, we obtain the optimal values of x_1 to x_n at which y is maximized or minimized, as the case may be. C_1 to C_n and a_{11} to a_{jn} are all coefficients that are known upfront. So when we get the optimal values of x_1 to x_n, we can determine the value of y^+ right away.

7.2.2 Advantages of LP Method Over Lagrange Multiplier Method

Immediately, some of us feel that the Lagrange multiplier method can be used with the Kuhn–Tucker conditions for solving the system represented by Eqs. 7.1–7.5. But the catch is if it were possible to solve such a system using the Lagrange multiplier method and the Kuhn Tucker conditions, why then were special optimization techniques like LP developed? There must have been some special reasons. What are these?

There are some key differences between this problem and the Lagrange multiplier formulation. Please remember that though we handled inequalities, (it was just one or 2), most were equalities for the problems handled through the Lagrange multipliers. *If everything is an inequality, the Kuhn–Tucker conditions get very messy.* So, essentially the Lagrange multiplier method is used for solving problems in which we have primarily equality constraints. For one or a few inequality constraints, the KTC formulation is very advantageous. The LP on the other hand is basically designed for handling inequality constraints.

7.2.3 Key Differences with Respect to the Lagrange Multiplier Method

1. In LP, almost all the constraints are inequalities.
2. j can be greater than n where j-number of constraints and n-number of variables. Not possible with Lagrange multipliers.

This is possible because they are inequalities and inequalities only reduce the feasible domain; these prohibit certain solutions. But equality constraints have to be exactly obeyed. This is the reason why the equality constraint is a lot more restrictive than the inequality constraint. Therefore, in the Lagrange multiplier method, if "m" is the number of equality constraints and "n" is the number of variables, then $m \leq n$. The advantage with Lagrange multiplier method is that the objective function can be as complicated as it can get and need not be a simple linear function as in this case. Even so, we can have any number of inequality constraints in the LP problem.

7.2.4 Why Is Everybody so Fond of Linear Programming?

The industrial engineering community has developed **very advanced commercial software for solving large LP problems**. Many business people use it for maximizing their profit and many analysts also use it. Therefore the LP is widely patronized. Furthermore, we are always looking at issues like, what happens if the cost of certain processes or items change. What happens if the selling price of this changes if the demand goes up. We want to evaluate the sensitivity of our optimal solutions to these changes. **Sensitivity analysis can be eminently and easily handled by LP**. So largely for these two reasons, namely, (i) the availability of commercial software and (ii) ease of doing sensitivity analysis or post-optimality analysis, LP is very popular.

For a slight degree of nonlinearity, for example, if a variable comes as x_3^2, one can assume this to be a variable say $x_{10} = x_3^2$ and continue working out the problem using LP. So a reasonable or slight amount of nonlinearity can be handled using the LP method.

7.2.5 Graphical Method of Solving an LP Problem

We now look at the graphical method of solving an LP problem. Though the graphical method has limited scope as only two-variable problems can be handled, all the features of an LP problem can be elegantly brought out using the graphical solution itself.

Example 7.1 A furniture company can make tables or chairs or both. The amount of wood required for making one table is 30 kg while that for a chair is 18 kg. The total quantity of wood available is 300 kg. The labor required for making a table is 20 man hours, while that for a chair is 8 man hours. The total number of man hours available is 160 (say for one day). The profit from a table is Rs. 1500, while that from the chair is Rs. 800. Determine the optimal product mix of the company using the graphical method of LP.

Solution

First we formulate the LP problem and then plot the constraints. Let x_1 be the number of tables and x_2 be the number of chairs.

$$\text{Maximize: } y = 1500x_1 + 800x_2 \tag{7.6}$$

Subject to:

(i) Material constraint:

$$30x_1 + 18x_2 \leq 300 \quad or \quad 3x_1 + 1.8x_2 \leq 30 \tag{7.7}$$

Fig. 7.1 Example 7.1 with
the labor constraint plotted

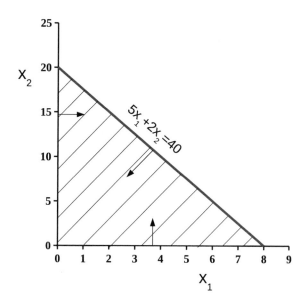

It is not that after making x_1 tables and x_2 chairs, the 300 kg of wood need to be finished, there could be some left too, which is why we are using the less than equal to sign rather than the equality sign. (In this sense, it is different from the mass balance or energy balance equation in thermal sciences). There can be residual wood at the end of the day. (ii) Labor constraint:

$$20x_1 + 8x_2 \leq 160 \ or \ 5x_1 + 2x_2 \leq 40 \tag{7.8}$$

(iii) Non-negativity constraint:

$$x_1 \geq 0 \ \text{and} \ x_2 \geq 0 \tag{7.9}$$

We first plot the labor constraint $5x_1 + 2x_2 \leq 40$. When $x_1 = 0$, $x_2 = 20$; when $x_2 = 0$, $x_1 = 8$. So we get a straight line that satisfies the equation $5x_1 + 2x_2 = 40$. The feasible region lies below this line. Since $x_1 \geq 0$ and $x_2 \geq 0$, we insert arrows pointing inwards as shown in Fig. 7.1. So the solution can only be in the first quadrant and these constraints only reduce the feasible region. Let us now plot the material constraint given by $3x_1 + 1.8x_2 = 30$. The feasible region is indicated by the shaded area in Fig. 7.2. Any point within the feasible region will satisfy all the constraints. But each point will result in a different value of y. So first, we have identified the region in which y_{max} is expected to lie, because y_{max} should first be a feasible solution. Now getting to y_{max} is not so straightforward. We have to assume some value of y, draw the straight line representing the profit function $y = 1500x_1 + 800x_2$ and keep moving the straight line till it just moves out of the feasible region. So when the line just moves out of the feasible region, the points it cuts in the feasible

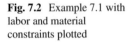
Fig. 7.2 Example 7.1 with
labor and material
constraints plotted

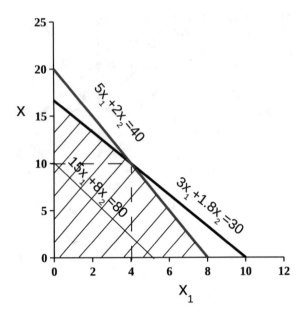

region is/are the optimal solutions to the problem. So let us take one iso-objective line $y = 1500x_1 + 800x_2 = 8000$. To plot this line, we need two points. These are $(5.33, 0)$ and $(0, 10)$. This line is shown in Fig. 7.2. Any point on this line will give combinations of x_1 and x_2 such that $y = $ Rs. 8000.

We can have $x_1 = x_2 = 0$, where we have all the material and labor but are not making anything, which is a trivial solution. We can also have a line that goes out of the board itself such that the optimum is unbounded. So both x_1 and x_2 are ∞ and y is also ∞. But this tells us that we have not put in the constraints. So within the feasible region, we have got to find the solution.

There is a theorem which states that even though there are infinite solutions within the feasible region, the corners of the polygon that are formed by the coordinate axes and by the plotting of all the constraints, qualify as candidates for the optimum solution. This can be proved. One reason for this is that the objective function is convex. To evaluate y only at the vertices of this polygon, which, in this example, are 4 in number.

- The first solution $(0, 0)$ is trivial and can be discarded.
- The vertex on the x_2 axis $(0, 16)$ indicates that all objects made are chairs.
- The vertex on the x_1 axis $(8, 0)$ corresponds to only tables being made and no chairs being produced.
- The fourth point is a mix of tables and chairs.

Each of these will result in a value of y and now we can determine the product mix that results in the maximum value for y. The y values for the 4 vertices are given in the accompanying table (Table 7.1).

Table 7.1 Graphical solution for Example 7.1

(x_1, x_2)	y
$(0, 0)$	0
$(0, 16)$	12800
$(8, 0)$	12000
$(4, 10)$	14000

However, if it so happens that if the iso-objective line, which is $1500x_1 + 800x_2$, becomes parallel to one of the constraints, then instead of these 2 lines intersecting at a point, we will get an **optimal line** instead of an optimal point. So all points on it will be **alternate optima**, all of which will give the same value of y.

Example 7.2 Solve Example 7.1 using slack variables

Solution

This is how we solve a two-variable LP problem using the graphical method. This can also be solved by introducing what are called **slack variables**. The slack variables are introduced to convert the inequalities to equalities. Let us consider the same problem.

$$\text{Maximize: } y = 1500x_1 + 800x_2 \tag{7.10}$$

Subject to:

$$\text{Material constraint : } 30x_1 + 18x_2 \leq 300 \tag{7.11}$$
$$\text{Labor constraint : } 20x_1 + 8x_2 \leq 160 \tag{7.12}$$
$$\text{Non-negativity constraints: } x_1 \geq 0 \text{ and } x_2 \geq 0 \tag{7.13}$$

The two constraints are rewritten using slack variables as

$$30x_1 + 18x_2 + s_1 = 300 \tag{7.14}$$
$$20x_1 + 8x_2 + s_2 = 160 \tag{7.15}$$

s_1 and s_2 are called slack variables and have to be positive. Why should they be positive? Because after the optimal solution is determined, s_1 tells us the amount of wood which is available and has been wasted in the process. s_2 refers to the number of man hours that are available. Neither of them can be negative and can be, at best, 0. Hence, they are *not unrestricted* in sign.

Already we had 2 variables and we have now converted the same into a four-variable problem. How do we solve this system? With 2 equations and 4 unknowns, there is no hope of solving this. One way is to make 2 variables 0 at a time and determine the resulting y. Since s_1 and s_2 also have to be positive, it is equivalent to representing x_1, x_2, s_1 and s_2 on a four-dimensional plane and then trying to find out the feasible region. We then use the same logic that iso-objective line cuts the polygon at one of the points and emerges as the optimal solution.

How many combinations are there if 2 variables are made 0 at a time? There are 6. The number of combinations is given by $n!/m!(n-m)!$, where n is the total number of variables including the slack variables and m is the number of constraints. In this example, $n = 4$ and $m = 2$. Hence the number of combinations is $4!/(2!2!) = 6$. So we need to evaluate 6 combinations.

$$30x_1 + 18x_2 + s_1 = 300 \tag{7.16}$$
$$20x_1 + 8x_2 + s_2 = 160 \tag{7.17}$$

1. $x_1 = x_2 = 0$; trivial solution to be discarded.
2. $x_1 = s_1 = 0$; $x_2 = 16$; $s_1 = 12$; y = 12800.
3. $x_1 = s_2 = 0$; $x_2 = 20$; $s_1 = -60$; Not allowed as we cannot have a surplus of −60 man hours.
4. $x_2 = s_1 = 0$; $x_1 = 10$; $s_2 = -40$; Not allowed as we cannot have a surplus of −40 man hours.
5. $x_2 = s_2 = 0$; $x_1 = 8$; y = 12000.
6. $s_1 = s_2 = 0$; $x_1 = 4$; $x_2 = 10$; y = 14000.

Hence, the optimal solution for this problem is $x_1 = 4$, $x_2 = 10$, y = 14000, which is the same as we got when we used the graphical method.

If one does not like the graphical method, we can do it this way. This has been modified and a very powerful technique called the **Simplex method** has been developed. It can be used for any number of variables and any number of constraints so long as the objective function and the constraints are written as a linear combination of the variables. We have demonstrated how this algorithm works for a two-variable problem. We can extend it to several variables and the Simplex does this elegantly.

So it is possible to introduce slack variables, which represent the difference between the left-hand side and the right-hand side of the constraints. If we have a material constraint, it represents the amount of material left. If we have a labor constraint, the slack variable represents the amount of man hours left. In this case, it so happens that both s_1 and s_2 are 0 at the optimal solution. This means that both the labor constraint and the material constraint are active and binding in this problem. They operate as equalities here. We may also get an optimal solution where one of these constraints is not binding.

So, if we encounter multi-variable problems, where the objective function and the constraints are linear combinations of the variables, this method can be used to obtain the optimal solution and then, one can do the post-optimality analysis.

Example 7.3 An ethylene refining plant receives 500 kg/h of 50% pure ethylene and refines it into 2 types of output 1 and 2. Type 1 has a purity of 90% while type 2 has a purity of 70%. The raw material cost is Rs. 40/kg and the selling price of type 1 is Rs. 200/kg, while that of type 2 is Rs. 120/kg. Packaging facilities allow a maximum of 200 kg/h of type 1 and 225 kg/h of type 2. The transportation cost of type 1 is Rs. 8/kg while that of type 2 is Rs. 16/kg. Total transportation cost should not exceed Rs. 4000. Set up the optimization problem to maximize the profit and solve it using the graphical method of LP. Declare the mass flow rates of types 1 and 2 as x_1 and x_2.

Solution

First we have to plot all the constraints, identify the feasible region, and plot one iso-objective line. Then, we move the iso-objective line till it cuts the feasible region at the farthest point because as we move away from the origin, each iso-objective line represents a higher profit. The highest profit, subject to the constraints, is what we are seeking.

$$\text{Maximize } y = 192x_1 + 104x_2 - (500 \times 40) \tag{7.18}$$
$$\text{subject to: } x_1 \leq 200 \text{ and } x_2 \leq 225 \tag{7.19}$$

There are 2 constraints in this problem. The selling price for type 1 is Rs. 200/kg and that for type 2 is Rs. 120/kg. But the transportation cost for type 1 is Rs. 8/kg while that for type 2 is Rs. 16/kg. That is why the profit in terms of x_1 and x_2 is written as $192x_1 + 104x_2$ by subtracting the transportation cost from the selling price. The raw material cost of 500×40 is further subtracted from this to get the actual profit. The constraints involved in the packaging are taken care of by the conditions $x_1 \leq 200$ and $x_2 \leq 225$.

$$\text{Mass balance: } 0.9x_1 + 0.7x_2 \leq 0.5 \times 500 \text{ or } 0.9x_1 + 0.7x_2 \leq 250 \tag{7.20}$$

$$\text{Transportation cost: } 8x_1 + 16x_2 \leq 4000 \tag{7.21}$$
$$\text{Non-negativity constraints: } x_1 \text{ and } x_2 \geq 0 \tag{7.22}$$

When we first plot the non-negativity constraints (Eq. 7.22) we get the feasible region indicated by the shaded region in Fig. 7.3.

We then plot the mass balance constraint (see Fig. 7.4) and finally plot the transportation constraint (Fig. 7.5).

The shaded region gives us the final feasible region. We saw that there is a property for linear programming problem, where we will have to search only at the vertices. There are 6 vertices from A–F as seen in Fig. 7.5. So we evaluate the objective function

Fig. 7.3 Example 7.3 with
bounds for the two variables
plotted

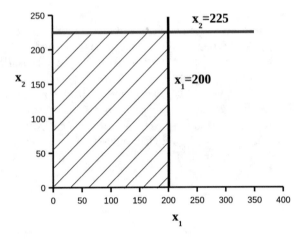

Fig. 7.4 Example 7.3 with
bounds on the variables and
the mass balance constraint
plotted

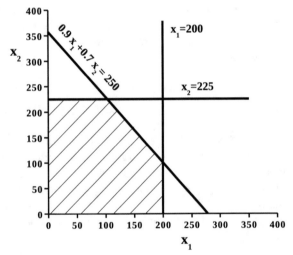

Fig. 7.5 Example 7.3 with
all the constraints plotted

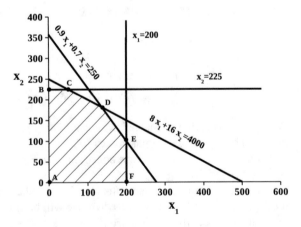

Table 7.2 Graphical solution of example 7.3

Points	x_1	x_2	y
A	0	0	−20000
B	0	225	3400
C	50	225	13000
D	136.36	181.81	25090
E	200	100	28800
F	200	0	18400

at each of these 6 vertices. One of these is a trivial solution $x_1 = 0$ and $x_2 = 0$. Profit becomes negative in this case because we bought 500 kg of raw material and did nothing with it! For the sake of the record, we take this as a solution and show that profit is negative. We then evaluate y for all the other solutions and invariably, the solution will be that which is farthest from the origin.

We can assume the profit y to be 20000 and find 2 values of (x_1, x_2), so that y can be plotted and it can be moved to represent different iso-objective lines that are parallel to each other. One such profit line y will be such that it just escapes the feasible region. Since this is a convex polygon, when y just escapes the feasible region, it will touch at one point or it can touch at several points, if one of the constraints is parallel to the iso-objective line, in which case, we will get alternate optima, as already discussed. But in this case, it does not happen to be so. In this case, E seems the farthest and hence E must be the solution (See Fig. 7.5). The exact solution can be found by working out y at A, B, C, D, E, and F (see Table 7.2).

$$y_A = -20000, \quad y_B = -3400, \quad y_C = 13000, \quad y_D = 25090,$$
$$y_E = 28800, \quad y_F = 18400$$

Now we know that y is a maximum at E and this is the optimum solution to the problem. This is how the graphical method is used to find the optima.

At the optimum, $x_1^+ = 200$ kg/h and $x_2^+ = 100$ kg/h, $y^+ = $ Rs. 28,800.

7.2.5.1 Sensitivity Analysis

At this point, it is pertinent to mention the importance of sensitivity analysis or post-optimality analysis in LP. The sensitivity of the solution can be studied with respect to (a) change in the unit profit/cost in the objective function or (b) change in the constraint. Let us see both, in the context of the above example.

1. If the raw material cost changes to Rs. 45/kg, even then, the solution remains the same (i.e, x_1 and x_2 remain the same).

   However $y_{new}^+ = y_{old}^+ - 500 \times 5 = 28800 - 2500 = $ Rs.26300.

Table 7.3 Sensitivity analysis with Example 7.3

Points	x_1	x_2	y_{new}
A	0	0	-20000
B	0	225	3400
C^+	112.5	225	25000
D^+	96.59	232.9	22766.88
E	200	100	28800
F	200	0	18400

2. If the selling price of x_1 goes down to Rs. 160/kg and the selling price of x_2 goes up to Rs. 140/kg and other things are the same, what is the optimum solution? Even in this case, the feasible region still remains the same, as the constraints are not affected and the optimum solution will again lie at one of the vertices of the polygon ABCDEFA. But now, the objective function y_{new} has now changed to

$$y_{new} = 152x_1 + 124x_2 - 500 \times 40 \qquad (7.23)$$

and this has to be evaluated at A, B, C, D, E, and F in order to determine y_{new}^+ Hence, the optimum solution now changes from E to D and y_{new} = Rs. 24545.3.

3. If the selling prices of x_1 and x_2 remain at Rs. 200/kg and Rs. 120/kg, respectively, and all the constraints remain the same except for the transportation cost which has only a maximum value of Rs.3000, what is the optimum solution? Here, only the transportation constraint has been changed to $8x_1 + 16x_2 \leq 3000$. Therefore points A, B, E, and F remain the same as before, but we need to evaluate C^+ and D^+, which will be different from C and D we obtained earlier (results are given in Table 7.3). For this case, the solution remains as E and the sensitivity is given by

$$\Delta y / \Delta \psi = 0. \qquad (7.24)$$

So, the additional relaxation of Rs. 500 for the transportation constraint has no bearing on the final solution.

7.2.6   Simplex Method

As discussed earlier, the simplex method may be considered as a variant of solving the LP problem with the slack variable method. In real-world problems in several fields such as management sciences, operation research, sociology, and often in thermal engineering, the number of unknown variables are way more than two. Thus, the graphical method cannot be used in such scenarios. In the graphical method, we demonstrated that one of the corner points of the feasible region is the optimum solution. This fundamental property is made use of in an iterative approach in the simplex method.

The standard form of the Linear Programming problem is given by
Maximize or minimize,

$$Y = C_1x_1 + C_1x_1 + \cdots + C_nx_n \tag{7.25}$$

Subject to

$$a_{11}x_1 + a_{12}x_2 + \cdots + a_{1n}x_n = b_1 \tag{7.26}$$
$$a_{21}x_1 + a_{22}x_2 + \cdots + a_{2n}x_n = b_2 \tag{7.27}$$
$$\vdots$$
$$a_{m1}x_1 + a_{m2}x_2 + \cdots + a_{mn}x_n = b_m \tag{7.28}$$
$$x_1, x_2, ...x_n \geq 0 \tag{7.29}$$

Slack variables are used for handling inequality constraints.
For example,

$$x_1 \leq 8 \tag{7.30}$$
$$x_1 + s_1 = 8 \tag{7.31}$$

where s_1 is a nonnegative quantity known as a slack variable.

Before getting into the methodology, it is instructive to get familiar with the terminology, such as basic variable, basic feasible solution, and basic solution.

- **Basic variable**: A variable x_1 is said to be a basic variable if it appears with a unit coefficient in the equation and zero in all other equations. This can be achieved by applying elementary row operations, thus converting a given variable to a basic variable. Such operations are known as pivot operations.
- **Basic feasible solution**: It is the solution in which the values of the basic variables are non-negative.
- **Basic solution**: It is the solution obtained by setting the non-basic variables to zero and solving for the basic variables.

The pivot operations can be explained with an example as follows.
Consider,

$$x_1 - 2x_2 + x_3 - 4x_4 + 2x_5 = 2 \tag{7.32}$$
$$x_1 - x_2 - x_3 - 3x_4 - x_5 = 4 \tag{7.33}$$

The coefficients of these equations can be represented as follows:

$$\begin{bmatrix} 1 & -2 & 1 & -4 & 2 & | & 2 \\ 1 & -1 & -1 & -3 & -1 & | & 4 \\ 1 & -2 & 1 & -4 & 2 & | & 2 \\ 0 & 1 & -2 & 1 & -3 & | & 2 \end{bmatrix}$$

$$R_2 \implies R_2 - R_1$$

$$R_1 \implies R_1 + 2R_2$$

$$\begin{bmatrix} 1 & 0 & -3 & -2 & -4 & | & 6 \\ 0 & 1 & -2 & 1 & -3 & | & 2 \end{bmatrix}$$

Thus the set of equations can be written as

$$x_1 - 0x_2 - 3x_3 - 2x_4 - 4x_5 = 6 \tag{7.34}$$
$$0x_1 + x_2 - 2x_3 + x_4 - 3x_5 = 2 \tag{7.35}$$

In the first equation, x_1 is the basic variable and in the latter, x_2 is the basic variable. These equations are said to be in canonical form with the basic variables x_1 and x_2. The basic solution is given by

$$x_3 = x_4 = x_5 = 0$$
$$x_1 = 6$$
$$x_2 = 2$$

The simplex method is now illustrated with an example

Example 7.4 Solve the following optimization problem using the simplex method.

Maximize,
$$Y = 5x_1 + 2x_2 + 3x_3 - x_4 + x_5$$

Subject to,

$$x_1 + 2x_2 + 2x_3 + x_4 = 8$$
$$3x_1 + 4x_2 + x_3 + x_5 = 7$$
$$x_1, x_2, x_3, x_4, x_5 \geq 0$$

Solution

Here, we can observe that the constraints are in their canonical form, eliminating the need for pivot operations. There x_4 and x_5 are the basic variables in Eqs. 1 and 2, respectively. The basic solution is given by

$$x_2 = x_3 = 0 \tag{7.36}$$
$$x_4 = 8 \tag{7.37}$$
$$x_5 = 7 \tag{7.38}$$
$$Y = -8 + 7 = -1 \tag{7.39}$$

The next step is to improve the basic solution. Let us consider x_1.
Let $x_1 = 1$ and $x_2 = x_3 = 0$.
Thus,

$$x_4 = 8 - x_1 = 8 - 1 = 7 \tag{7.40}$$

and,

$$x_5 = 7 - 3x_1 = 7 - 3 = 4 \tag{7.41}$$

So,

$$Y_{new} = 5 - 7 + 4 = 2 \tag{7.42}$$
$$Y_{new} - Y = 2 - (-1) = 3 \tag{7.43}$$

Therefore, the relative profit of x_1 is 3, which means that, by increasing the variable x_1 by one unit, the total change in the profit is 3 units. The question now to ask is by how much can x_1 increase? x_1 can be increased to the extent that it does not violate the two constraints. So taking both the constraints into consideration, we have

$$x_1 + x_4 = 8 \tag{7.44}$$
$$3x_1 + x_5 = 7 \tag{7.45}$$

In the first constraint, the maximum value that x_1 can take is 8, and in the second one the maximum value is 7/3. Thus, we take the minimum of the two, $x_1 = 7/3$. The minimum is chosen so that the nonnegative constraint is not violated. If x_1 is taken to be 8, x_5 would be equal to -17, which violates the non-negative constraint. This rule is called the minimum ratio rule. Thus,

$$x_2 = x_3 = x_5 = 0 \tag{7.46}$$
$$x_4 = 8 - \frac{7}{3} = \frac{17}{3} \tag{7.47}$$
$$Y = 5 \times \frac{7}{3} - \frac{17}{3} + 0 = 6 \tag{7.48}$$

Now the constraints are as follows.

Table 7.4 Simplex tableau no.1 for Example 7.4

C_B	Basis\\C_J	5	2	3	−1	1	Constants
		x_1	x_2	x_3	x_4	x_5	
−1	x_4	1	2	2	1	0	8
1	x_5	3	4	1	0	1	7
$\bar{C}$ row		3	0	4	0	0	$Y = -1$

Table 7.5 Minimum ratio rule for Example 7.4

Row	Row variable	Upper limit
1	x_4	$8/2 = 8$
2	x_5	$7/1 = 7$

This table is known as the *Simplex Tableau*. $\bar{C}$ is the relative profit which is given by the following equation:

$\bar{C} = C_j$ − (inner product of C_B and column vector of x_i in the canonical system)

$$C_1 = 5 - (-1\ \ 1)\begin{pmatrix} 1 \\ 3 \end{pmatrix} = 3 \tag{7.49}$$

$$C_2 = 2 - (-1\ \ 1)\begin{pmatrix} 2 \\ 4 \end{pmatrix} = 0 \tag{7.50}$$

$$C_3 = 3 - (-1\ \ 1)\begin{pmatrix} 2 \\ 1 \end{pmatrix} = 4 \tag{7.51}$$

$$C_4 = -1 - (-1\ \ 1)\begin{pmatrix} 1 \\ 0 \end{pmatrix} = 0 \tag{7.52}$$

$$C_5 = 1 - (-1\ \ 1)\begin{pmatrix} 0 \\ 1 \end{pmatrix} = 0 \tag{7.53}$$

Here, C_j is the coefficient of the variables in the objective function. We can observe that x_3 has the highest relative profit. This implies that by increasing the value of x_3 by one unit, the objective function increases by 4 units, which is the highest among all the variables. So, x_3 is chosen to be the basic variable. Thereafter, to check to what extent x_3 can be incremented and to find the variable which would leave the basis, the *minimum ratio rule* is applied as follows (Table 7.4).

Here, the constants of the equation are divided by the corresponding coefficients of x_3. In Table 7.5, Row 1 has the minimum ratio, thus x_4 exists and x_3 enters the basis column. Now the constraints are to be converted into the canonical form with the basic variables as (x_3, x_5). So the following pivot operations are done:

Table 7.6 Simplex tableau no.2 for Example 7.4

C_B	Basis\C_J	5	2	3	-1	1	Constants
		x_1	x_2	x_3	x_4	x_5	
3	x_3	1/2	1	1	1/2	0	4
1	x_5	5/2	3	0	$-1/2$	1	3
$\overline{C}$ row		1	-4	0	-2	0	$Y = 15$

$$R_1 \implies \frac{R_1}{2}$$

$$R_2 \implies R_2 - \frac{R_1}{2}$$

The constraints in the canonical form then become

$$\left(\frac{1}{2}\right) x_1 + x_2 + x_3 + \left(\frac{1}{4}\right) x_4 = 4 \tag{7.54}$$

$$\left(\frac{5}{2}\right) x_1 + 3x_2 - \left(\frac{1}{2}\right) x_4 + x_5 = 3 \tag{7.55}$$

Thus the basic solution is

$$x_1 = x_2 = x_4 = 0$$
$$x_3 = 4, \ x_5 = 3 \ and \ Y = 15$$

The relative profits are

$$C_1 = 5 - (3 \ 1) \begin{pmatrix} 1/2 \\ 5/2 \end{pmatrix} = 1 \tag{7.56}$$

$$C_2 = 2 - (3 \ 1) \begin{pmatrix} 1 \\ 3 \end{pmatrix} = -4 \tag{7.57}$$

$$C_3 = 3 - (3 \ 1) \begin{pmatrix} 1 \\ 0 \end{pmatrix} = 0 \tag{7.58}$$

$$C_4 = -1 - (3 \ 1) \begin{pmatrix} 1/2 \\ -1/2 \end{pmatrix} = -2 \tag{7.59}$$

$$C_5 = 1 - (3 \ 1) \begin{pmatrix} 0 \\ 1 \end{pmatrix} = 0 \tag{7.60}$$

In the second iteration, it can be seen that the maximum relative profit is for the variable x_1. Hence, x_1 enters the basis. To determine which variable leaves the solution, the *minimum ratio rule* is again applied. It can be observed that variable x_5 leaves the basis (Tables 7.6, 7.7).

Table 7.7 Minimum ratio rule for Example 7.4

Row	Row variable	Upper limit
1	x_3	$4/(1/2) = 8$
2	x_5	$3/(5/2) = 6/5$

Table 7.8 Simplex tableau no.3 for Example 7.4

C_B	Basis\\C_J	5	2	3	-1	1	Constants
		x_1	x_2	x_3	x_4	x_5	
3	x_3	0	2/5	1	3/5	$-1/5$	17/5
5	x_1	1	6/5	0	$-1/5$	2/5	6/5
$\bar{C}$ row		0	$-26/5$	0	$-9/5$	$-2/5$	$Y = 81/5$

The next step is to carry out pivot operations to convert the equation in the canonical form with the basic variables (x_3, x_1).

$$R_2 \implies \frac{2}{5} \times R_2$$
$$R_1 \implies R_1 - \frac{R_2}{5}$$

In Table 7.8, the elements of the $\bar{C}$ row are all either 0 or negative, thus confirming no further improvement is possible in this problem.

Final solution:
$$x_2 = x_4 = x_5 = 0,$$
$$x_1^+ = 6/5, \quad x_3^+ = 17/5 \ and \ Y^+ = 81/5$$

Example 7.5 A furniture company can make tables or chairs or both. The amount of wood required for making one table is 30 kg while that for a chair is 18 kg. The total quantity of wood available is 300 kg. The labor required for making a table is 20 man hours, while that for a chair is 8 man hours. The total number of man hours available is 160 (say for one day). The profit from a table is Rs. 1500, while that from the chair is Rs. 800. Determine the optimal product mix of the company using the simplex method.

Solution

We now solve the above problem using the simplex method. The slack variables are introduced as discussed in Example 7.2 to convert the inequalities to equalities. Let us consider the same problem.

Table 7.9 Simplex tableau no.1 for Example 7.5

C_B	Basis\C_J	1500	800	0	0	Constants
		x_1	x_2	s_1	s_2	
−1	s_1	30	18	1	0	300
1	s_2	20	8	0	1	160
$\overline{C}$ row		1500	800	0	0	$Y = 0$

Table 7.10 Minimum ratio rule for Example 7.5

Row	Row variable	Upper limit
1	s_1	300/30 = 10
2	s_2	160/20 = 8

Table 7.11 Simplex tableau no.2 for Example 7.5

C_B	Basis\C_J	5	2	3	−1	Constants	Upper limit
		x_1	x_2	x_3	x_4		
0	s_1	0	6	1	−3/2	60	60/6 = 10
1500	x_1	1	2/5	0	1/20	8	8/(2/5) = 20
$\overline{C}$ row		0	200	0	−75	$Y = 12{,}000$	

Maximize: $Y = 1500x_1 + 800x_2$
Subject to

$$\text{Material constraint}: \quad 30x_1 + 18x_2 \leq 300 \qquad (7.61)$$
$$\text{Labor constraint}: \quad 20x_1 + 8x_2 \leq 160 \qquad (7.62)$$
$$\text{Nonnegative constraints}: \quad x_1 \geq 0 \; and \; x_2 \geq 0 \qquad (7.63)$$

The two constraints are rewritten using slack variables as

$$30x_1 + 18x_2 + s_1 = 300 \qquad (7.64)$$
$$20x_1 + 8x_2 + s_2 = 160 \qquad (7.65)$$

s_1 and s_2 are slack variables and have to be positive as discussed in the previous example. The next step to identify the basic variables. s_1 and s_2 are the basic variable because they have unit coefficient in one equation and zero in all the other equations. Let us formulate the simplex tableau.

We can observe that the maximum relative profit of x_1 is the highest of all variables. Hence, x_1 enters the basis column. Now, the next step is to find the variable which leaves the basis, and for which the following table is made (Tables 7.9–7.11).

Table 7.12 Simplex tableau no.3 for Example 7.5

C_B	Basis\C_J	1500	800	0	0	Constants
		x_1	x_2	s_1	s_2	
800	x_2	0	1	1/6	$-1/4$	10
1500	x_1	1	0	$-1/15$	3/20	4
$\overline{C}$ row		0	0	$-100/3$	-25	$Y = 14{,}000$

Here, s_2 has the minimum ratio, and hence it leaves the basis column. The next step is to convert the equations into their canonical form with basic variables (s_1, x_1).

$$R_2 \implies \frac{R_2}{20}$$
$$R_1 \implies R_1 - \frac{3}{2} \times R_2$$

After calculating the relative profit for each of the variables, we can see that the variable x_2 has the maximum value. In the eighth column of the table, ratios are calculated and we can see that the minimum is for s_1. Hence, s_1 leaves and x_2 enters the basis column. The next step is to convert the equations into their canonical form with basic variables (x_2, x_1) (Table 7.12).

After the second iteration, we can notice that the relative profits of all the variables as smaller than or equal to zero. This implies that by increasing the value of any of the variables by one unit, the value of the objective function would not increase. Hence, we have arrived at the optimum value of the objective function.

Hence, the final solution is, $x_1^+ = 4 \quad x_2^+ = 10$ and $Y^+ = 14,000$.

Minimization Problem:

In the case of the minimization problem, we can follow the same algorithm with just one modification. In the $\overline{C}$ row, the variable with the highest negative value is chosen and the iteration is stopped when all the values are greater than or equal to zero. An alternative approach for a minimization problem is to convert the objective function into a maximization problem by multiplying it with minus one.

$$\text{Minimize, } y = 40x_1 + 36x_2 \tag{7.66}$$
$$\text{Maximize, } y' = -40x_1 - 36x_2 \tag{7.67}$$

7.2.7 Integer Programming

The optimization problem in which some or all of the variables are restricted to integer (or discrete) values is commonly referred to as Pure integer programming (IP) or Mixed-integer problems. IP problems are inherently nonlinear since the different functions of the problem are defined only at the discrete values of the variables. However, for the purpose of developing the solution procedure, after removing the integer restrictions on the variables concerned, the resulting problem is an LP then we can treat the IP problem as being an LP. Otherwise, it is classified as a nonlinear problem.

Most algorithms of IP are based on the continuous version of the IP model. IP algorithms not based on the continuous version are generally known to be unpredictable.

There are mainly two methods to solve IP problems.

1. **Cutting plane methods**:

   We first solve for the continuous version of the problem and then add special "secondary" constraints, which represent necessary conditions for ensuring the variables are integers. In this way, the continuous solution space is gradually modified till its continuous optimum extreme points satisfy the integer conditions. The added constraints cut certain portions of the solution that do not contain feasible integer points.

   e.g., Gomory's fractional cut algorithm.

2. **Searching method**:

   This method is based on the simple idea of enumerating all the feasible integer points. The basic idea is to develop "clever tests" that consider only a "small" portion of that feasible integer points explicitly but automatically account for the remaining points implicitly.

   e.g., Branch and bound method.

Cutting plane method/Gomory's fractional cut algorithm:

Consider the following IP problem,

$$\text{Maximize:} \quad Y = 2x_1 + 5x_2 \tag{7.68}$$
$$\text{Subject to:} \quad 3x_1 + 6x_2 \leq 16 \tag{7.69}$$
$$\text{Non-negativity constraints:} \ x_1 \geq 0 \ \text{and} \ x_2 \geq 0 \tag{7.70}$$

We can first attempt to solve it as an LP problem using the graphical method.

From Fig. 7.6, it is seen that OABO is the feasible domain and the values of the objective functions at the corner points are,

Thus, the solution to the linear programming problem is $x_1^+ = 0$ and $x_2^+ = 8/3$. However, we see that of x_2^+ is not an integer (Table 7.13).

Fig. 7.6 Feasible region OABO for the above IP problem

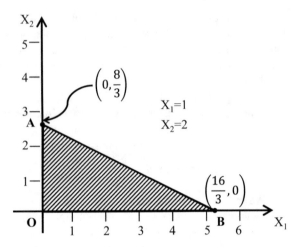

Table 7.13 Value of Y at corner points

x_1	x_2	Y
0	0	0
16/3	0	32/3
0	8/3	40/3

Fig. 7.7 Plot showing possible integer values for the variables in the feasible region for the above IP problem

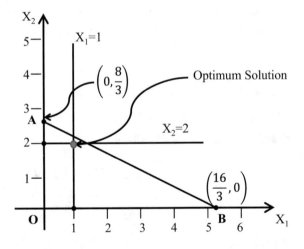

The nearest integer value to x_2 is 2. So, we draw a line $x_2 = 2$ on the graph and in turn cut the domain into two parts. Now we arrive at an integer solution with $x_1 = 0$ and $x_2 = 2$. In this case, the value of $Y = 10$. But this might not be the maximum value of the objective function. Now, we check what other integer values can x_1 take without leaving the feasible domain (Fig. 7.7).

Let us assume that $x_1 = 1$. The constraint then turns out to be

$$(3 \times 1) + (6 \times 2) = 3 + 12 = 15 \le 16$$

We can see that the constraint is not violated.
Now let us take the next possible integer for $x_1 = 2$. The constraint is

$$(3 \times 2) + (6 \times 2) = 6 + 12 = 18 > 16$$

Here, the constraint is violated. Thus, the optimum value of the objective function is at $(1, 2)$ and is 12.

Example 7.6 Solve the following integer programming problem using Gomory's fractional cut algorithm.
Maximize
$$Y = 3x_1 + 5x_2$$

subject to,

$$x_1 + 4x_2 \le 9$$
$$2x_1 + 3x_2 \le 11$$
$$x_1, \ x_2 \ge 0$$

x_1 and x_2 are integers.

Solution

$$x_1 + 4x_2 + x_3 = 9 \qquad (7.71)$$
$$2x_1 + 3x_2 + x_4 = 11 \qquad (7.72)$$

We now use the tableau to solve the problem.

$$x_1^+ = 17/5 \ \ x_2^+ = 7/5 \ and \ Y^+ = 86/5$$

Since x_1^+ and x_2^+ are not integers, the above is not an acceptable solution and so we need to work on this problem further.

$$1x_2 + \frac{2}{5}x_3 - \frac{1}{5} = \frac{7}{5} \qquad (7.73)$$

$$(0 + 1)x_2 + \left(0 + \frac{2}{5}\right)x_3 + \left(-1 + \frac{4}{5}\right)x_4 = 1 + \frac{2}{5} \qquad (7.74)$$

Table 7.14 Tableau no.1 for Example 7.6

C_B	Basis\C_J	3	5	0	0	Constants
		x_1	x_2	x_3	x_4	
0	x_3	1	4	1	0	9
0	x_4	2	3	0	1	11
$\overline{C}$ row		3	5	0	0	

Table 7.15 Tableau no.2 for Example 7.6

C_B	Basis\C_J	3	5	0	0	Constants	Upper limit
		x_1	x_2	x_3	x_4		
5	x_2	1/4	1	1/4	0	9/4	9
0	x_4	5/4	0	−3/4	1	17/4	17/5
$\overline{C}$ row		7/4	0	−5/4	0		

Table 7.16 Tableau no.3 for Example 7.6

C_B	Basis\C_J	3	5	0	0	Constants
		x_1	x_2	x_3	x_4	
3	x_1	1	0	−3/5	4/5	17/5
5	x_2	0	1	2/5	−1/5	7/5
$\overline{C}$ row		0	0	−1/5	−7/5	$Y = 86/5$

$$\frac{2}{5} + \frac{4}{5}x_4 = \frac{2}{5} + (1 - x_2 + x_4) \tag{7.75}$$

For the above to be true, the following arguments are in order. Since x_2, x_3 and x_4 are all integers, $(1 - x_2 + x_4)$ has to be positive as $\frac{2}{5}x_3 + \frac{4}{5}x_4$ will have to be greater than $\frac{2}{5}$ for all values of x_3 and x_4. Hence to satisfy the constraint of being integers, the following additional condition is proposed (Tables 7.14–7.17).

$$\frac{2}{5} + \frac{4}{5}x_4 \geq \frac{2}{5} \tag{7.76}$$

$$-\frac{2}{5} - \frac{4}{5}x_4 \leq -\frac{2}{5} \tag{7.77}$$

$$-\frac{2}{5} - \frac{4}{5}x_4 + x_5 = -\frac{2}{5} \tag{7.78}$$

Please note that in the above tableau the "cut" is implemented as a constraint in Row 3

Table 7.17 Tableau no.4 for Example 7.6

C_B	Basis\C_J	3	5	0	0	0	Constants
		x_1	x_2	x_3	x_4	x_5	
3	x_1	1	0	$-3/5$	4/5	0	17/5
5	x_2	0	1	2/5	$-1/5$	0	7/5
0	x_5	0	0	$-2/5$	$-4/5$	1	$-2/5$
$\overline{C}$ row		0	0	$-1/5$	$-7/5$	0	

Table 7.18 Tableau no.5 for Example 7.6

C_B	Basis\C_J	3	5	0	0	0	Constants
		x_1	x_2	x_3	x_4	x_5	
3	x_1	1	0	0	2	$-3/2$	4
5	x_2	0	1	0	-1	1	1
0	x_5	0	0	1	2	$-5/2$	1
$\overline{C}$ row		0	0	0	-1	$-1/2$	$Y = 17$

$$x_3 = \frac{-1/5}{-2/5} = \frac{1}{2} \tag{7.79}$$

$$x_4 = \frac{-7/5}{-4/5} = \frac{7}{4} \tag{7.80}$$

We now continue with the same approach as the simplex method. The final Tableau becomes Table 7.18.

From the above tableau, it is seen that all entries in $\overline{C}$ row are either negative or zero, implying that no further improvement in the objective function is possible. Furthermore, x_1^+ and x_2^+ are now integers, thereby satisfying the integer constraint.

Hence, the final solution is $x_1^+ = 4$ $x_2^+ = 1$ and $Y^+ = 17$.

7.3 Dynamic Programming (DP)

Dynamic programming is an optimization technique applicable to a class of problems, where a large or complex problem can be broken down into stages. We keep evaluating the optimum for each stage and keep proceeding. The belief is that, if we divide a complex problem into stages and optimize each of the sub-stages, and when we reach the end, we will get the overall optimum. This can be proved for certain classes of problems and this technique was developed by Prof. Richard Bellman, a Professor of Applied Mathematics in the University of Southern California. He figured out the *Bellman optimality theorem*, the Bellman condition, and so on.

DP is used in process industries or if we want to lay gas pipelines between two cities such that they pass through intermediate cities. There are various ways of going from one city to another through the intermediate cities. We then ask the question, what will be the optimum path for going from one city to another? Suppose we are going from city 1 to city 2 and in between we have cities A, B, and C, what is the shortest path from 1 to C, C to B, B to A, and from A to 2. Suboptimal solutions at each stage are eliminated. This way there is a computational gain.

Optimization is the process of getting the optimum without working out all the possibilities. If all the possibilities have to be worked out, it is called the exhaustive search. It will no longer be called dynamic programming (DP). There is a computational gain here compared to the exhaustive search. For problems involving many such networks, it may be extremely complicated for us to do an exhaustive search. In view of this, dynamic programming is also used extensively by computer engineers and scientists.

The other example where DP can be possibly used is the transportation of goods from one place to another. Of course, there are some constraints that the goods have to pass through some intermediate cities and there are only some fixed paths and so on. The complete DP algorithm can be mathematically derived and one can have an algorithmic approach. But here we just take a simple problem and using a common-sense solution, with out going into the numerical evaluation of the objective function, we will make the technique look clear.

Dynamic programming is also used in the **Duckworth–Lewis method** for prescribing target scores for the team playing second in a rain-interrupted limited-overs cricket match. This proposal appeared in a paper in the Journal of Operations Research Society in 1997. Basically, there are some control variables and they are optimally divided. If there are 50 overs and 10 wickets in a one-day international match, how much does the second team need to score to win? If we have 40 overs and 10 wickets or 40 overs and 6 wickets, the situation is different. Duckworth and Lewis proposed an exponential distribution for each of these. Taking the past data from rain-interrupted matches and prepared a lookup table. They figured out what the constants of these curves should be. This table is available and is now used to determine the target for the team playing second, in rain-interrupted matches.

Example 7.7 Determine the cheapest cost of transporting goods from city A to city B. The constraint is that the path must involve one of three C's (C_1, C_2, C_3), one of the three D's, and one of the three E's. All the constants are given in Fig. 7.8. Using DP, determine the optimal path to minimize the total cost.

Solution

There are totally 27 paths available. We have to pass through one of the nodes at each level. The point to bear in mind, is that, if we are able to calculate the costs of all

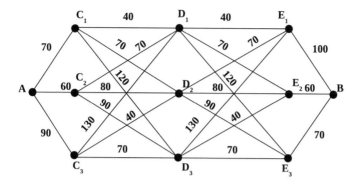

Fig. 7.8 Paths represented by nodes and associated costs for example 7.7

27 paths comfortably and cheaply, there is no need for dynamic programming. But imagine that instead of 3 choices, we have 6 or 7 choices at each level and instead of 3 intermediate levels, there are 7 or 8 intermediate stages. If the objective function also involves formidable calculations, then it is worthwhile to develop an algorithm for this. The problem has to be subdivided into stages.

Cost from A to D:

We have to find out the cost from A to any D first, because the final optimal path must pass through either D_1 or D_2 or D_3. First, we get the total cost to reach any D from A. There are 9 paths to reach D. We will come out with a tabular column as shown in Table 7.19. The path from A to any D can be accomplished by starting from any A and passing through one of the three Cs and reaching any D. There are 3 choices for C and 3 choices for D. So there are 9 possibilities. $A - C_1 - D_1$, $A - C_1 - D_2$, $A - C_1 - D_3$, $A - C_2 - D_1$, $A - C_2 - D_2$, $A - C_2 - D_3$, $A - C_3 - D_1$, $A - C_3 - D_2$, $A - C_3 - D_3$. We evaluate the 9 entries.

Now the optimal path to D_1, optimal path to D_2 and the optimal path to D_3 are all indicated by boldface. At this stage, it is too premature to look at the minimum of these three and proceed from there, because we do not know what the story from D to E is. So we should not try to get premature convergence by going in for a local optimum. But one thing which is sure is, if we have to reach the end, it has to be through a D. So to reach each of these D's, we first figure out what optimum path is. Once we determine the optimal path to each of these Ds, when we go from D to E, we use these optimal paths alone for calculation. Suboptimal paths are left out. This way a complex is broken down in to stages.

Till the calculation of these 9 entries, no algorithm or dynamic programming was done. Now, when the lowest in every column is already noted (with boldface), we have used the algorithm already. There are two things to be noted. We are (i) looking at the minimum cost to D_1, D_2 and D_3 (ii) but not looking at the minimum cost in the table and removing the other 8 entries. That may result in a suboptimal solution and these are 2 cardinal points in dynamic programming.

Table 7.19 Cost from A to D through C in Example 7.7

Cost through	Cost to D_1	Cost to D_2	Cost to D_3
C_1	**110**	140	190
C_2	130	140	**150**
C_3	220	**130**	160

Table 7.20 Cost for A-E in Example 7.7

Cost through	Cost to E_1	Cost to E_2	Cost to E_3
D_1	**150**	**180**	230
D_2	200	210	**220**
D_3	280	190	**220**

Cost from A to E:

How do we reach E_1, E_2 or E_3? It has to be either through D_1, D_2 or D_3. When we reach from D_1 to E_1, the other paths like $A - C_2 - D_1 - E_1$ or $A - C_3 - D_1 - E_1$ are not evaluated because we have already eliminated them. If we calculate those again, it becomes an exhaustive search. Up to D_1, anyway, we know what the optimal path is! We again indicate with boldface, the lowest in every column of the Table as shown in Table 7.20. Through E_1, if we want to calculate the total cost to B, it is 250. Through E_2 it is 240. Through E_3, it is 290. Therefore, the optimal path is $A - C_1 - D_1 - E_2 - B$.

The control variables in the above problem were basically the cost associated with each stage. But in the cricket match, there are two control variables: the number of wickets and the number of overs. So it is a lot more difficult and so Duckworth and Lewis must have written a program and solved it. So in principle, one can have any number of variables. Here, it is simple with just one variable.

Computational gain:

What is the big deal? How many evaluations did we do? We did 9 at the first stage, 9 more at the second and 3 at the last stage and so a total of 21 evaluations. An exhaustive search would have required 27 evaluations. So the computational gain is 6/27. There is a reduction of 28.52% in our effort because we applied dynamic programming.

At each and every stage, we identify the optimal solution and in the subsequent stage, we carry over from the optimal solution left behind in the previous stage and there is no distinction between the start and the finish. We could have started from B and found the optimal path from B to E, B to D, B to C, and B to A, we will get the same solution. It does not matter whether we start from the left or the right. At each and every stage, we proceed with whatever optimum we get and discard the suboptimal solutions at that stage. But we do not look at the overall minimum and discard all the others, as that may actually mislead us.

The example, we considered above, closely resembles the classical "stage coach problem". This program is said to originate from a traveling salesman who moves from one small town A to another small town B, through several intermediate towns. His objective was not to pay more than what is necessary for transport. Between small towns, he is supposed to have used "stage coaches". Dynamic programming is also applicable in other problems involving reliability, manpower planning, optimum layout of gas pipelines, machine replacement problem, and so on. Dynamic programming is extensively treated in operations research texts (see, for example, Srinivasan 2010).

Problems

7.1 Use the graphical method of linear programming to solve the LP problem given below and obtain the optimal solution.

$$\text{Maximize } y = 2x_1 + 3x_2$$
$$\text{Subject to } x_1 \leq 6$$
$$x_2 \leq 4$$
$$1.5x_1 + 2.4x_2 \leq 12$$
$$x_1, x_2 \geq 0$$

7.2 Consider the LP problem given above. However, the objective function is now $y = x_1 + 4x_2$. Other constraints remain the same. Use the graphical method of LP to solve this problem. Comment on your result.

7.3 A raw bauxite ore consisting of 50% of Aluminum (Al) is processed in a plant and the output consists of two different grades of Al, one consisting of 70% purity and the other 90%.

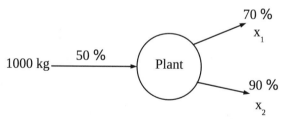

The cost of raw bauxite is Rs. 10/kg and the selling price of processed Al is Rs.20/kg (70%) while that of the other grade is Rs. 25/kg (90%). Assume that 1000 kg of bauxite is fed into the plant at a time. Let x_1 and x_2 (in kg) be the outputs, respectively, of the 70 and 90% grades. The design of the plant gives rise to the constraint $5x_1 + 3x_2 \leq 3000$. Determine x_1 and x_2 for maximum profit by using the graphical method of linear programming. Assume dumping of Al is allowed during the processing.

7.4 Revisit exercise Problem 1 of Chap. #4. A turbine receives 5 kg/s steam and can produce superheated steam or electricity. The prices are Rs. 4/kWhr electricity;

Rs. 0.15/kg low-pressure steam, 0.25/kg high-pressure steam. Assume that each kg/s into the generator can produce 0.025 kWhr/s electricity.

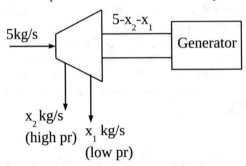

To prevent overheating of the generator, the mass flow into the generator should be less than 3 kg/s. To prevent unequal loading on the shaft, the extraction rates should be such that $2x_1+3x_2 \leq 10$. The design of the bleed outlets allows the constraint $6x_1 + 5x_2 \leq 20$. Find x_1 and x_2 for maximum profit by using the graphical method of linear programming.

7.5 A typical gas turbine engine for a military aircraft consists of a combustor and an afterburner for increased thrust. The total energy at the exit has to be maximized for maximum work output. The aircraft is flying with a speed of 600 m/s and intakes air at a rate of 10 kg/s. (C_p for air $= 1000$ J/kgK). The temperature at the inlet of the engine is 250 K and heating value of the fuel is 45000 kJ/kg. Due to limitations in the spraying of fuel, $4x_1+3x_2 \leq 1.5$. The combustor can withstand more heat than the afterburner and this is employed in the design of the spray rates of fuel by the constraint: $x_1 \geq x_2 + 0.1$. Further, due to limitations in fuel storage and distribution, $2x_1+5x_2 \leq 1$.

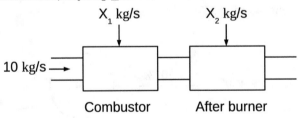

Determine the optimum values of x_1 and x_2 for maximum energy at the exit using the graphical method of linear programming.

7.6 Solve problem no. 7.1 given above, using the method of slack variables. Confirm that the solution you obtain is the same as that obtained using the graphical method.

7.7 Solve the gas turbine with afterburner problem (problem no. 7.5 above) using the method of slack variables. Confirm that the solution you obtain is the same as that obtained using the graphical method. Comment on the values of the slack variables in the final solution.

7.8 The first stage of a space shuttle consists of a central liquid engine consisting of three nozzles and two solid straps on boosters as shown.

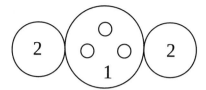

For convenience, assume complete expansion in the nozzles, i.e., pressure at exit = ambient pressure.

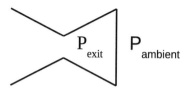

The exit gas velocities are $v_{e1} = 4000\,\text{m/s}$; $v_{e2} = 1500\,\text{m/s}$; based on the design of the nozzles, $\dot{m}_1$ and $\dot{m}_2$ are the mass flow rates (in kg/s) for each of the central and external nozzles. Optimize $\dot{m}_1$ and $\dot{m}_2$ for maximum total thrust. The constraints are

(a) Due to a limit on the maximum propellant weight that can be carried, $5\dot{m}_1 + \dot{m}_2 \le 6500$.

(b) The total fuel that can be carried in the main central tank is 750 tons and this has to burn for 8 min.

(c) The total fuel that can be carried in the external casings is 1000 tons and this has to burn for 2 min.

(d) The nozzle geometry allows the constraint: $4\dot{m}_1 + \dot{m}_2 \le 6000$.

Determine the optimum thrust using (a) the graphical method of linear programming and (b) the method of slack variables.

7.9 Revisit exercise problem 7.1 and solve it using the simplex method.

7.10 Revisit exercise problem 7.3 and solve it using the simplex method.

7.11 Revisit exercise problem 7.4 and solve it using the simplex method.

7.12 Consider a container truck that is climbing a "ghat" (hill) road. There are three sections on the ghat road denoted by B-C, C-D, and D-E. The fuel consumed in each section varies with the time taken to cover the particular section and is given in the accompanying Table 7.21.

Please note that section CD is the toughest part of the ghat road that "drinks" or "guzzles" so much fuel. The hill climb needs to be completed within a total time of 34s. Using dynamic programming, determine the optimal time to be spent by the truck in the three sections so as to minimize the total fuel consumption. What is the computational gain achieved by using dynamic programming for this problem? (adapted with modifications from Stoecker 1989).

7.13 A person wants to go from city A to city J. The various options and distances(in terms of km) are given in the schematic and Table 7.22. Using DP, obtain the minimum distance to be traveled (Fig. 7.9).

Table 7.21 Time and fuel consumption for various sections of the hill climbing (Problem 7.12)

Section	Time, t (s)	Fuel consumption, (g)
B-C	10	60
	11	51
	12	43.5
	13	37.5
C-D	10	91.5
	11	78
	12	67.5
	13	57
D-E	10	73.5
	11	61.5
	12	52.5
	13	45

Table 7.22 Path and distance for various combinations in (Problem 7.13)

Path	A-B	A-C	A-D	B-E	B-F	B-G	C-E	C-F
Distance	7	7	9	6	5	8	9	7
Path	C-G	D-E	D-F	D-G	E-H	E-I	F-H	F-I
Distance	10	6	5	7	8	9	8	6
Path	G-H	G-I	H-J	I-J				
Distance	9	7	12	13				

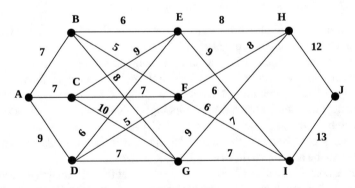

Fig. 7.9 Schematic of the network for Problem 7.13

References

Srinivasan, G. (2010). *Operations Research- Principles and Applications*. New Delhi, India: Prentice Hall India.

Stoecker, W. F. (1989). *Design of Thermal Systems*. Singapore: Mc Graw Hill.

Chapter 8
Nontraditional Optimization Techniques

8.1 Introduction

In this chapter, we will look at two optimization techniques, namely, (a) Genetic algorithms (GA) and (b) Simulated annealing (SA), both of which are unconventional, yet powerful. Both use only the information on the objective function and not any auxiliary information like derivatives. GA and SA are both search techniques that also employ probabilistic laws in the search process and have gained immense popularity in the recent past.

8.2 Genetic Algorithms (GA)

A very powerful nontraditional optimization technique is the class of algorithms that mimics the process of evolution. We look at the features of biological systems and see if these can be implemented in engineering optimization. Anything that looks at the process of evolution comes under the class of **evolutionary optimization** techniques. There are several of them, of which the most important is the genetic algorithms or GA. This is relatively new, developed in 1975, and is gaining popularity in the past 20 years or so.

Genetic algorithms are basically non-calculus-based search algorithms. Every technique has a search logic and this varies. Previously, for a two-dimensional search, we searched around a particular point in the east, west, north, south, northeast, northwest, southeast, and southwest direction. We then proceeded in one direction. That was "hill climbing". Or based on a two-point test, we eliminated a portion of the interval. If it was a two-variable problem, we worked with one variable at a time. So we can have an efficient, single-variable search technique, which could be broadly based on either a hill-climbing or a region elimination method.

© The Author(s) 2021
C. Balaji, *Thermal System Design and Optimization*,
https://doi.org/10.1007/978-3-030-59046-8_8

8.2.1 Principle Behind GA

Genetic algorithm is based on the mechanics of natural selection and natural genetics. The central idea is the *survival of the fittest* concept, which stems from the Darwinian theory, wherein, the fittest will survive and procreate such that successive generations become better and better progressively. This is true if we look at any parameter. For example, the average life expectancy of an Indian male is now 68 years. It was only 40 years at the time of independence. The life expectancy in Japan is about 88 years. The probability that a child born in Germany today will live for 100 years is more than 0.5.

If we consider the field of medicine, we cannot say that all the diseases have been conquered. Several diseases have been conquered, but new diseases have come and medical research is going on at a frenetic pace and intensity. However, it cannot be denied that a lot more people live longer "with" and "in spite" of diseases. So, outside of Darwinian evolution, there has been intervention by man too. Successive generations are becoming better, i.e., in several metrics.

So, basically, genetic algorithms simulate or mimic the process of evolution. The key idea here is that evolution is an optimizing process. If evolution is an optimizing process and successive generations are becoming better and better, each generation is like an iteration in numerical methods. So if we have 5 generations, it is like having 5 iterations in a numerical method and with each iteration, there is a progressive improvement in the objective function. This means that more or less, GA is like a hill-climbing technique. For example, if y is a function of x_1, x_2, x_3, we start with some combination of values of x_1, x_2, x_3, and we get a value of y. As we apply the genetic algorithm, with successive iterations, y keeps increasing (for a maximization problem). Based on a suitable convergence criterion, we stop the iterations (generations).

8.2.2 Origin of GA

Prof. John Holland, along with his colleagues and students developed this method at the University of Michigan around 1975. Prof. David Goldberg, an illustrious student of Holland, is the author of the book "Genetic Algorithms in search, optimization and machine learning" (1989). Goldberg is a civil engineer who did his Ph.D. with Prof. Holland and is now an independent consultant. He also worked with Prof. Holland on genetic algorithms and has contributed significantly to the spectacular growth of this field along with a few other key scientists.

8.2.3 Robustness of GA

The central theme of research on genetic algorithms has been robustness. If we say something is robust, it means it can survive even in hostile environments. Robustness in optimization refers to a fine balance between efficiency and efficacy that is essential for survival in varied environments. The robustness of a species or the robustness of human beings, as a race, can mean many things. How high a temperature can a man withstand? How long can we go on without food and water? How long can we survive under hostile conditions? Equivalently from an optimization point of view, we are looking at how many different classes of problems are amenable to genetic algorithms.

We do not claim that for all the optimization problems, genetic algorithms will solve it better than other optimization techniques. But the point is that for a wide class of problems, genetic algorithms work very well and that is what we mean by robustness. For a specific two-variable problem, which is continuous and differentiable, where the first and second derivatives can be obtained, the steepest descent or the conjugate gradient method will work very fast and will defeat genetic algorithms hands down. There is a class of problems where specialized techniques will be superior to GA.

However, if we apply the conjugate gradient method for a function that has a lot of local minima and maxima and is oscillating wildly, the conjugate gradient will just get choked. In the class of problems, where getting the derivatives is messy and is computationally expensive, GA can be very potent and useful. Needless to say, in a large majority of exceedingly complex problems, GA would outperform traditional methods.

8.2.4 Features of Biological Systems that Are Outstanding

1. Expensive redesigns can be substantially reduced or even eliminated if systems can be made more robust. This is easy to understand because even for the slightest deviation from design conditions, if a system does not work, it is not robust. If we have a very specialized air conditioner, and the ambient temperature goes above 40 °C, it could possibly switch off. This has actually led to the development of "tropicalized compressors". For example, an air conditioner that works in Europe may not work in Saudi Arabia. The maximum outside temperature in Saudi Arabia can reach 50 °C and so the compressor has to be redesigned. The same will work in Europe, though this design may not be required there. So we say that the air conditioner is robust if it can work in different types of climatic conditions. So systems should not be finicky and fail if the design conditions change even slightly.

2. The adjustability, robustness, and efficiency of biological systems are all amazing.

3. Self-repair, self-guidance, and reproduction in biological systems are also amazing. If we take animals in the wild like tigers and elephants, they do not have
   doctors and hospitals, but they do get injured and carry out self-repair. So if an
   animal species gets diseased or injured, some will die, while the best will survive.
   So the ones that survive are already immune to this disease. The weaklings are
   taken care of or consumed by the disease. In human beings too, it used to be the
   same case but now we have started intervening medically. For example, if we get
   a fracture, we do not remove that part and send it to a service center. Along with
   our daily functions, if we just use a sling and take pain killers, the fracture gets
   healed automatically in due course of time, while the other processes go on as
   usual. So self-repair is an outstanding feature of biological systems. Self-repair
   is almost impossible in artificial systems.

8.2.5   Conclusions About Living and Artificial Systems

1. If robustness is the objective, compared to man-made designs, natural designs
   are far superior.
2. Biological examples or evolution can provide wonderful insights on survival and
   adaptation and help get the idea of how species survive and adapt to changing
   circumstances and how we can incorporate it in engineering problems. This has
   opened up a new field called **biomimetics**, wherein we try to mimic features of
   biological systems in engineering design.

8.2.6   Why Would We Want to Use Genetic Algorithms?

For example, we consider a one-variable problem in x, as shown in Fig. 8.1. Let us
look at a function f(x) that has 2 local minima and one global minimum. The two local
minima are 1 and 3 while the global minimum is 2. A robust optimization algorithm
is one, which, regardless of the starting point, will always converge to point 2, the
global minimum.

Let us look at it this way. If the initial guess is in the A region, traditional algorithms
like the steepest ascent/descent or the conjugate gradient method will converge the
solution to 1. They will declare that 1 is the solution because if an iterate goes to the
right of 1, the algorithm will push it to 1 and when the iterate goes to the left of 1, it
will push it to 1. Therefore, the algorithm will aver that 1 is the true solution. If we
start in the B region, traditional optimization algorithms will converge to 3.

However, if we use an evolutionary optimization technique or a global optimization technique, regardless of whether we start in region A or B, we get the final
answer as 2. If we start in region C, the regular algorithm and GA will both give
the same answer. However, if one knows this much, then he/she may not need an
optimization tool. A robust optimization technique is one that gives us the solution

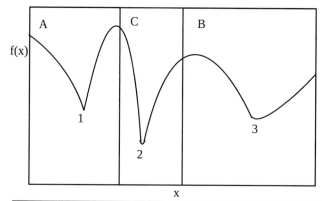

Initial guess	Traditional optimization	Genetic Algorithm
A	Converges to 1	Converges to 2
B	Converges to 3	Converges to 2
C	Converges to 2	Converges to 2

Fig. 8.1 Concept of local and global minima in an optimization problem

regardless of the starting point. So the GA is one algorithm that can be used to get the correct solution if f(x) looks as shown here.

Several engineering problems may have local minima like this and just one global optimum and we want to determine the global optimum. Now, this picture should give us additional ideas. If we look at whether the genetic algorithm will be fast or slow compared to the conjugate gradient method, we see that there is a disadvantage of GA, that it converges slowly. The advantage is that it converges to the global optimum. Can we combine the speed with the robustness? How do we do this?

We start initially with genetic algorithms and zero in on region C. Once we get to this region, we quickly go to the conjugate gradient or some other faster traditional technique. So this is called a **hybrid optimization technique**. So, here we start with GA and switch later to a gradient-based method. This is routinely used nowadays for complex optimization problems.

8.2.7 What Are the Mathematical Examples for Multi-Modal Functions?

The Rastrigin function is one such example. The function is given here (Eq. 8.1). It really tortures y and has several peaks and valleys but the minimum of the function is 0. The global minimum occurs at (0, 0).

$$Ras(x_1, x_2) = 20 + x_1^2 + x_2^2 - 10(cos(2\pi x_1) + cos(2\pi x_2)) \qquad (8.1)$$

$$Ban(x_1, x_2) = (1 - x_1)^2 + 100(x_2 - x_1^2)^2 \qquad (8.2)$$

The Rosenbrock function is called the Banana function, $Ban(x_1, x_2)$ (Eq. 8.2). The global minimum occurs at $(1, 1)$. If somebody is interested in working on new optimization methods, one needs to first test his/her algorithm against standard functions like this. We have to benchmark and find out on a particular computer, how many iterations this function takes using the new algorithm and what level of accuracy it reaches. First, it should reach the global optimum, then the speed and time are tested. These are all considered standard.

8.2.8 Efficiency of the Optimization Technique Versus the Kind of Problem

An optimization problem falls under one of the three categories listed below.

1. Unimodal problems are those that have got only one peak or valley.
2. A combinatorial problem is one that permits only a finite number of combinations of the variables under question. For example, there are 5 machines and each can do 4 types of operations. There are various combinations of which machine will do which job, so that we maximize the profit or minimize the cost as the case may be.
3. Multimodal problems are mathematical or engineering problems that have several peaks and valleys.

Now if we look at unimodal problems, the conjugate gradient or the steepest ascent/descent method will be very efficient. In Fig. 8.2, we see that the efficiency is almost close to 1. But if we try to apply the same to a multimodal function, it may not converge or may converge very slowly. That means we have to restart with different initial points. So the efficiency is very low. Even when the exhaustive search or the random walk algorithm is used for a combinatorial problem, the efficiency is very low. This is because, for a combinatorial type of problem, we may try to exhaustively search for the optima. Now, if we look at GA, its efficiency is more or less the same for all classes of problems and looks, as shown here. So the GA has an efficiency that is far greater compared to what a traditional optimization technique or exhaustive search method has for multimodal and combinatorial problems. However, for a specific unimodal problem, the efficiency of GA will be lower than a sophisticated traditional technique developed for it.

While the exhaustive search or the random walk works with a uniform or systematically low efficiency, for a class of problems, GA works with a reasonably high efficiency for a wide range of problems. This is known as robustness. We cannot claim that for all the problems, the efficiency of GA is the highest. Such a statement would be far from the truth!

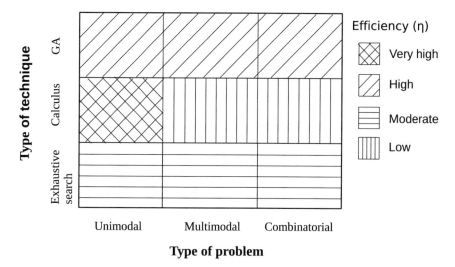

Fig. 8.2 Efficiency of various classes of optimization algorithms for three kinds of problems

8.2.9 Philosophy of Optimization

Now we have to look at the philosophy of optimization. What is the goal of optimization? We keep improving the performance to reach a or some optimal points. We see that as y improves, we keep going in that direction.

The implications are as follows:

- We seek improvement to approach some optimal point. There is always an optimal point and the goal is to reach that. That is the belief!
- What Prof. Goldberg alleges is that this is a highly calculus-oriented view of optimization. We have learned so much of calculus that we feel that dy/dx must be equal to 0 such that there is a peak and on both sides, y should drop sharply. He says it never happens in nature or any engineering problem.
- He alleges that it is not a natural emphasis.

According to human perception,

- Goodness is invariably judged relative to the competition.
- The best scientist, tennis player or cricketer, poet, we cannot define a maximum y in these cases! When somebody is outstanding it means that compared to others, he or she is very good and is far better compared to the others. But after 20 years, someone else may come who is better than him/her. We cannot simply specify criterion and say he/she satisfies all this, he/she is the best.
- Convergence to the best is beside the point as no one knows the definition of best or what the criteria for it are.
- So the definition of best in most of the situations is "far better compared to others", but there is no objectivity here.

- Can we say the best human being has arrived? How do we define such a person? Someone who has the maximum money or has solved a 200 year problem or propose a radically new theory?
- Doing better relative to the others is the key. This also justifies, in a lighter vein, why many universities in the world have a relative grading in place.

Therefore, we have to look at whether the optimization algorithm goes in the right direction. After it has reached some particular level, we just stop working on the problem. We do not try to get one optimum solution. So from the perspective of a GA analyst, the priorities of optimization have to be relooked. The most important goal of optimization is improvement. We look at how the objective function changes. The key point now is that instead of optimizing, can we reach a satisfying level of performance quickly? For example, with the design variable like temperature, pressure, and so on, can we quickly reach a good level of efficiency for the power plant? That is the goal. So the attainment of the optimum is not a critical requirement for complex systems.

A parallel that can be drawn here is the Kaizen philosophy used extensively in Japan. In Japanese, Kaizen means "continuous improvement" is that we do not have a separate set of supervisors. Usually, someone makes a product and someone else checks it. They got rid of it and said that the person who makes it also checks it. He/she reports efficiency and there is a self-correction. We cannot see rapid progress in Kaizen but there is continuous improvement. But because there is incremental progress over a period of time, we get substantial progress. The Japanese carmaker Toyota uses this, and this automobile giant is known for its quality consciousness.

8.2.10 Key Differences Between GA and Traditional Optimization Methods

- **GA employs the coding of the parameter set** and not the parameters themselves. Generally, the variables x_1 to x_n are replaced by their binary equivalent such that we convert them into 0s and 1s. While it is possible to write a genetic algorithm code without this binary representation too, the most popular of them all is the binary representation.
- **The GA searches for a population of points and not a single point**. Instead of taking a single value of x_1 and x_2 and taking y, we see how y changes with different sets of x_1 and x_2. If we have a two-variable problem, we take a, b, c, and d which represent different (x_1, x_2) values. a, b, c, d are known as candidates. For a, b, c, and d, we now calculate y(a), y(b), y(c), and y(d). We then find which among them is maximum and rank them. We convert x_1 and x_2 into 0 and 1. We mix and match the bits from "better" parents. The "better" parents are those for which y is higher (for a maximization problem). We now get new values of a, b, c, and d. The

"children" are now born, who then become parents. The strategy involved is that among all these parents, only the fittest can survive and reproduce. The number of members in a population, (the sample size) is kept fixed (in a constant population size approach). So if we have 4 or 5 solutions, we look at how these solutions evolve generation after generation. After a certain stage, we can take the average of these points and say that is the optimum. We will reach a stage where there is little difference between members of the species. That is the way it should be. If there is continuous improvement, all members should be equally strong. That is what GA strives to achieve.

- **GA uses the objective function information** and does not use the information of the derivative or second derivative.
- **GA uses stochastic transition rules** and not deterministic rules.

8.2.11 The Basic GA

- A population of 2n–4n trial solutions is used, where n is the number of variables. For a two-variable problem, we typically start with 4–8 solutions.
- Each solution is represented by a string of binary variables corresponding to chromosomes in genetics. So, each solution is represented as 0s and 1s. While this is not the only way of coding, it is the most popular one.
- The string length can be made long enough to accomplish any desired accuracy of the fitness and thus, any desired accuracy is attainable.
- The numerical value of y mimics the concept of "fitness"in genetics. The fittest members of the species will survive and procreate. This comes from the Darwinian theory. Fitness in genetics is analogous to the objective function value in optimization problems.
- After the candidate solutions are selected, a new generation is produced by selecting, using stochastic principles, the fittest parents to produce children from among the trial solutions. There is no distinction of father or mother in so far as the selection of parents is concerned in GA.
- In each child, we mix and match the bits from 2 parents and produce new children. Then we again convert the 0s and 1s back into y. We now calculate y for the new values of the variables.
- We rank the solution in the ascending or descending order of fitness. For a maximization problem, we choose those values of design parameters that give the highest values of y.
- The random alteration of binary digits in a string may introduce beneficial or harmful effects. Sometimes what happens is that some solution is already having a high fitness, and because we always want 2 parents to mate and produce children, as defective chromosomes from the second parent may replace the good ones of the first parent. It may happen sometimes that this child may have poor fitness. So this constant mixing and matching may result in a drop in quality and we randomly do what is called **mutation**. That is, one in 20 bits, we change 1–0 and 0–1. This

is randomly or stochastically done so that the undesirable effects of this crossover are neutralized.

A basic flowchart for solving an optimization problem with GA is shown in Figure 8.3

8.2.12 Elitist Strategy

Genetic algorithms have some other techniques also to overcome this problem of losing strings of high quality. There is another strategy called the **elitist strategy**. What we do in this is that out of the n solutions, the best solution is left untouched and automatically goes to the next generation. So all the crossover and permutations–combinations are done for the (n-1) solutions while the "king" remains as it is. But "he" cannot remain king forever. When we do mixing and matching and the (n-1) parents recombine to produce new (n-1) children, the king is added to the list and ranked along with the others. If he is no longer king, i.e., he does not have the highest fitness, then he also joins the process of crossover and mutation. The king in a particular generation is not touched. This is the elitist strategy.

This is also used in yet another optimization technique known as **particle swarm optimization**. This is also very similar to the genetic algorithm and is an evolutionary technique. For example, when a group of birds is moving and searching for food, the group adjusts its orientation such that each of the birds is at the smallest possible distance from the leader and the leader is the one who is closest to the food. After some time, the position of the leader may change. This change of leadership is analogous to the elitist strategy in GA.

8.2.13 Some Doubts About GA?

(a) How do we represent a set of variables as a string?

Let us answer this through an example, $X = (x_1, x_2, x_3, \ldots x_n)$. If we have 4 variables, $x_1 = 15$, $x_2 = 4$, $x_3 = 21$, $x_4 = 7$, we know that in the binary system, 15 can be represented as 01111, 4 as 00100, and so on. So $X = (01111, 00100, 10101, 00111)$. Now we remove the comma separating the values of the variables from X. In our computer program, we know that the last 5 bits represent x_4, the next 5 bits represent x_3, and so on. So each X is a design vector, which is a combination of the design variables that are represented as combinations of 0 and 1.

(b) How do we handle decimals?

Suppose we want to represent 4.87, we multiply this by 100 which gives 487. We need 9 bits to represent 487 because $2^8 = 256$ and $111111111 = 511$. So 487 being smaller than 511, 9 bits are required to represent it. But we have to remember that

Fig. 8.3 Genetic Algorithm
flowchart

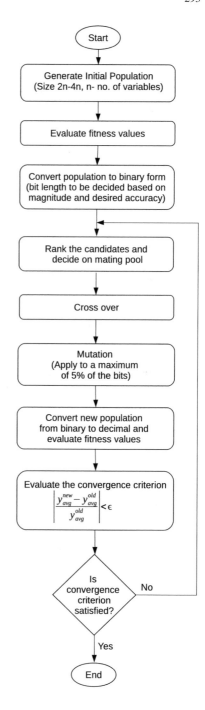

when we interpret the final value of these 9 bits, we have to divide it by 100. So if the accuracy we seek keeps increasing, then the number of bits required will also increase dramatically. The values of the variables alone do not decide the string length and the accuracy too influences it. If we know for a fact that a variable x lies between 1 and 5, and we need 2-digit accuracy, then 9 bits are required or are sufficient for its representation.

8.2.14 The Main Features of GA

GA has evolved a lot since it first made its appearance in 1975. The main features of GA are

1. GA is conceptually easy to understand.
2. GA works very well in exceedingly complex problems and in problems, where getting derivatives of the objective function, is very cumbersome.
3. Lots of approaches and possibilities exist and so considerable subjectivity exists in implementation.
4. It is easy to parallelize. That is, when we want the value of y for 4 different cases, each can be sent to a different processor on a server and this speeds up the process. However, GA is not massively parallel as crossover, reproduction, mutation, and so on cannot be parallelized.
5. It is not easy to make GA work very efficiently though it is easy to make it work.

In summary, GA is a contemporary powerful tool available to us to solve complex optimization problems. However, not all the features of GA can be established rigorously by mathematics. Even so, GA is certainly not a helter-skelter technique that has no philosophy or logic behind it.

Example 8.1 The cost of engines plus fuel for a cargo ship (in lakhs of rupees per year for 100 tons of cargo carried) varies with speed and is given by $0.2x^2$, where x is the speed of the ship in m/s. The fixed cost of hull and crew (again in the same units) is given by $450/x$. Using genetic algorithms, perform 2 iterations to determine the optimal operating speed of the ship. The original interval of uncertainty can be taken as $0.5 \leq x \leq 25.5$ m/s and 1.

Solution:

The largest value of x is given as 25.5 m/s and one decimal accuracy is required. So the number we are looking at is 255. The largest number with 8 bits is 255 and so, 8 bits are enough to represent x here. $y = 0.2x^2 + 450/x$; $0.5 \leq x \leq 25.5$ m/s We will work with 4n solutions, where n is the number of variables. In this case, n = 1. The number of design solutions = 4n = 4.

We randomly take 4 values of x that are uniformly spread over the interval $0.5 \leq x \leq 25.5$. Here, when we are using 8 bits and the maximum value is 255, we must remember after all the operations are done and we are reconverting from binary to decimal, we have to divide by 10 to get the value of x.

Now we convert these values to binary and check for bias. We have 16 ones and 16 zeros, which is very good. We can also generate the initial population by using a random number table. A typical random number is given at the end of the book. In this table, we proceed from the top to the bottom and once a column is over, we proceed from left to right. We now use the first random number. If this is less than 0.5, we generate "0" in the binary form of the first candidate solution. Else it is 1. We then go to the next random number and apply the same rule and proceed till we generate 32 bits.

Next we calculate the value of y by substituting the values of x in the equation $y = 0.2x^2 + 450/x$. We determine $\Sigma y = 386.44$ and average fitness $\bar{y} = 98.61$. Please note that the original GA is for maximization, while we are trying to do minimization, in this example.

For a minimization problem, the fittest is string 2 whose y value is 64.95 and relative fitness is 0.168. Now we want to know the count so that we can decide on the mating pool. We must keep total count as 4n so that we have a constant number of members in the mating pool. Now we do the single-point crossover though we can do a uniform crossover bit by bit. We now generate the new population by deciding on (i) the mate for a particular parent and (ii) the exact point of crossover. We need to do both of these randomly (or stochastically).

After these operations with the mates and crossover sites are done, as shown in Table 8.1, we generate the new population and evaluate the fitness (y_i) of all the candidates (Table 8.2). We take note of the one with the lowest y_i, meaning the one with the lowest cost.

The beauty of the GA is that in just one iteration, Σy has come down from 386.44 to 315.89 and the average fitness has come down from 98.6 to 78.97. This means the cost has come down dramatically. This may sound controversial but we want the average fitness to decrease as we are looking at a minimization problem. The most important thing in genetic algorithm is that the variance in y or the difference

Table 8.1 Initial population and mating pool for Example 8.1

String No	Initial population, x	Initial population (binary)	y_i	$y_i / \sum y_i$	Count	Mating pool
1	4.2	00101010	110.67	0.286	1	01100101
2	10.1	01100101	64.95	0.168	2	01100101
3	16.4	10100100	81.23	0.21	1	00101010
4	23.5	11101011	129.59	0.335	0	10100100
Σ			386.44			

Table 8.2 First iteration for Example 8.1

Mating pool	Mate	Crossover site	New population	New population (decimal)	y_i	$y_i / \sum y_i$	count
0110 | 0101	4	4	01100100	10	65	0.2	2
01100 | 101	3	5	01100010	9.8	65.12	0.206	1
00101 | 010	2	5	00101101	4.5	104.05	0.329	0
1010 | 0100	1	4	10100101	16.5	81.72	0.259	1
$\sum$					315.89		

Table 8.3 Second iteration for Example 8.1

Mating pool	Mate	Crossover site	New population	New population (decimal)	y_i	$y_i / \sum y_i$	Count
011001 | 00	3	6	01100110	10.2	65	0.24	1
0110 | 0100	4	4	01100101	10.1	64.95	0.23	2
011000 | 10	1	6	01100000	9.6	65.31	0.24	1
1010 | 0101	2	4	10100100	16.4	81.23	0.29	0
$\sum$					276.5		

between the minimum and maximum values of y will dramatically come down, with generations. Initially, this variation was 64.64 and now it is 39.05. It will come down even further and we have to do one more iteration according to the question. The results of the second iteration are shown in Table 8.3. The average fitness has come down further to 69.125. Furthermore, the variance in the fitness across the population comes down with iterations, which is the hallmark of evolution.

We can have a double crossover also in this stage if we want and in 3 or 4 iterations, we will get an average fitness that is very close to the correct answer. We may again say that we can do an exhaustive search, but for multimodal problems, it will not work. However, GA searches in the whole solution space and even if a solution is very weak, it is not discarded. How does this happen? This can come in the form of mutations or some patterns of strings that can be picked up and so on. In the tournament selection , we get a good mating pool compared to just ranking the trial solution where we compare the fitness of two candidates at a time, just as we do in cricket or tennis tournament in ascending or descending order. This way, some diversity in the population is preserved and we can avoid premature convergence.

MATLAB code for Example 8.1

```
1  clear;
2  close all;
3  clc;
4
5  a=0.5;    % Initial value in the range
6  b=25.5;   % Final value in the range
7  ps=20;    % Population size (an even number)
8  nb=8;     % Number of bits reqd. to represent
9  itr=20;   % Number of iterations (generations)
10
11 % Initialize fitness
12 y=zeros(1,ps);
13
14 %Initial population randomly generated and
15 %rounded to 2 decimal accuracy
16 x=a+(b-a)*rand(1,ps);
17 x=round(x,1);
18
19 count=0;
20 while count<itr  % Starting the iteration
21     count=count+1;
22
23     y=0.2*x.^2+450./x; %Calculating the Fitness
24
25     %Converting x,y from decimal to binary
26     x_bin=de2bi(floor(x*10),nb);
27     x_bin2=x_bin;
28
29
30     %Generating the mating pool
31
32     % Noting the value and index for the maximum
33     [Vmax, j1] = max(y);
34     % Noting the value and index for the minimum
35     [Vmin, j2] = min(y);
36     xmin_bin = x_bin(j2,:); % Noting the minimum value
37     xmin = bi2de(xmin_bin)/10;
38
39     % Count 2 for minimum fitness and
40     x_bin2(j1,:)=x_bin(j2,:);
41
42    % eliminating the one with minimum fitness
43     %Select Mates
44     m=ones(2,ps/2);
45     k=1:ps;
46
47     m(1,1)=j1;
48     m(1,2)=j2;
49     r=ceil((ps-2)*(rand(1)));
```

```
50      m(2,1)=k(r);
51      k(r)=[];
52      r=ceil((ps-3)*(rand(1)));
53      m(2,2)=k(r);
54      k(r)=[];
55
56      if ps>4
57          for i=1:(ps-4)/2
58              r=ceil((ps-2-2*i)*(rand(1)));
59              m(1,2+i)=k(r);
60              k(r)=[];
61              r=ceil((ps-3-2*i)*(rand(1)));
62              m(2,2+i)=k(r);
63              k(r)=[];
64          end
65      end
66
67      %Generate random cross-over site
68      cr=ceil(rand(1,ps/2)*(nb-1));
69
70      %Cross-over and generate a new population
71      for i=1:ps/2
72          x_bin(2*i-1,1:cr(i))=x_bin2(m(1,i),1:cr(i));
73          x_bin(2*i-1,cr(i)+1:nb)=x_bin2(m(2,i),cr(i)+1:nb);
74          x_bin(2*i,1:cr(i))=x_bin2(m(2,i),1:cr(i));
75          x_bin(2*i,cr(i)+1:nb)=x_bin2(m(1,i),cr(i)+1:nb);
76
77      end
78
79      % Converting the new population to decimal
80      %for calculating the fitness
81      x=bi2de(x_bin)/10;
82
83      % Print
84      prt = ['Itr = ',num2str(count),...
85          ', minVel = ',num2str(xmin),...
86          ', minY = ',num2str(Vmin)];
87      disp(prt)
88
89  end
```

The output of the program is

> *Itr = 1, minVel = 11.1, minY = 65.1905*
> *Itr = 2, minVel = 10.9, minY = 65.0464*
> *Itr = 3, minVel = 10.9, minY = 65.0464*
>
> ⋮   ⋮   ⋮
>
> *Itr = 19, minVel = 10.4, minY = 64.9012*
> *Itr = 20, minVel = 10.4, minY = 64.9012*

Example 8.2 Revisit Example 8.1 and solve it using genetic algorithm with elitist strategy and perform two iterations to determine the optimal operating speed of the ship.

Solution:

In the elitist strategy, the best solution is left untouched and it carries forward to the next generation. Therefore in this case we begin with a population size of 4n+1, i.e., 5 as it is more compatible to perform the crossover operation in an even number of strings.

We randomly take 5 values of x that are uniformly spread over the interval $0.5 \leq x \leq 25.5$ (Table 8.4).

Next, we calculate the value of y by substituting the values of x in the equation $y = 0.2x^2 + 450/x$. We determine $\Sigma y = 456.82$ and average fitness $\bar{y} = 91.36$. Please note that the original GA is for maximization, while we are trying to do minimization, in this example.

For a minimization problem, the fittest is string 2 whose y value is 65.68 and relative fitness is 0.144. Now we want to know the count so that we can decide on the

Table 8.4 Initial population and mating pool for Example 8.2

String No	Initial population, x	Initial population (binary)	y_i	$y_i / \sum y_i$	Count	Mating pool
1	9.4	01011101	65.68	0.143	2	01011101
2	3.5	00100011	131.02	0.287	0	01011100
3	14.7	10010011	73.83	0.162	1	10010011
4	22.4	11100000	120.44	0.264	1	11100000
5	9.2	01011100	65.84	0.144	1	**01011101**
Σ			456.82			

mating pool. We must keep the total count as 4n+1 so that we have a constant number of members in the mating pool. Since the fittest string is given a count 2, only one fittest string is used for the crossover operation. The other fittest string directly goes to the next generation without any crossover.

Now we do the single-point crossover though we can do a uniform crossover bit by bit. We now generate the new population by deciding on (i) the mate for a particular parent and (ii) the exact point of crossover. We need to do both of these randomly (or stochastically).

After these operations with the mates and crossover sites are done, as shown in Table 8.5, we generate the new population and evaluate the fitness (y_i) of all the candidates (Table 8.5). We take note of the one with the lowest y_i, meaning the one with the lowest cost.

The beauty of the GA is that in just one iteration, Σy has come down from 456.82 to 411.57 and the average fitness has come down from 91.36 to 82.31. This means the cost has come down dramatically. This may sound controversial but we want the average fitness to decrease as we are looking at a minimization problem. It will come down even further and we have to do one more iteration according to the question. The results of the second iteration are shown in Table 8.6. The most important thing in genetic algorithm is that the variance in y or the difference between the minimum

Table 8.5 First iteration for Example 8.2

Mating pool	Mate	Crossover site	New population	New population (decimal)	y_i	$y_i / \sum y_i$	Count
0101 | 1101	4	4	11101101	23.7	131.32	0.32	0
010 | 11100	3	3	01010011	8.3	67.99	0.165	1
100 | 10011	2	3	10011100	15.6	77.52	0.188	1
1110 | 0000	1	4	01010000	8.0	69.05	0.168	1
01011101	–	–	01011101	9.3	65.68	0.16	2
Σ					411.57		

Table 8.6 Second iteration for Example 8.2

Mating pool	Mate	Crossover site	New population	New population (decimal)	y_i	$y_i / \sum y_i$	count
0 | 1011101	2	1	01010011	8.3	67.99	0.2	1
0 | 1010011	1	1	01011101	9.3	65.68	0.194	2
010 | 10000	4	3	01011100	9.2	65.84	0.195	1
100 | 11100	3	3	10010000	14.4	72.72	0.21	0
01011101	–	–	01011101	9.3	65.68	0.194	1
Σ					337.93		

and maximum values of y will dramatically come down, with generations. Initially, this variation was 65.34 and now it is 7.04. The average fitness has come down further to 67.58. Furthermore, the variance in the fitness across the population comes down with iterations, which is the hallmark of evolution.

MATLAB code for Example 8.2

```
1   %Matlab code for Genetic Algortihm
2   %Example #8.1
3
4   clear;
5   close all;
6   clc;
7
8   a=0.5;          % Initial value in the range
9   b=25.5;         % Final value in the range
10  ps=21;          % Population size (an odd number)
11  nb=8;           % Number of bits reqd. to represent
12  itr=20;         % Number of iterations
13
14  % Initialize fitness
15  y=zeros(1,ps);
16
17  %Initial population randomly generated and
18  %rounded to 1 decimal accuracy
19  x=a+(b-a)*rand(1,ps);
20  x=round(x,1);
21
22  count=0;
23  while count<itr                % Starting the iteration
24      count=count+1;
25
26      y=0.2*x.^2+450./x;         %Calculating the Fitness
27      %Converting x,y from decimal to binary
28      x_bin=de2bi(floor(x*10),nb);
29      x_bin2=x_bin;
30
31      [Y,idx] = sort(y,'descend');
32      for i=1:ps
33          x_bin2(i,:) = x_bin(idx(i),:);
34      end
35
36      % Count 2 for minimum fitness and
37      % eliminating the one with maximum fitness
38      x_bin2(1,:) = x_bin2(ps,:);
39      %Generating the mating pool
40      Ymin = Y(ps);
41      xmin = bi2de(x_bin2(ps,:))/10;
42
```

```
43        %Select  Mates
44        m=ones(2,(ps-1)/2);
45        k=1:ps-1;
46
47            for  i=1:(ps-1)/2
48                r=ceil((ps-1+2-2*i)*(rand(1)));
49                m(1,i)=k(r);
50                k(r)=[];
51                r=ceil((ps-1+1-2*i)*(rand(1)));
52                m(2,i)=k(r);
53                k(r)=[];
54            end
55
56
57        %Generate  random  cross-over  site
58        cr=ceil(rand(1,(ps-1)/2)*(nb-1))
59
60        %Cross-over  and  generate  a  new  population
61        for  i=1:(ps-1)/2
62        x_bin(2*i-1,1:cr(i))=x_bin2(m(1,i),1:cr(i));
63        x_bin(2*i-1,cr(i)+1:nb)=x_bin2(m(2,i),cr(i)+1:nb);
64        x_bin(2*i,1:cr(i))=x_bin2(m(2,i),1:cr(i));
65        x_bin(2*i,cr(i)+1:nb)=x_bin2(m(1,i),cr(i)+1:nb);
66        end
67        x_bin(ps,:)  =  x_bin2(ps,:);
68        % Converting  the  new  population  to  decimal
69        % for  calculating  the  fitness
70        x=bi2de(x_bin)/10;
71
72        % Print
73        prt = ['Itr = ',num2str(count),...
74            ', v_min = ',num2str(xmin),...
75            ', Y_min = ',num2str(Ymin)];
76        disp(prt)
77
78  end
```

The output of the program is:

```
Itr = 1, minVel = 10.1, minY = 64.9565
Itr = 2, minVel = 10.7, minY = 64.9541
Itr = 3, minVel = 10.7, minY = 64.9541
  ⋮               ⋮               ⋮
Itr = 19, minVel = 10.4, minY = 64.9012
Itr = 20, minVel = 10.4, minY = 64.9012
```

Example 8.3 Consider the problem of maximization of the volume of a rect-angular box to be made of sheet metal with a total area of $2\,m^2$. The three dimensions of the box are length-x, breadth-y, and height-z (all in m).

(a) Convert this to a two-variable unconstrained optimization problem in say, x and y.
(b) We would like to use genetic algorithm to solve the above problem, as a two-variable maximization problem in x and y

Perform three iterations of the genetic algorithm with an initial population size of 4.

You may assume that $0.1 \le x$ or $y \le 2.55\,m$. Two decimal accuracies are desired on the dimensions. Decide appropriate strategies for crossover. No mutation is required. Report maximum, minimum, and average fitness values for the initial population (which may be randomly chosen) and the populations at the end of the first, second, and third iterations. Make maximum use of the random number tables.

Solution:

The dimensions of the box are length-x, breadth-y, and height-z.
  Given the area of the box, $A = 2(xy + yz + zx) = 2m^2$.
  Volume of the box, $V = xyz$.
  Our objective is to maximize the volume for the given area by converting this into a two-variable unconstrained optimization problem in x and y.
  From the equation for area, we get

$$z = \frac{1 - xy}{x + y} \tag{8.3}$$

Therefore, we substitute z in the volume expression so as to convert it into a two-variable unconstrained optimization problem in x and y.

$$Maximize\ V = xy\left(\frac{1 - xy}{x + y}\right) \tag{8.4}$$

The largest value of x or y is given as $2.55\,m$ and two decimal accuracies are required. So the number we are looking at is 255. The largest number with 8 bits is 255 and so, 8 bits are enough to represent x or y here in the range $0.01 \le x$ or $y \le 2.55\,m$ We will work with 4n solutions, where n is the number of variables. In this case, $n = 2$. Number of design solutions $= 4n = 8$, i.e., 4 for each variable.

  We randomly take 4 values each of x and y that are uniformly spread over the interval $0.01 \le x$ or $y \le 2.55$. Here, when we are using 8 bits and the maximum value is 255, we must remember after all the operations are done and we are reconverting

Table 8.7 Initial population and mating pool for Example 8.3

S. No.	Initial population		Binary		Volume	Count	Mating pool	
	x	y	x	y			x	y
1	1.55	0.28	10011011	00011100	0.1342	1	10011011	00011100
2	0.90	0.48	01011010	00110000	0.1778	2	01011010	00110000
3	0.77	1.15	01001101	01110011	0.0528	1	01001101	01110011
4	0.40	2.14	00101000	11010110	0.0485	0	01011010	00110000

Max. fitness = 0.1778; Min. fitness = 0.0485; Avg. fitness = 0.1026

Table 8.8 First iteration for Example 8.3

Mating pool		Mate	Crossover	New population		Decimal		Volume	Count
x	y			x	y	x	y		
10011011	00011100	2	6	10011010	00010000	1.54	0.16	0.1092	1
01011010	00110000	1	6	01011011	00111100	0.91	0.60	0.1642	1
01001101	01110011	4	4	01001110	01110000	0.78	1.12	0.0581	0
01011010	00110000	3	4	01011001	00110011	0.89	0.51	0.1771	2

Max. fitness = 0.1771; Min. fitness = 0.0581; Avg, fitness = 0.1271

from binary to decimal, we have to divide by 100 to get the values of x and y (Table 8.7).

Next, we calculate the value of V by substituting the values of x and y. We note the maximum fitness to be 0.1778, minimum fitness to be 0.0485, and average fitness to be 0.1026. We shall look at how these 3 values change across the generations(iterations). Please note that the original GA is for maximization, while we are trying to do minimization, in this example.

For a maximization problem, the fittest is string 2 whose V value is 0.1778. Now we want to know the count so that we can decide on the mating pool. We must keep the total count as 4n+1 so that we have a constant number of members in the mating pool. Since the fittest string is given a count 2, only one fittest string is used for the crossover operation. The other fittest string directly goes to the next generation without any crossover. The string with the minimum fitness is given a count 0 as it needs to be eliminated.

Now we do the single-point crossover though we can do a uniform crossover bit by bit. We now generate the new population by deciding on (i) the mate for a particular parent and (ii) the exact point of crossover. We need to do both of these randomly (or stochastically). In this problem, since it involves two variables, we have the option of choosing the mates and crossover sites differently for both the variables. But we chose the mates and crossover sites the same for both the variables.

After these operations with the mates and crossover sites are done, as shown in Table 8.8, we generate the new population and evaluate the volume of all the candidates (Table 8.8). We note the maximum fitness to be 0.1771, minimum fitness

Table 8.9 Second iteration for Example 8.3

Mating pool		Mate	Crossover	New population		Decimal		Volume	Count
x	y			x	y	x	y		
10011010	00010000	3	7	10011011	00010001	1.55	0.17	0.1128	0
01011011	00111100	4	5	01011001	00111011	0.88	0.50	0.1786	1
01011001	00110011	1	7	01011000	00110010	0.89	0.59	0.1685	1
01011001	00110011	2	5	01011011	00110100	0.91	0.52	0.1743	2

Max. fitness = 0.1786; Min. fitness = 0.1128; Avg, fitness = 0.1585

Table 8.10 Third iteration for Example 8.3

Mating pool		Mate	Cross-over	New population		Decimal		Volume	Count
x	y			x	y	x	y		
01011011	00110100	3	6	01011000	00110110	0.88	0.54	0.1756	1
01011001	00111011	4	4	01011011	00110000	0.91	0.48	0.1770	1
01011000	00110010	1	6	01011000	00110010	0.88	0.50	0.1786	2
01011011	00110100	2	4	01011001	00111011	0.89	0.59	0.1685	0

Max. fitness = 0.1786; Min. fitness = 0.1685; Avg; fitness = 0.1749

to be 0.0581, and average fitness to be 0.1271. The maximum fitness is slightly reduced whereas the minimum fitness and the average fitness have increased.

We follow the similar procedure for iteration 2 and 3 by allotting the count 2 to the candidate with maximum fitness and count 0 to the candidate with minimum fitness and proceed with the mating and crossover. The iterations 2 and 3 are shown in Table 8.9 and Table 8.10, respectively.

As we move from the initial population to the end of the third iteration, we observe that the maximum fitness has increased from 0.1778 to 0.1786, whereas the minimum fitness has increased from 0.0485 to 0.1685 and the average fitness has increased from 0.1026 to 0.1749. The beauty of the GA is that in three iterations, we are still able to observe that the solution is approaching the maximum. The MATLAB code and output for this problem are shown below. We observe that in 20 iterations, the maximum value for the Volume has been obtained.

The implementation of genetic algorithm can be made even robust by implementing mutation, multiple-point crossover, and selecting different mates for the variables x and y. In this problem, only a simpler version of the genetic algorithm has been implemented but still, we are able to obtain the maximum quite quickly which shows the robustness of the genetic algorithm.

MATLAB code for Example 8.3

```
 1  clear;
 2  close all;
 3  clc;
 4
 5  a=0.1;          % Initial value in the range
 6  b=2.55;         % Final value in the range
 7  ps=20;          % Population size (an even number)
 8  nb=8;           % Number of bits reqd. to represent
 9  itr=50;         % Number of iterations
10
11  % Initialize fitness
12  V=zeros(1,ps);
13  Vmax=zeros(1,itr);
14  Vmin=zeros(1,itr);
15  Vavg=zeros(1,itr);
16
17  %Initial population randomly generated
18  % and rounded to 2 decimal accuracy
19  x=a+(b-a)*rand(1,ps);
20  x=round(x,2);
21  y=a+(b-a)*rand(1,ps);
22  y=round(y,2);
23
24  count=0;
25  while count<itr             % Starting the iteration
26      count=count+1;
27      %Calculating the volume(fitness)
28      for i=1:ps
29          V(i)=x(i)*y(i)*(1-x(i)*y(i))/(x(i)+y(i));
30          % Eliminating samples with negative fitness
31          if V(i)<=0
32              %and regenerating new samples
33              label=1;
34              while label==1
35                  p=rand(1);
36                  if p>0.5
37                      x(i)=a+(b-a)*rand(1);
38                      x=round(x,2);
39                  else
40                      y(i)=a+(b-a)*rand(1);
41                      y=round(y,2);
42                  end
43                  V(i)=x(i)*y(i)*(1-x(i)*y(i))/(x(i)+y(i));
44                  if V(i)>0
45                      label=0;
46                  end
47              end
48          end
49      end
```

```
50
51      %Converting  x,y  from  decimal  to  binary
52      x_bin=de2bi(floor(x*100),nb);
53      y_bin=de2bi(floor(y*100),nb);
54
55      x_bin2=x_bin;
56      y_bin2=y_bin;
57
58      % Noting  the  value  and  index  for  the  maximum
59      [V_max,  j1]  =  max(V);
60      % Noting  the  value  and  index  for  the  minimum
61      [V_min,  j2]  =  min(V);
62
63      xmax  =  bi2de(x_bin(j1,:))/100;
64      ymax  =  bi2de(y_bin(j1,:))/100;
65
66      % Count  2  for  maximum  fitness  and
67      %eliminating  the  one  with  minimum  fitness
68      x_bin2(j2,:)=x_bin(j1,:);
69      y_bin2(j2,:)=y_bin(j1,:);
70
71      %Select  Mates
72      m=ones(2,ps/2);
73      k=1:ps;
74
75      % Making  sure  the  samples  with
76      % maximum  fitness  do  no  mate
77      if  j1>j2
78          k(j1)=[];
79          k(j2)=[];
80      else
81          k(j1)=[];
82
83          if  j2>1
84              k(j2-1)=[];
85          else
86              k(j2)=[];
87          end
88      end
89
90      m(1,1)=j1;
91      m(1,2)=j2;
92      r=ceil((ps-2)*(rand(1)));
93      m(2,1)=k(r);
94      k(r)=[];
95      r=ceil((ps-3)*(rand(1)));
96      m(2,2)=k(r);
97      k(r)=[];
98
99      if  ps>4
100         for  i=1:(ps-4)/2
```

```
101          r=ceil((ps−2−2*i)*(rand(1)));
102          m(1,2+i)=k(r);
103          k(r)=[];
104          r=ceil((ps−3−2*i)*(rand(1)));
105          m(2,2+i)=k(r);
106          k(r)=[];
107       end
108    end
109
110    %Generate random cross−over site
111    cr=ceil(rand(1,ps/2)*(nb−1));
112
113    %Cross−over and generate a new population
114    for i=1:ps/2
115       x_bin(2*i−1,1:cr(i))=x_bin2(m(1,i),1:cr(i));
116       x_bin(2*i−1,cr(i)+1:nb)=x_bin2(m(2,i),cr(i)+1:nb);
117       x_bin(2*i,1:cr(i))=x_bin2(m(2,i),1:cr(i));
118       x_bin(2*i,cr(i)+1:nb)=x_bin2(m(1,i),cr(i)+1:nb);
119       y_bin(2*i−1,1:cr(i))=y_bin2(m(1,i),1:cr(i));
120       y_bin(2*i−1,cr(i)+1:nb)=y_bin2(m(2,i),cr(i)+1:nb);
121       y_bin(2*i,1:cr(i))=y_bin2(m(2,i),1:cr(i));
122       y_bin(2*i,cr(i)+1:nb)=y_bin2(m(1,i),cr(i)+1:nb);
123    end
124
125    % Converting the new population to decimal and calculate
             the fitness
126    x=bi2de(x_bin)/100;
127    y=bi2de(y_bin)/100;
128
129    %Recording maximum fitness in every iteration
130    Vmax(count)=V_max;
131    %Recording maximum fitness in every iteration
132    Vmin(count)=V_min;
133    %Recording maximum fitness in every iteration
134    Vavg(count)=mean(V);
135
136    % Print
137    prt = ['Itr = ',num2str(count),...
138          ', x_max = ',num2str(xmax),...
139          ', y_max = ',num2str(ymax),...
140          ', Vol_max = ',num2str(max(V))];
141    disp(prt)
142
143 end
144 plot(1:itr,Vmax)
145 xlabel('Iterations')
146 ylabel('Vmax')
147 title('Vmax vs. Number of Iterations')
148 grid on
149 disp(x)
150 disp(y)
```

The output of the program is:

$Itr = 1, x_{max} = 0.5, y_{max} = 0.42, V_{max} = 0.18033$
$Itr = 2, x_{max} = 0.55, y_{max} = 0.4, V_{max} = 0.18063$
$Itr = 3, x_{max} = 0.55, y_{max} = 0.75, V_{max} = 0.18642$

$\vdots \qquad \vdots \qquad \vdots \qquad \vdots$

$Itr = 19, x_{max} = 0.53, y_{max} = 0.56, V_{max} = 0.19148$
$Itr = 20, x_{max} = 0.56, y_{max} = 0.57, V_{max} = 0.19237$

8.2.15 A Research Example

Suppose we have assorted electronic equipment in the cabinet of a desktop, each of which is generating some heat, we want to look at the optimum positions of the heat source such that maximum cooling is achieved. When is maximum cooling achieved? When the temperature is the lowest and the Nusselt number is the highest. But as we can see from Fig.8.4, there are infinite possibilities for these 4 devices to be kept in this space. Let us see how this problem is tackled using GA.

There is airflow here and it is a very simplified configuration of actual electronic equipment. There are 4 heat-generating devices and we want to minimize the maximum temperature. We can solve this problem using a CFD software or develop our own code. Upon doing this, we can get the temperature distribution around each of these chips or heat sources. What we then do is basically take some arbitrary positions of the devices, get the CFD solution, and obtain the maximum temperatures.

Fig. 8.4 Schematic of electronic equipment in a desktop cabinet

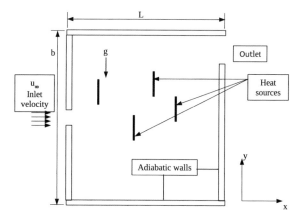

Then we use the GA and generate 4 new positions. The new configurations are again solved for the maximum temperatures using CFD. We use GA again and get 4 new positions. We kept repeating this till convergence is obtained (Refer to Madadi and Balaji, 2008, for a detailed discussion on this).

8.3   Simulated Annealing (SA)

We now look at another nontraditional optimization technique called "Simulated Annealing". It is non-traditional in the sense that we do not use calculus and neither do we use the method of dividing into intervals and eliminating portions of it repeatedly. However, it is also a search technique and we use some probabilistic laws and hence it is a *stochastic optimization technique*. Apart from that, we draw upon certain principles used in metallurgy, e.g., the process of annealing–slow cooling of an object in air, in solving the problem and hence the name "Simulated Annealing".

Some features of simulated annealing are as follows:

- It is similar to genetic algorithms in one sense that it is also based on probabilistic rules.
- It was developed by Kirkpatrick, Gelatt, and Vecchi in 1983 and was published in the Journal Science.
- It is an offshoot of the metropolis algorithm, which is a very powerful sampling technique in statistics.
- Simulated annealing is a global optimization procedure like genetic algorithms, which ensures that there is no premature convergence.
- It is very robust but not very efficient in the sense that it does not converge quickly.
- It works very well for discrete optimization problems as well as for exceedingly complex problems.

8.3.1   Basic Philosophy

Consider the cooling process of molten metals through annealing. At high temperatures, the atoms in the molten state can move freely with respect to one another. However, as the temperature reduces, the movement gets restricted. So analogously, during the initial iterations, the samples are free to move anywhere in the domain. For a single-variable problem in x, x can move anywhere in the domain. This is similar to the situation at the start of annealing, where the atoms have the probability of being in any state.

However, as the energy becomes low, the probability of attaining a higher energy state also becomes low. If the energy is high, the system has an equal probability of attaining any of the states. In short, it means that the freedom gets reduced as the energy level goes down. Similarly in the initial iterations, the freedom is very high. If we have variables x_1 and x_2, they can move here and there initially. But as the iterations proceed, the conditions for accepting a particular sample, i.e., when we are proceeding from $(x_1, x_2)_i$ to $(x_1, x_2)_j$, the conditions for accepting the latter $(x_1, x_2)_{i+1}$ become stricter and stricter, when the iteration is not proceeding in the direction of decreasing objective function, for a minimization problem.

At the end of annealing, the atoms begin to get ordered and finally form crystals with minimum potential energy. If the cooling takes place very fast, it may not reach the final state of minimum potential energy. The crucial parameter here is the cooling rate which decides if eventually we reach the state of minimum potential energy. In view of this, if the cooling rate can be tweaked, fine-tuned, or controlled in such a way that we get the optimum end product in metallurgy. Analogously, the convergence rate of the optimization algorithm is controlled in such a way that we reach global convergence.

If the cooling rate is not properly controlled in metallurgy, instead of reaching the crystalline state, finally one may reach a polycrystalline state that may be at a higher energy state than the crystalline state. Analogously, for the optimization problem, we may get a solution that has converged prematurely. The final solution may well be an optimum but unfortunately it could turn out to be a local optimum. There is no guarantee that this is the global optimum.

The key to achieving the absolute minimum state is slow cooling. This slow cooling is known as annealing in metallurgy and simulated annealing mimics this process. Achieving the global optimum is akin to reaching the minimum energy state in the end.

8.3.2 Key Point in Simulated Annealing (SA)

The key point is that the cooling is controlled by a parameter called the temperature, that is, closely related to the concept of Boltzmann probability distribution. According to the Boltzmann distribution, a system in thermal equilibrium at a temperature T has its energy distributed probabilistically according to

$$P(E) = e^{-\frac{E}{kT}} \tag{8.5}$$

In Eq. 8.5, k is *Boltzmann constant*, 1.38×10^{-23} J/K. Equation 8.5 suggests that as T increases, the system has a uniform probability of being in any energy state. If T is very high, it means that the expression reduces to the exponent raised to the power of minus of a very low quantity, which makes it close to 1, i.e., $P(E) = e^{-very\ small\ quantity} \approx 1$, for all E.

While Eq. 8.5 suggests that when T increases, the system has a uniform probability of being at any energy state, it also tells us that when T decreases, the value of the expression $P(E) = e^{-\frac{E}{kT}}$ takes on a very small value and hence the system has a small probability of being at a higher energy state (see Fig. 8.5). How does it work?

Fig. 8.5 Plot of P(E) versus
E for different temperatures

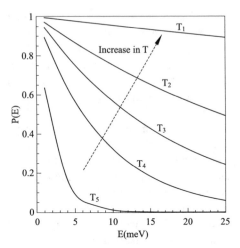

The probability is exponential to the power of a negative quantity. So for small values of (-E/kT), P(E) will be $e^{-0.1}$, $e^{-0.2}$, $e^{-0.05}$, and so on, there is a chance of getting different values of P(E). However, once we have reached e^{-4}, e^{-5}, and so on, P(E) is almost 0. Therefore by controlling T and assuming that the search process follows Eq. 8.5 (the Boltzmann probability distribution), the convergence of the algorithm can be controlled. We use the Boltzmann distribution like condition, to decide if the next sample will be accepted or not.

What is the key difference between this and a traditional algorithm? If we are looking for a search algorithm, conventional thinking says that when Y(x) has to be minimized, when we go from X^0 to X^1 and from Y^0 to Y^1, we accept X^1 only if $Y^1 < Y^0$. In simulated annealing too, there is no problem with this. X is the design vector with variables x_1 to x_n. So what we are doing in simulated annealing is that if we are seeking a minimum and Y^1 decreases compared to Y^0, there can be no doubt in our mind that X^1 has to be accepted.

But the beauty is that if X^1 is such that $Y^1 > Y^0$, we do not reject X^1 right away. Instead, we reject it with a probability. We get this probability from the Boltzmann distribution. We recast Eq. 8.5 as $P(E) = e^{-\frac{\Delta E}{kT}}$, where ΔE is the change in energy in the metallurgical context and is the change in objective function $Y^1 - Y^0$, in the context of optimization. Furthermore, the *Boltzmann constant k* can be made 1 for our algorithm and there are several ways to represent the "temperature" in the Boltzmann distribution. This temperature could be the average of Y for say 4 values of X.

How the system proceeds is like this. Initially, if we want to start, what we do is, if we have only one X, we take 4 arbitrary values of X and obtain the 4 values of Y. We take the average. If we recall, GA also proceeded the same way. Now we assume that $\bar{Y} = T$ and generate a new sample X^1 from X^0. There are several ways of doing this. Either we use the random number table or a Gaussian distribution. Now we compare Y^1 and Y^0. If Y^1 decreases, we accept X^1. But if Y^1 increases, we apply the probability criterion and we will get a number between 0 and 1. We generate another random number r between 0 and 1. We compare r with P. If r is less than or equal to P, we accept the new sample.

The major departure in SA is that even if the objective function becomes worse, i.e., if for a maximization problem, Y decreases or for a minimization problem, Y increases, initially we allow such an aberration. But as we proceed, what happens is that this T will come down as generations proceed because T represents the average Y. When T comes down, it has a smaller probability of being at a higher state. Therefore, compared to the random number r that we are generating, this probability P which is also varying from 0 to 1 will be such that if the function decreases for a maximization or if the function increases for a minimization, as the iterations proceed, when the temperature T comes down, it will become more and more difficult for us to accept the sample, if ΔY increases for a minimization problem. The algorithm will proceed along the route of "conventional thinking" only after the solution space has been initially thoroughly searched.

In 1955, Metropolis suggested a way to implement the Boltzmann probability distribution in "simulated" thermodynamic systems. This can also be used in optimization problems too. The Metropolis-Hastings algorithm is basically a sampling algorithm that helps us get samples of the design vector $X(x_1, x_2, \ldots x_n)$.

Let us say the current point is $X^{(t)}$ and the value of the function at that point is $E(t) = f(X^{(t)})$. The probability of the next point being at $X^{(t+1)}$ depends on $\delta E = E(t + 1) - E(t) = $ change in objective function. For a minimization problem, applying the Boltzmann probability distribution, we get $P[E(t + 1)] = min[1, e^{-\frac{\Delta E}{kT}}]$. If $\Delta E \leq 0$, Y^1 is less than Y^0, the probability is 1 and hence $X^{(t+1)}$ always gets accepted for the minimization problem. We are not questioning conventional thinking. While the acceptance is straightforward, the rejection is not so immediate. We reject only with a probability and this probability is such that the rejection becomes stricter and stricter as iterations proceed.

The random number r will always be between 0 and 1. But P will be such that it becomes very close to 0 as the iterations keep progressing. The criterion is the same. We generate a random number r. If $r \leq P$, we accept it, else the sample is rejected. P will also vary between 0 and 1. So now if we consider the condition as stated before that $r \leq P$ for the sample to be accepted, in the first few iterations, when P is close to 1, the chance of X^1 being accepted is higher as the criterion $r \leq P$ may be easily satisfied. But as P reduces and becomes almost 0 as the iterations proceed, while r remains a random value between 0 and 1, the probability that X^1 will be accepted when $Y^1 \geq Y^0$ for a minimization problem becomes lesser and lesser.

8.3.3 Steps Involved in SA

So unlike GA, SA is generally a point by point method (SA for multiple points is also available and can be considered to be a variant of the classic SA algorithm).

Fig. 8.6 Illustration of the
iteration process in a
two-variable minimization
problem with SA

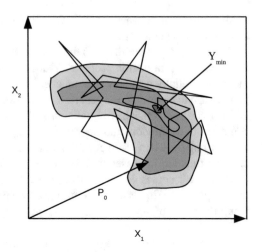

- Begin with an initial point and a high temperature T. The initial temperature could be the mean of a few values of Y calculated from randomly chosen values of x.
- A second point is created in the vicinity of the initial point by using a sampling technique and ΔE is calculated.
- If ΔE $(Y_{i+1} - Y_i)$ is negative, the new point is accepted right away.
- If ΔE is positive (for a minimization problem), the point is accepted with a probability of $e^{-\frac{\Delta E}{kT}}$, k=1. This completes one iteration.
- In the next generation, again a new point is chosen, but now the temperature T is reduced. This is where the cooling rate is controlled. Usually $T_{new} = \frac{T_{old}}{2}$.
- At every temperature, a number of points (4 or 6) are evaluated before reducing the temperature.
- The stopping criterion is either $\Delta E \leq \epsilon_1$ or $T \leq \epsilon_2$ or $|X^{i+1} - X^i| \leq \epsilon_3$.

The progress of iterations in a typical two-variable minimization problem in x_1 and x_2 is shown in Fig. 8.6. We can see that the initial search covers almost every possible subregion of the solution space, so that premature convergence is avoided.

A typical flowchart for solving an optimization problem with SA is shown in Fig. 8.7.

Consider the cargo ship problem, we discussed while learning GA.

$$y = 0.2x^2 + \frac{450}{x}; \ 0.5 \leq x \leq 25.5 \, \text{m/s}.$$

Now let us take the first sample as $x_0 = \frac{(0.5+25.5)}{2} = 13 \, \text{m/s}$. This is the mean such that we do not want samples to exceed 25.5 m/s and we do not want them to fall below 0.5 m/s. 99% of the time we can do this, if we follow a Gaussian or a normal distribution whose mean = 13 m/s and whose $3\sigma = \frac{(25.5-0.5)}{2} = 12.5 \, \text{m/s}$ (see Fig. 8.8).

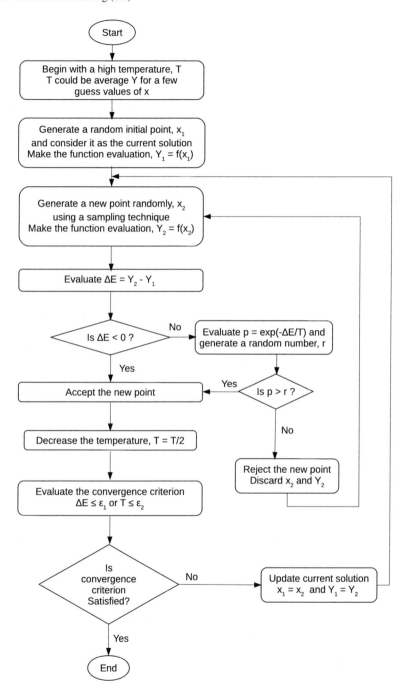

Fig. 8.7 Simulated Annealing flowchart for a minimization problem

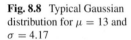

Fig. 8.8 Typical Gaussian
distribution for $\mu = 13$ and
$\sigma = 4.17$

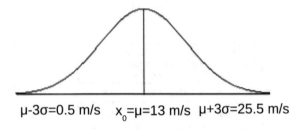

$\mu\text{-}3\sigma\text{=}0.5$ m/s $x_0\text{=}\mu\text{=}13$ m/s $\mu\text{+}3\sigma\text{=}25.5$ m/s

So we start with $x_0 = \mu = 13\,\text{m/s}$. How do we generate x_1 now? The normal
distribution function can be written as $f = \frac{1}{\sqrt{2\pi}\times\sigma} \times e^{\frac{-(x-\mu)^2}{2\sigma^2}}$. What is the ordinate
of this distribution? When $x = \mu$, $f = \frac{1}{\sqrt{2\pi}\times\sigma}$. This is the maximum probability. So
what we do is first generate a random number. We will take it as f and since we know
μ and σ, we will determine the new x. The problem with this procedure is sometimes
it may give some meaningless values and since f here is between 0 and $\frac{1}{\sqrt{2\pi}\times\sigma}$ while
we actually want f to be between 0 and 1. We need to use the normalized standard
distribution.

We have a sample which is the mean. We want the next sample either to the
left or right of this. So what we do here is generate the random number, divide it
by $\sqrt{2\pi} \times \sigma$ and then equate to the probability distribution function 'f'. The new
problem is that since it is $(x - \mu)^2$, when we generate the random value, it will
always go to one side of μ. So what we do to overcome this is that we generate
another random number between 0 and 1, say k. So if $k > 0.5$, we use $+\Delta x$ and if
$k < 0.5$, we use $-\Delta x$.

In view of the above, it is evident that a lot of tweaking is required to make the
method work well. That said, this leads to the variety or diversity we are introducing
in the population. So if there is a treacherous function that goes up and down, the SA
will not leave it, it will be able to catch it. The Golden section search, for example, will
work only for unimodal functions. SA and GA methods can work for any function.
They may be slow but they will get to the global optimum. In any case, these are
infinitely superior to exhaustive search.

So we have to use 3 sets of random numbers for simulated annealing. The first
set of random numbers is for sampling, we can stick to rows 1 and 2 of the random
numbers given in Table 8.4. Row 3 of the Table can be used for generating r. Row
4 or row 5 can be used for generating the value k, which decides whether we take
$+\Delta x$ or $-\Delta x$.

What is the cooling schedule mentioned in step 6? $T^{i+1} = \frac{T^i}{2}$ is the cooling schedule. If T is high, convergence is slow, while if T is low, it may lead to premature
convergence. Hence T and n govern convergence. To calculate the initial T, we draw a
few random values of x and calculate the average of f(x). Now we set T = average f(x).

Example 8.4 Consider the cargo ship problem. We would like to solve it using SA. Perform 4 iterations of the SA with an initial interval of uncertainty $0.5 \leq x \leq 25.5$. Use the random number table provided to you.

Solution:

$$y = 0.2x^2 + \frac{450}{x}; \quad 0.5 \leq x \leq 25.5 \, \text{m/s}.$$

Calculate the initial cooling rate T_0. For this, we need to use 4 values of x and the corresponding y. We already did this for GA (refer to Example 8.1), we will use the same 4 values. We got y = 98.61; Hence $T_0 = 98.61$.

$$\mu = \frac{0.5 + 25.5}{2} = 13 \, \text{m/s}$$

$$3\sigma = \frac{25.5 - 0.5}{2}; \sigma = 4.17 \, \text{m/s}$$

We now draw a Gaussian around $\mu = 13$ m/s. The objective of writing the Gaussian distribution is that we have to solve this equation to obtain x_{new}.

$$\frac{Rand}{\sqrt{2\pi} \times \sigma} = \frac{1}{\sqrt{2\pi} \times \sigma} e^{-\frac{(x_{new} - \mu)^2}{2\sigma^2}}$$

$$x_{new} = \mu \pm \sqrt{-2\sigma^2 \log(Rand)} = x_{old} \pm \sqrt{-2\sigma^2 \log(Rand)}$$

The random numbers are selected from the random number table in Table 8.11. Now, the iterations of SA follow as

Iteration 1

$$x_{old} = \mu = 13m/s, \, y_{old} = 68.42, \, T_0 = 98.61$$

Generate a set of 3 random numbers

r_1	r_2	r_3
0.0012	0.8989	0.5788

Table 8.11 A random number table

0.001213	0.898980	0.578800	0.676216	0.050106
0.499629	0.282693	0.730594	0.701195	0.182840
0.108501	0.386183	0.769105	0.683348	0.551702
0.557434	0.799824	0.456790	0.216310	0.876167
0.092645	0.589628	0.332164	0.031858	0.611683
0.762627	0.696237	0.170288	0.054759	0.915126
0.032722	0.299315	0.308614	0.833586	0.517813
0.352862	0.574100	0.265936	0.859031	0.433081
0.941875	0.240002	0.655595	0.385079	0.908297
0.199044	0.936553	0.888098	0.817720	0.369820
0.339548	0.543258	0.624006	0.091330	0.416789
0.155062	0.582447	0.858532	0.887525	0.337294
0.751033	0.239493	0.535597	0.333813	0.493837
0.634536	0.199621	0.650020	0.745795	0.791130
0.227241	0.191479	0.406443	0.081288	0.734352
0.721023	0.222878	0.072814	0.641837	0.442675
0.789616	0.052303	0.106994	0.558774	0.141519
0.760869	0.120791	0.277380	0.657266	0.792691
0.805480	0.826543	0.294530	0.208524	0.429894
0.585186	0.986111	0.344882	0.343580	0.115375

$$\text{Here, } r_2 > 0.5, \text{ so } x_{new} = x_{old} + \sqrt{-2\sigma^2 \log(r_1)}$$

$$x_{new} = 13 + \sqrt{-2 \times 4.17^2 \log(0.0012)}$$

$$x_{new} = 28.28, \text{ which is out of range.}$$

Generate a set of 3 random numbers

r_1	r_2	r_3
0.4996	0.2827	0.7306

$$\text{Here, } r_2 < 0.5, \text{ so } x_{new} = x_{old} - \sqrt{-2\sigma^2 \log(r_1)}$$

$$x_{new} = 13 - \sqrt{-2 \times 4.17^2 \log(0.4996)}$$

$$x_{new} = 8.087, \ y_{new} = 68.72, \ \Delta y = 0.3$$

$$P = e^{-\frac{\Delta y}{T}} = 0.9969$$

$$A = min(1, P) = 0.9969 > r_3, \text{ so } x_{new} = 8.087 \text{ is accepted.}$$

Iteration 2

$$x_{old} = \mu = 8.087, \ y_{old} = 68.72, \ T_1 = \frac{98.61}{2} = 49.31$$

Generate a set of 3 random numbers

r_1	r_2	r_3
0.1085	0.3862	0.7691

Here, $r_2 < 0.5$, so $x_{new} = x_{old} - \sqrt{-2\sigma^2 \log(r_1)}$

$$x_{new} = 8.087 - \sqrt{-2 \times 4.17^2 \log(0.1085)}$$

$$x_{new} = -0.7017, \text{ which is out of range.}$$

Generate a set of 3 random numbers

r_1	r_2	r_3
0.5574	0.7998	0.4568

Here, $r_2 > 0.5$, so $x_{new} = x_{old} + \sqrt{-2\sigma^2 \log(r_1)}$

$$x_{new} = 8.087 + \sqrt{-2 \times 4.17^2 \log(0.5574)}$$

$$x_{new} = 12.59, \ y_{new} = 67.44, \ \Delta y = -1.28$$

$$P = e^{-\frac{\Delta y}{T}} = 1.026$$

$A = min(1, P) = 1 > r_3$, so $x_{new} = 12.59$ is accepted.

Iteration 3

$$x_{old} = \mu = 12.59, \ y_{old} = 67.44, \ T_2 = \frac{49.31}{2} = 24.65$$

Generate a set of 3 random numbers

r_1	r_2	r_3
0.0926	0.5896	0.3322

Here, $r_2 > 0.5$, so $x_{new} = x_{old} + \sqrt{-2\sigma^2 \log(r_1)}$

$$x_{new} = 12.59 + \sqrt{-2 \times 4.17^2 \log(0.0926)}$$

$$x_{new} = 21.68, \ y_{new} = 114.76, \ \Delta y = 47.32$$

$$P = e^{-\frac{\Delta y}{T}} = 0.1466$$

$$A = min(1, P) = 0.1466 < r_3, \text{ so } x_{new} = 21.68 \text{ is rejected.}$$

Generate a set of 3 random numbers

r_1	r_2	r_3
0.7626	0.6962	0.1703

Here, $r_2 > 0.5$, so $x_{new} = x_{old} + \sqrt{-2\sigma^2 \log(r_1)}$

$$x_{new} = 12.59 + \sqrt{-2 \times 4.17^2 \log(0.7626)}$$

$$x_{new} = 15.66, \ y_{new} = 77.78, \ \Delta y = 10.34$$

$$P^{-\frac{\Delta y}{T}} = 0.6574$$

$$A = min(1, P) = 0.6574 > r_3, \text{ so } x_{new} = 15.66 \text{ is accepted.}$$

$$x_{old} = \mu = 15.66, \ y_{old} = 77.78, \ T_3 = \frac{24.65}{2} = 12.33$$

Generate a set of 3 random numbers

r_1	r_2	r_3
0.0327	0.2993	0.3086

Here, $r_2 < 0.5$, so $x_{new} = x_{old} - \sqrt{-2\sigma^2 \log(r_1)}$

$$x_{new} = 15.66 - \sqrt{-2 \times 4.17^2 \log(0.0327)}$$

$$x_{new} = 4.75, \ y_{new} = 99.25, \ \Delta y = 21.47$$

$$P = e^{-\frac{\Delta y}{T}} = 0.1753$$

$$A = min(1, P) = 0.1753 < r_3, \text{ so } x_{new} = 4.75 \text{ is rejected.}$$

Generate a set of 3 random numbers

r_1	r_2	r_3
0.3528	0.5741	0.2659

$$\text{Here, } r_2 > 0.5, \text{ so } x_{new} = x_{old} + \sqrt{-2\sigma^2 \log(r_1)}$$

$$x_{new} = 15.66 + \sqrt{-2 \times 4.17^2 \log(0.3528)}$$

$$x_{new} = 21.67, \ y_{new} = 114.68, \ \Delta y = 36.90$$

$$P = e^{-\frac{\Delta y}{T}} = 0.0501$$

$$A = min(1, P) = 0.0501 < r_3, \text{ so } x_{new} = 21.67 \text{ is rejected.}$$

Generate a set of 3 random numbers

r_1	r_2	r_3
0.9418	0.2400	0.6556

$$\text{Here, } r_2 < 0.5, \text{ so } x_{new} = x_{old} - \sqrt{-2\sigma^2 \log(r_1)}$$

$$x_{new} = 15.66 - \sqrt{-2 \times 4.17^2 \log(0.9418)}$$

$$x_{new} = 14.22, \ y_{new} = 72.08, \ \Delta y = -5.7$$

$$P = e^{-\frac{\Delta y}{T}} = 1.58$$

$$A = min(1, P) = 1 > r_3, \text{ so } x_{new} = 14.22 \text{ is accepted.}$$

So this is how simulated annealing works and after some time, by the zigzag path, it will cover the whole solution space well, so that the global optimum is not missed. If the objective function is computationally expensive, for example, if a CFD solution or something as complex is required for getting each value, we can develop a neural network. We can run it for certain combinations of x, validate it and train it. After we get the optimum, we can substitute the values of x back into our original forward model or governing equations and check if the values predicted by the neural network are the same, as got by substituting in the equations. This completes the loop. The following points are in order.

- We are suggesting that the cooling rate algorithm for this problem can be $\frac{T}{2}$ at every stage. We can also have different rates. Ultimately the reduction in cooling rate has to be decided based on our problem. We do not want to reduce it by 4 times or 8 times or 10 times because the rejection will become very strict and though it may accelerate our convergence, the latter could become premature.
- At the start, we calculate the average of y at 4 points and use it as T.
- Additionally, the new value of x becomes the mean of the distribution for that iteration. That is the way all sampling algorithms work. When we start with 13, the mean is around 13. If x reduces to 8, the mean is around 8. Next, if x becomes 8.6, the mean is around 8.6. The new value of x becomes μ automatically.

MATLAB code for Example 8.4

```
 1  clear;
 2  close all;
 3  clc;
 4
 5  a = 0.5;            % Range of v
 6  b = 25.50;
 7  cs = 0.75;          % Cooling Schedule
 8  itr=100;            % No. of iterations
 9
10  % New sample selected as v2 = v1 (+ or −) rand * sigma
11  mu = (b+a)/2;       % mu
12  sig = (b−a)/6;      % sigma
13  v = a+(b−a)*rand(1,4);
14  y = 0.2*v.^2 + 450./v;
15  T = mean(y);
16  count=0;
17  vi = zeros(1,itr);
18  Yi = zeros(1,itr);
19  vi(1) = mu;
20  Yi(1) = 0.2*vi(1)^2 + 450/vi(1);
21
22  while count<itr
23
24      count = count +1;
25      label = 0;
26
27      while label==0 % To make sure that the sample is in the
                        range
28
29          if rand(1) > 0.5
30              vi(count+1)=vi(count) + sig * rand(1);
31          else
32              vi(count+1)=vi(count) − sig * rand(1);
33          end
34          if vi(count+1)>a && vi(count+1)<b
35              label=1;
36          end
37      end
38
39      Yi(count+1) = 0.2*vi(count+1)^2 + 450/vi(count+1);
40
41      if Yi(count+1)<Yi(count) % Accept if Ynew < Yold
42          T = cs*T;  % Reducing T using cooling schedule
43      else
44
45      % Calculating probability of rejection
46          P = exp((Yi(count)−Yi(count+1))/T);
47          if P<rand(1)
48              vi(count+1)=vi(count);
```

```
49                    Yi(count+1)=Yi(count);
50            else
51                    T = cs*T;
52            end
53        end
54
55        % Print
56        prt = ['Itr = ',num2str(count) ,...
57               ', v = ',num2str(vi(count)) ,...
58               ', Y = ',num2str(Yi(count))];
59        disp(prt)
60  end
```

The output of the program is

> Itr = 1, v = 12.5, Y = 67.25
> Itr = 2, v = 8.8582, Y = 66.494
> Itr = 3, v = 10.7258, Y = 64.9635
> Itr = 4, v = 10.1376, Y = 64.9434
> Itr = 5, v = 14.1872, Y = 71.9739
>
> ⋮   ⋮   ⋮
>
> Itr = 99, v = 10.3879, Y = 64.9013
> Itr = 100, v = 10.3879, Y = 64.9013

Example 8.5 Minimize the following function $y = 2x^2 + \frac{7}{x^3} - 4$, in the interval $0.5 \le x \le 5.5m$. Perform three iterations of the SA with a starting value of $x = 3m$.

Solution:
$$y = 2x^2 + \frac{7}{x_3} - 4; \ 0.5 \le x \le 5.5 \, m.$$

Calculate the initial cooling rate T_0. For this, we need to use 4 values of x and the corresponding y (shown in Table 8.12). The mean value of y is 19.28 which is the initial cooling rate (T_0).

Table 8.12 Calculating the initial cooling rate

x	0.8	2.5	3.5	4.5
y	10.95	8.95	20.66	36.58

$$\mu = \frac{0.5+5.5}{2} = 3$$

$$3\sigma = \frac{5.5-0.5}{2}; \sigma = 0.833$$

Now, the iterations of SA follow as

Iteration 1

$$x_{old} = \mu = 3, \ y_{old} = 14.26, \ T_0 = 19.28$$

Generate a set of 3 random numbers

r_1	r_2	r_3
0.0012	0.8989	0.5788

Here, $r_2 > 0.5$, so $x_{new} = x_{old} + \sqrt{-2\sigma^2 \log(r_1)}$

$$x_{new} = 3 + \sqrt{-2 \times 0.833^2 \log(0.0012)}$$

$$x_{new} = 6.05, \text{ which is out of range.}$$

Generate a set of 3 random numbers

r_1	r_2	r_3
0.4996	0.2827	0.7306

Here, $r_2 < 0.5$, so $x_{new} = x_{old} - \sqrt{-2\sigma^2 \log(r_1)}$

$$x_{new} = 3 - \sqrt{-2 \times 0.833^2 \log(0.4996)}$$
$$x_{new} = 2.018, \ y_{new} = 4.996, \ \Delta y = -9.264$$

$$P = e^{-\frac{\Delta y}{T}} = 1.62$$

$$A = min(1, P) = 1 > r_3, \text{ so } x_{new} = 2.018 \text{ is accepted.}$$

Iteration 2

$$x_{old} = 2.018, \ y_{old} = 4.996, \ T_1 = \frac{19.28}{2} = 9.64$$

Generate a set of 3 random numbers

r_1	r_2	r_3
0.1085	0.3862	0.7691

Here, $r_2 < 0.5$, so $x_{new} = x_{old} - \sqrt{-2\sigma^2 \log(r_1)}$

$$x_{new} = 2.018 - \sqrt{-2 \times 0.833^2 \log(0.1085)}$$

$$x_{new} = 0.2624, \text{ which is out of range.}$$

Generate a set of 3 random numbers

r_1	r_2	r_3
0.5574	0.7998	0.4568

Here, $r_2 > 0.5$, so $x_{new} = x_{old} + \sqrt{-2\sigma^2 \log(r_1)}$

$$x_{new} = 2.018 + \sqrt{-2 \times 0.833^2 \log(0.5574)}$$

$$x_{new} = 2.918, \ y_{new} = 13.31, \ \Delta y = 8.314$$

$$P = e^{-\frac{\Delta y}{T}} = 0.4221$$

$A = min(1, P) = 0.4221 < r_3$, so $x_{new} = 2.918$ is rejected.

Generate a set of 3 random numbers

r_1	r_2	r_3
0.0926	0.5896	0.3322

Here, $r_2 > 0.5$, so $x_{new} = x_{old} + \sqrt{-2\sigma^2 \log(r_1)}$

$$x_{new} = 2.018 + \sqrt{-2 \times 0.833^2 \log(0.0.0926)}$$

$$x_{new} = 3.835, \ y_{new} = 25.54, \ \Delta y = 20.54$$

$$P = e^{-\frac{\Delta y}{T}} = 0.1187$$

$A = min(1, P) = 0.1187 < r_3$, so $x_{new} = 3.835$ is rejected.

Generate a set of 3 random numbers

r_1	r_2	r_3
0.7626	0.6962	0.1703

Here, $r_2 > 0.5$, so $x_{new} = x_{old} + \sqrt{-2\sigma^2 \log(r_1)}$

$$x_{new} = 2.018 + \sqrt{-2 \times 0.833^2 \log(0.7626)}$$

$$x_{new} = 2.631, \ y_{new} = 10.23, \ \Delta y = 5.234$$

$$P = e^{-\frac{\Delta y}{T}} = 0.5810$$

$$A = min(1, P) = 0.5810 > r_3, \text{ so } x_{new} = 2.631 \text{ is accepted.}$$

Iteration 3

$$x_{old} = 2.631, \ y_{old} = 10.23, \ T_2 = \frac{9.64}{2} = 4.82$$

Generate a set of 3 random numbers

r_1	r_2	r_3
0.0327	0.2993	0.3086

$$\text{Here, } r_2 < 0.5, \text{ so } x_{new} = x_{old} - \sqrt{-2\sigma^2 \log(r_1)}$$

$$x_{new} = 2.631 - \sqrt{-2 \times 0.833^2 \log(0.0327)}$$

$$x_{new} = 0.4532, \text{ which is out of range.}$$

Generate a set of 3 random numbers

r_1	r_2	r_3
0.3528	0.5741	0.2659

$$\text{Here, } r_2 > 0.5, \text{ so } x_{new} = x_{old} + \sqrt{-2\sigma^2 \log(r_1)}$$

$$x_{new} = 2.631 + \sqrt{-2 \times 0.833^2 \log(0.3528)}$$

$$x_{new} = 3.83, \ y_{new} = 25.46, \ \Delta y = 15.23$$

$$P = e^{-\frac{\Delta y}{T}} = 0.0424$$

$$A = min(1, P) = 0.0424 < r_3, \text{ so } x_{new} = 3.83 \text{ is rejected.}$$

Generate a set of 3 random numbers

r_1	r_2	r_3
0.9418	0.2400	0.6556

$$\text{Here, } r_2 < 0.5, \text{ so } x_{new} = x_{old} - \sqrt{-2\sigma^2 \log(r_1)}$$

$$x_{new} = 2.631 - \sqrt{-2 \times 0.833^2 \log(0.9418)}$$

$$x_{new} = 2.34, \, y_{new} = 7.49, \, \Delta y = -2.74$$

$$P = e^{-\frac{\Delta y}{T}} = 1.76$$

$A = min(1, P) = 1 > r_3$, so $x_{new} = 2.34$ is accepted.

MATLAB code for Example 8.5

```matlab
clear;
close all;
clc;

a = 0.5;          % Range of x
b = 5.5;
cs = 0.5;         % Cooling Schedule
itr=100;          % No. of iterations

% New sample selected as x2 = x1 (+ or -) rand * sigma
mu = (b+a)/2;     % mu
sig = (b-a)/6;    % sigma
x = a+(b-a)*rand(1,4);
y = 2*x.^2 + 7./x.^3-4;
 T = mean(y);
count=0;
xi = zeros(1,itr);
Yi = zeros(1,itr);
xi(1) = mu;
Yi(1) = 2*xi(1)^2 + 7/xi(1)^3-4;

while count<itr
    count = count +1;
    label = 0;
    while label==0 % To make sure that the sample is in the
        range
        if rand(1) > 0.5
            xi(count+1)=xi(count) + sig * rand(1);
        else
            xi(count+1)=xi(count) - sig * rand(1);
        end
        if xi(count+1)>a && xi(count+1)<b
            label=1;
        end
    end
    Yi(count+1) = 2*xi(count+1)^2 + 7/xi(count+1)^3-4;
    if Yi(count+1)<Yi(count) % Accept if Ynew < Yold
        T = cs*T;  % Reducing T using cooling schedule
```

```
38        else
39        % Calculating  probability  for  rejection
40             P = exp((Yi(count)-Yi(count+1))/T);
41             if P<rand(1)
42                  xi(count+1)=xi(count);
43                  Yi(count+1)=Yi(count);
44             else
45                  T = cs*T;
46             end
47        end
48
49        % Print
50        prt = ['Itr = ',num2str(count),...
51             ', x = ',num2str(xi(count)),...
52             ', Y = ',num2str(Yi(count))];
53        disp(prt)
54   end
```

The output the program is:

Itr = 1, x = 3, Y = 14.2593
Itr = 2, x = 3.1091, Y = 15.5665
Itr = 3, x = 3.874, Y = 26.1368
Itr = 4, x = 4.2115, Y = 31.5674
Itr = 5, x = 4.2115, Y = 31.5674

$\vdots$ $\vdots$ $\vdots$

Itr = 99, x = 1.3912, Y = 2.4706
Itr = 100, x = 1.3912, Y = 2.4706

Example 8.6 Consider a thermal system, whose heat dissipation rate is given by $Q = 2.5 + 6.2v^{0.8}$, where Q is in kW and v is the velocity in m/s of the fluid being used as the medium for accomplishing the heat transfer. The accompanying pumping power is given by $P = 1.3 + 0.04v^{1.8}$, again in kW with v in m/s (in both the expressions, the constants ensure that both Q and P are in kW). It is desired to maximize the performance parameter Q/P in the range $3 \le v \le 12$ m/s. Perform 5 iterations of Simulated Annealing.

Solution:

$$Maximize \ y = \frac{Q}{P} = \frac{2.5 + 6.2v^{0.8}}{1.3 + 0.04v^{1.8}}; \ 3 \le v \le 12 \, \text{m/s}.$$

Table 8.13 Calculating the initial cooling rate

v	4	6	8	10
y	11.93	12.36	11.78	10.88

Calculate the initial cooling rate T_0. For this, we need to use 4 values of v and the corresponding y (shown in Table 8.13). The mean value of y is 11.74 which is the initial cooling rate (T_0).

Since this is a maximization problem, the probability of accepting the sample is calculated as $P = e^{\frac{\Delta y}{T}}$.

$$\mu = \frac{12+3}{2} = 7.5$$

$$3\sigma = \frac{12-3}{2}; \sigma = 1.5$$

Now, the iterations of SA follow as

Iteration 1

$$v_{old} = \mu = 7.5, \ y_{old} = 11.97, \ T_0 = 11.74$$

Generate a set of 3 random numbers

r_1	r_2	r_3
0.0012	0.8989	0.5788

Here, $r_2 > 0.5$, so $v_{new} = v_{old} + \sqrt{-2\sigma^2 \log(r_1)}$

$$v_{new} = 7.5 + \sqrt{-2 \times 1.5^2 \log(0.0012)}$$

$$v_{new} = 12.99, \text{ which is out of range.}$$

Generate a set of 3 random numbers

r_1	r_2	r_3
0.4996	0.2827	0.7306

Here, $r_2 < 0.5$, so $v_{new} = v_{old} - \sqrt{-2\sigma^2 \log(r_1)}$

$$v_{new} = 7.5 - \sqrt{-2 \times 1.5^2 \log(0.4996)}$$

$$v_{new} = 5.73, \ y_{new} = 12.37, \ \Delta y = 0.4$$

$$P = e^{\frac{\Delta y}{T}} = 1.03$$

$$A = min(1, P) = 1 > r_3, \text{ so } v_{new} = 5.73m/s \text{ is accepted.}$$

Iteration 2

$$v_{old} = 5.73, \ y_{old} = 12.37, \ T_1 = \frac{11.74}{2} = 5.87$$

Generate a set of 3 random numbers

r_1	r_2	r_3
0.1085	0.3862	0.7691

Here, $r_2 < 0.5$, so $v_{new} = v_{old} - \sqrt{-2\sigma^2 \log(r_1)}$

$$v_{new} = 5.73 - \sqrt{-2 \times 1.5^2 \log(0.1085)}$$

$$v_{new} = 2.56, \text{ which is out of range.}$$

Generate a set of 3 random numbers

r_1	r_2	r_3
0.5574	0.7998	0.4568

Here, $r_2 > 0.5$, so $v_{new} = v_{old} + \sqrt{-2\sigma^2 \log(r_1)}$

$$v_{new} = 5.73 + \sqrt{-2 \times 1.5^2 \log(0.5574)}$$

$$v_{new} = 7.53, \ y_{new} = 12.03, \ \Delta y = -0.34$$

$$P = e^{\frac{\Delta y}{T}} = 0.9437$$

$$A = min(1, P) = 0.9437 > r_3, \text{ so } v_{new} = 7.53 \text{ is accepted.}$$

Iteration 3

$$v_{old} = 7.35, \ y_{old} = 12.03, \ T_2 = \frac{5.87}{2} = 2.935$$

Generate a set of 3 random numbers

r_1	r_2	r_3
0.0926	0.5896	0.3322

$$\text{Here, } r_2 > 0.5, \text{ so } v_{new} = v_{old} + \sqrt{-2\sigma^2 \log(r_1)}$$

$$v_{new} = 7.35 + \sqrt{-2 \times 1.5^2 \log(0.0926)}$$

$$v_{new} = 10.62, \, y_{new} = 10.58, \, \Delta y = -1.45$$

$$P = e^{\frac{\Delta y}{T}} = 0.6102$$

$$A = min(1, P) = 0.6102 > r_3, \text{ so } v_{new} = 10.62m/s \text{ is accepted.}$$

Iteration 4

$$v_{old} = 10.62, \, y_{old} = 10.58, \, T_2 = \frac{2.935}{2} = 1.467$$

Generate a set of 3 random numbers

r_1	r_2	r_3
0.7626	0.6962	0.1703

$$\text{Here, } r_2 > 0.5, \text{ so } v_{new} = v_{old} + \sqrt{-2\sigma^2 \log(r_1)}$$

$$v_{new} = 10.62 + \sqrt{-2 \times 1.5^2 \log(0.7626)}$$

$$v_{new} = 11.72, \, y_{new} = 10.07, \, \Delta y = -0.51$$

$$P = e^{\frac{\Delta y}{T}} = 0.7064$$

$$A = min(1, P) = 1 > r_3, \text{ so } v_{new} = 11.72m/s \text{ is accepted.}$$

Iteration 5

$$v_{old} = 11.72, \, y_{old} = 10.07, \, T_4 = \frac{1.467}{2} = 0.734$$

Generate a set of 3 random numbers

r_1	r_2	r_3
0.0327	0.2993	0.3086

$$\text{Here, } r_2 < 0.5, \text{ so } v_{new} = v_{old} - \sqrt{-2\sigma^2 \log(r_1)}$$

$$v_{new} = 11.72 - \sqrt{-2 \times 1.5^2 \log(0.0327)}$$

$$v_{new} = 7.79, \, y_{new} = 11.86, \, \Delta y = 1.79$$

$$P = e^{\frac{\Delta y}{T}} = 11.46$$

$A = min(1, P) = 1 > r_3$, so $v_{new} = 7.79$ is accepted.

8.4 Hybrid Optimization Techniques

In some cases, instead of proceeding with a single algorithm till the end, switching to a different algorithm after a certain number of iterations will ensure faster convergence and better accuracy of the solution. This idea will become clear after going through Examples 8.7 and 8.8. In Example 8.7, we start with GA and then switch to the golden section search. In Example 8.8, we start with GA and then switch to simulated annealing.

> **Example 8.7** Consider the cargo ship problem. Solve it using hybrid optimization technique by starting with genetic algorithm and proceeding till 20 iterations With the results obtained from the genetic algorithm, switch to golden section search method and perform 7 iterations.

Solution:

Using genetic algorithm, the population (20) after 20th iteration is shown in Table 8.14. This table is the output of the MATLAB code presented earlier.

From this population we have, $x_{min} = 9.6$ and $x_{max} = 12.6$.

So we use golden section search method to find the optimum in the range [9.6, 12.6]. The iterations using golden section search method are as follows:

Table 8.14 Population and fitness values after 20th iteration

S. No.	Population	Fitness value	S. No.	Population	Fitness value
1	10.0	65.00	**11**	**12.6**	**67.47**
2	10.4	64.90	12	11.0	65.11
3	10.4	64.90	13	10.4	64.90
4	10.4	64.90	14	12.0	66.30
5	10.8	64.99	15	10.8	64.99
6	9.8	65.13	16	10.7	64.95
7	10.4	64.90	**17**	**9.6**	**65.31**
8	12.0	66.30	18	10.5	64.91
9	10.2	64.93	19	10.5	64.91
10	10.5	64.91	20	10.8	64.99

Iteration 1

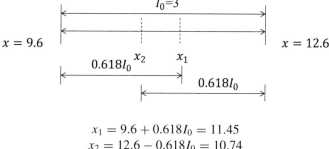

$$x_1 = 9.6 + 0.618 I_0 = 11.45$$
$$x_2 = 12.6 - 0.618 I_0 = 10.74$$
$$y(x_1) = 65.53, \ y(x_2) = 64.97, \ y(x_1) > y(x_2)$$

Hence, the region to the right of x_1 can be eliminated.

Iteration 2

$$x_3 = 9.6 + 0.618 I_1 = 10.74$$
$$x_4 = 11.45 - 0.618 I_1 = 10.31$$
$$y(x_3) = 64.97, \ y(x_4) = 64.91, \ y(x_4) < y(x_3)$$

Hence, the region to the right of x_3 can be eliminated.

Iteration 3

$$x_5 = 9.6 + 0.618 I_2 = 10.30$$
$$x_6 = 10.74 - 0.618 I_2 = 10.04$$
$$y(x_5) = 64.91, \ y(x_6) = 64.98, \ y(x_6) > y(x_5)$$

Hence, the region to the left of x_6 can be eliminated.

Iteration 4

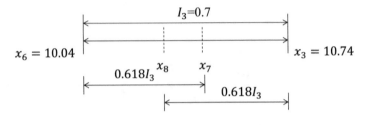

$$x_7 = 10.04 + 0.618 I_3 = 10.47$$
$$x_8 = 10.47 - 0.618 I_3 = 10.31$$
$$y(x_7) = 64.904, \ y(x_8) = 64.906, \ y(x_8) > y(x_7)$$

Hence, the region to the left of x_8 can be eliminated.

Iteration 5

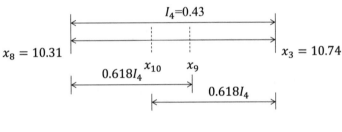

$$x_9 = 10.31 + 0.618 I_4 = 10.57$$
$$x_{10} = 10.74 - 0.618 I_4 = 10.47$$
$$y(x_9) = 64.918, \ y(x_{10}) = 64.904, \ y(x_9) > y(x_{10})$$

Hence, the region to the right of x_9 can be eliminated.

Iteration 6

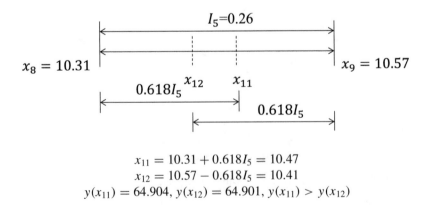

$$x_{11} = 10.31 + 0.618 I_5 = 10.47$$
$$x_{12} = 10.57 - 0.618 I_5 = 10.41$$
$$y(x_{11}) = 64.904, \ y(x_{12}) = 64.901, \ y(x_{11}) > y(x_{12})$$

Hence, the region to the right of x_{11} can be eliminated.

Iteration 7

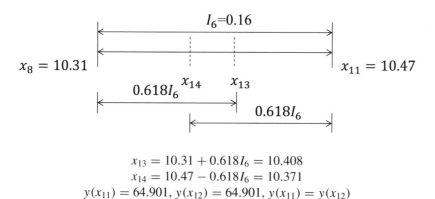

$$x_{13} = 10.31 + 0.618I_6 = 10.408$$
$$x_{14} = 10.47 - 0.618I_6 = 10.371$$
$$y(x_{11}) = 64.901, \ y(x_{12}) = 64.901, \ y(x_{11}) = y(x_{12})$$

Hence, the optimum value of x is 10.4.

Example 8.8 Consider the cargo ship problem. Solve it using hybrid optimization technique by starting with genetic algorithm and proceeding till 20 iterations With the results obtained from genetic algorithm, switch to simulated annealing and perform 5 iterations.

Solution:

For simulated annealing (SA), the objective function is given by

$$y = 0.2x^2 + 450/x; \ 9.6 \leqslant x \leqslant 12.6$$

Calculate the initial cooling rate T_0. For this, we need to use 4 values of x and the corresponding y (shown in Table 8.15). The mean value of y is 65.37 which is the initial cooling rate (T_0).

$$\mu = \frac{9.6 + 12.6}{2} = 11.1, \ \sigma = \frac{12.6 - 9.6}{2} = 0.5$$

Table 8.15 Calculate the initial cooling rate

x	9.8	10.1	11.0	12.0
y	65.13	64.96	65.11	66.30

Now, the iterations of SA follow as

Iteration 1

$$x_{old} = \mu = 11.1, \; y_{old} = 65.18, \; T_0 = 65.37$$

Generate a set of 3 random numbers

r_1	r_2	r_3
0.0012	0.8989	0.5788

Here, $r_2 > 0.5$, so $x_{new} = x_{old} + \sqrt{-2\sigma^2 \log(r_1)}$

$$x_{new} = 12.93, \text{ which is out of range.}$$

Generate a set of 3 random numbers

r_1	r_2	r_3
0.4996	0.2827	0.7306

Here, $r_2 < 0.5$, so $x_{new} = x_{old} - \sqrt{-2\sigma^2 \log(r_1)}$

$$x_{new} = 10.51, \; y_{new} = 64.908, \; \Delta y = -0.272$$

$$P = e^{-\frac{\Delta y}{T}} = 1.004$$

$$A = min(1, P) = 1 > r_3, \text{ so } x_{new} = 10.51 \text{ is accepted.}$$

Iteration 2

$$x_{old} = \mu = 10.51, \; y_{old} = 64.908, \; T_1 = \frac{65.37}{2} = 32.685$$

Generate a set of 3 random numbers

r_1	r_2	r_3
0.1085	0.3862	0.7691

Here, $r_2 < 0.5$, so $x_{new} = x_{old} - \sqrt{-2\sigma^2 \log(r_1)}$

$$x_{new} = 9.456, \text{ which is out of range.}$$

Generate a set of 3 random numbers

r_1	r_2	r_3
0.5574	0.7998	0.4568

Here, $r_2 > 0.5$, so $x_{new} = x_{old} + \sqrt{-2\sigma^2 \log(r_1)}$

$$x_{new} = 11.05, \, y_{new} = 65.14, \, \Delta y = 0.236$$

$$P = e^{-\frac{\Delta y}{T}} = 0.9928$$

$$A = min(1, P) = 0.9928 > r_3, \text{ so } x_{new} = 11.05 \text{ is accepted.}$$

Iteration 3

$$x_{old} = \mu = 11.05, \, y_{old} = 65.14, \, T_2 = \frac{32.685}{2} = 16.343$$

Generate a set of 3 random numbers

r_1	r_2	r_3
0.0926	0.5896	0.3322

Here, $r_2 > 0.5$, so $x_{new} = x_{old} + \sqrt{-2\sigma^2 \log(r_1)}$

$$x_{new} = 12.14, \, y_{new} = 66.54, \, \Delta y = 1.4$$

$$P = e^{-\frac{\Delta y}{T}} = 0.9179$$

$$A = min(1, P) = 0.9179 > r_3, \text{ so } x_{new} = 12.14 \text{ is accepted.}$$

Iteration 4

$$x_{old} = \mu = 12.14, \, y_{old} = 66.54, \, T_3 = \frac{16.343}{2} = 8.171$$

Generate a set of 3 random numbers

r_1	r_2	r_3
0.7626	0.6962	0.1703

Here, $r_2 > 0.5$, so $x_{new} = x_{old} + \sqrt{-2\sigma^2 \log(r_1)}$

$$x_{new} = 12.51, \, y_{new} = 67.27, \, \Delta y = 0.73$$

$$P = e^{-\frac{\Delta y}{T}} = 0.9145$$

$$A = min(1, P) = 0.9145 > r_3, \text{ so } x_{new} = 12.51 \text{ accepted.}$$

Iteration 5

$$x_{old} = \mu = 12.51, \ y_{old} = 67.27, \ T_3 = \frac{8.171}{2} = 4.085$$

Generate a set of 3 random numbers

r_1	r_2	r_3
0.0327	0.2993	0.3080

Here, $r_2 < 0.5$, so $x_{new} = x_{old} - \sqrt{-2\sigma^2 \log(r_1)}$

$$x_{new} = 11.21, \ y_{new} = 65.27, \ \Delta y = -2$$

$$P = e^{-\frac{\Delta y}{T}} = 1.63$$

$$A = min(1, P) = 1 > r_3, \text{ so } x_{new} = 11.21 \text{ accepted.}$$

Problems

8.1 Consider the problem of minimization of convective heat loss from a cylindrical storage heater that makes use of the solar energy collected by a suitable system. The volume of the tank is 5 m^3 and is fixed. The radius of the tank is "r" and the height is "h".
(a) Convert this to a single-variable unconstrained optimization problem in radius "r".
(b) We would like to use Simulated Annealing (SA) to solve the above problem, as a single-variable minimization problem in "r".
Perform four iterations of the SA with a starting value of r=2 m. You may assume that $0.1 \le r \le 4$ m. Use random number tables. The "initial temperature T" (used in the algorithm that does not correspond to the physical temperature in this problem) may be taken to be the average of four objective function values for appropriately chosen values of the radius.

8.2 Revisit exercise Problem 8.1 and solve it using genetic algorithm for one variable. Perform three iterations of the GA with an initial population size of 4. You may assume that $0.1 \le r \le 2.55$ m.

Reference

Goldberg, D. E. (1989). *Algorithms, genetic in search, optimization and machine learning*. Boston, MA, USA: Addison-Wesley Longman Publishing.

Chapter 9
Inverse Problems

9.1 Introduction

An inverse problem, by definition, is one in which the effect is known (or measured) and the cause(s) need(s) to be identified. Let us take a simple example of a person suffering from fever and going to a doctor for treatment. Invariably, the first parameter the doctor checks is the patient's temperature (usually the oral temperature). If this quantity is more than 38 °C, then the patient has a fever. Fever is the effect and the doctor has to correctly identify the cause–or rather, has to identify the correct cause. The problem is "ill-posed", as there could be several causes for the fever. The fever could be because of a viral infection, bacterial infection, inflammation or allergy, or some very serious underlying disorder. In order to reduce the ill-posedness, the doctor either goes in for additional tests or simply starts treating the fever empirically, by guessing the cause based on his prior knowledge and applies midcourse corrections, if the patient does not feel better in, say, 3–5 days time.

Similarly, in thermal sciences, as in other branches of science and engineering, there are several situations where the correct cause needs to be identified from the effect. The "effect" is usually a temperature or heat flux distribution in thermal sciences. The cause we are seeking could be a thermophysical property like thermal conductivity, thermal diffusivity or emissivity, or a transport coefficient like heat or mass transfer coefficient or could even be the heat flux (in this case, the 'effect' is usually the measured temperature distribution).

A familiar example is the surface heat flux on a reentry vehicle that reenters the atmosphere. The velocities here are terrific and the kinetic energy of the fluid (because of the relative motion between stagnant air and the fast-moving vehicle) is "braked" and converted to an enthalpy rise on the outer surface of the vehicle. This phenomenon is frequently referred to as "aerodynamic heating" and results in a huge temperature rise on the surface. It is impossible to place a heat flux gauge on the outside surface of the reentry vehicle as the temperature is of the order of a few thousand Kelvin. In view of this, thermocouple measurements are made at locations, where the temperature is "measurable", and from these one has to correctly estimate

© The Author(s) 2021
C. Balaji, *Thermal System Design and Optimization*,
https://doi.org/10.1007/978-3-030-59046-8_9

the heat flux at the outer surface. This flux is a critical design parameter that decides the cooling strategies. For instance, the outer surface of a reentry vehicle can be coated with an ablating material with a thickness that ensures the heat flux at the outer surface, is "melted" or more correctly "sublimated" away, thereby protecting the inside of the vehicle.

Other important areas where inverse problems find applications in thermal sciences are as follows:

- Estimation of thermophysical properties.
- Computerized Axial Tomography (CAT) or CT scans, where mapping of the absorption coefficient in a tissue is diagnostic of the structure and functional status of the tissue.
- Identifying the distribution of the radiation source in applications like thermal control of spacecraft, condensers, and internal combustion engines.
- Retrieving the constituents like water vapor, liquid water, and ice in the atmosphere by remotely making measurements of infrared or microwave radiation from the earth's surface by placing sensors onboard a satellite orbiting the earth.

9.1.1 Parameter Estimation by Least Squares Minimization

From the foregoing discussion, it can be seen that parameter estimation is an important inverse problem in thermal sciences. The parameter estimation problem is invariably posed as an optimization problem, wherein minimization of the sum of the least squares of the residues is done. Mathematically, if $y_{data,i}$ is the measured data vector and $y_{sim,i}$ is the simulated or calculated vector of y values for assumed values of the parameters, then, we define S as

$$S = \sum_{i=1}^{N} (y_{data,i} - y_{sim,i})^2 \tag{9.1}$$

where N refers to the total number of measurements.

In Eq. 9.1, as aforesaid, $y_{sim,i}$ are the values of y that are calculated with assumed values of the parameters we are seeking. Let us explore this further with an example.

Example 9.1 Consider a thin aluminum foil coated with a paint of "high" emissivity ϵ with dimensions of 2 cm × 2 cm, 6 mm thickness suspended in an evacuated chamber (Fig. 9.1). The chamber is maintained at 373 K and the foil is initially at 303 K. The foil gets heated radiatively and its temperature response is given in Table 9.1. Propose a strategy for estimating the emissivity of the coating.

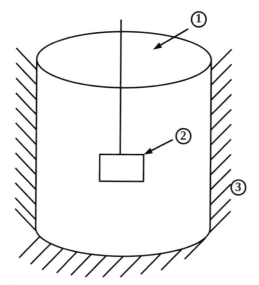

① Evacuated chamber ② Electrically heated foil
suspended in the enclosure

③ Outer isothermal wall

Fig. 9.1 Aluminum foil in an evacuated chamber

Table 9.1 Temperature versus time

t (s)	50	100	150	200	250	300
T (K)	313	323	331	337	344	348

Solution:
One can straightaway see that this is an inverse problem. In a direct (or forward problem), typically all the properties of the system will be known beforehand or "a priori" and the temperature response of the system will be sought. However, in this case, the experimentally measured temperature response is available and we have to estimate or retrieve the value of the emissivity. The first step would be to set up the mathematical model for the direct problem.

The following assumptions are to be made.

1. There is no heat loss from/to the aluminum foil other than that due to radiation from/to the walls of the enclosure.
2. The temperature of the enclosure remains constant throughout.
3. The properties of the foil do not change with temperature.
4. The foil is spatially isothermal (lumped capacitance formulation).

For the above assumptions, the heating of the foil can be mathematically represented as

$$mC_P \frac{dT}{dt} = -\varepsilon\sigma A(T^4 - T_\infty^4) \tag{9.2}$$

where

    m → mass of the foil, ρv, kg

    C_p → specific heat capacity of the foil, J/kgK

    A → surface area = 2(length × breadth), m^2

    ε → emissivity of the foil surface

    σ → Stefan Boltzmann constant, 5.6×10^{-8} $W/m^2 K^4$

    T_∞ → temperature of the inner surface of the evacuated enclosure, K

    T → instantaneous temperature of the foil, K

    t → time, s

In Eq. 9.2, the left-hand side represents the rate of change of enthalpy and the right-hand side represents the heat transfer by radiation. Since $T<T_\infty$ always, the minus sign on the RHS of Eq. 9.2 ensures that dT/dt is positive, thereby confirming that the foil is getting heated radiatively. Equation 9.2 is frequently referred to as the forward model, in the parlance of inverse problems.

Solution to the forward model

$$\frac{dT}{T^4 - T_\infty^4} = -\frac{\varepsilon\sigma A dt}{mC_p} \tag{9.3}$$

Initial condition: $T = T_i$ at t=0.

    The LHS can be integrated using partial fractions as follows:

$$\int_{T_i}^{T} \frac{dT}{T^4 - T_\infty^4} = \int_{T_i}^{T} \frac{dT}{(T^2 - T_\infty^2)(T^2 + T_\infty^2)} \tag{9.4}$$

$$= \frac{1}{2T_\infty^2} \int_{T_i}^{T} \left[\frac{1}{(T^2 - T_\infty^2)} - \frac{1}{(T^2 + T_\infty^2)} \right] dT \tag{9.5}$$

$$= \frac{1}{2T_\infty^2} \int_{T_i}^{T} \left\{ \left[\frac{1}{2T_\infty} \left(\frac{1}{(T - T_\infty)} - \frac{1}{(T + T_\infty)} \right) \right] - \frac{1}{T^2 + T_\infty^2} \right\} dT \tag{9.6}$$

$$= \frac{1}{4T_\infty^3} \int_{T_i}^{T} \frac{dT}{(T - T_\infty)} - \frac{1}{4T_\infty^3} \int_{T_i}^{T} \frac{dT}{(T + T_\infty)} - \frac{1}{2T_\infty^2} \int_{T_i}^{T} \frac{dT}{(T^2 + T_\infty^2)} \tag{9.7}$$

$$= \frac{1}{4T_\infty^3} \ln \left(\frac{T - T_\infty}{T + T_\infty} \right) \Big|_{T_i}^{T} - \frac{1}{2T_\infty^2} \frac{1}{T_\infty} \tan^{-1} \left(\frac{T}{T_\infty} \right) \Big|_{T_i}^{T} \tag{9.8}$$

$$= \frac{1}{4T_\infty^3} \left[\ln \left(\frac{T - T_\infty}{T + T_\infty} \right) - \ln \left(\frac{T_i - T_\infty}{T_i + T_\infty} \right) - 2\tan^{-1} \left(\frac{T}{T_\infty} \right) + 2\tan^{-1} \left(\frac{T_i}{T_\infty} \right) \right] \tag{9.9}$$

Table 9.2 Variation of the residual with emissivity for Example 9.1

ε	0.6	0.65	0.7	0.75	0.8	0.85	0.9	0.95
$S(\varepsilon)$	358.02	229.76	132.59	65.19	21.57	2.94	5.11	27.04

The RHS can be easily integrated as

$$-\frac{\varepsilon\sigma At}{mC_p} \tag{9.10}$$

Therefore, the solution to the forward model is

$$\frac{1}{4T_\infty^3}\left[\ln\left(\frac{T-T_\infty}{T+T_\infty}\right) - \ln\left(\frac{T_i-T_\infty}{T_i+T_\infty}\right) - 2\tan^{-1}\left(\frac{T}{T_\infty}\right) + 2\tan^{-1}\left(\frac{T_i}{T_\infty}\right)\right]$$

$$= -\frac{\varepsilon\sigma At}{mC_p} \tag{9.11}$$

Equation 9.11 is the solution to the forward problem or forward model. It can be seen that the solution is algebraically involved and is not explicit in "T". The rather complicated nature of the solution arises from the non-linearity associated with the $(T^4 - T_\infty^4)$ term. Equation 9.11 also suggests that it is not possible to apply the linear least squares directly to estimate the emissivity, as the resulting equation will be highly difficult to solve.

One possibility of solving the inverse problem is to substitute various values of ε in Eq. 9.11 and determine the temperatures T_i at various time instants given in the problem. With these, the following can be calculated.

$$S(\varepsilon) = \sum_{i=1}^{N}(T_{exp,i} - T_{calc,i})^2 \tag{9.12}$$

Upon doing such an exercise for ε ranging from 0.6 to 0.95 in steps of 0.05 (it is our prior belief that the coating will have its emissivity in the above range), we obtain $S(\varepsilon)$ as shown in Table 9.2.

Based on our discussion of optima in the previous chapters, it is clear that, at best, we can say that $0.8 \le \varepsilon \le 0.9$. The residuals are plotted against emissivity and is shown in Fig. 9.2. What we have essentially done now is an exhaustive, equal interval search. We can apply more sophisticated single variable searches like the Golden section search to get a better estimate of "ε" with the same number of functional evaluations, namely, 8. However, even with the inefficient exhaustive search method, as afore discussed, we can do a little better, by locally fitting a Lagrangian interpolation polynomial for $S(\varepsilon)$, by employing three values of ε where the residuals appear to

Fig. 9.2 Variation of residuals with emissivity for Example 9.1

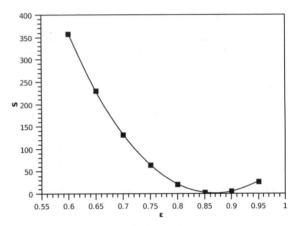

be heading towards a minimum. These happen to be 0.8, 0.85, and 0.9.

$$S = \frac{(\varepsilon - 0.85)(\varepsilon - 0.9)}{(0.8 - 0.85)(0.8 - 0.9)} \times 21.57 + \frac{(\varepsilon - 0.8)(\varepsilon - 0.9)}{(0.85 - 0.8)(0.85 - 0.9)} \times 2.94$$

$$+ \frac{(\varepsilon - 0.85)(\varepsilon - 0.8)}{(0.9 - 0.8)(0.9 - 0.85)} \times 5.11 \tag{9.13}$$

$$S = 6512\,\varepsilon^2 - 11235\,\varepsilon + 4841.89$$

Now we take $\frac{dS}{d\varepsilon}$ and equate it zero to make S stationary

$$\frac{dS}{d\varepsilon} = 13024\,\varepsilon - 11235 = 0 \tag{9.14}$$

$$\varepsilon = 0.86 \tag{9.15}$$

Therefore, the best estimate of ε with the level of computational intensity discussed above and our limited prior belief ($0.6 \le \varepsilon \le 0.95$) is $\varepsilon = 0.86$. Of course, we can employ the Gauss–Newton method or even the Levenberg–Marquardt method to get better estimates of ε.

It is instructive to mention that the above effort is for the estimation of just one parameter and the forward model is just an ordinary differential equation. The computational complexity of the inverse problem will dramatically increase if (i) the number of parameters to be estimated increases (ii) the direct problem becomes very involved, as, for example, if it becomes a CFD problem or an integrodifferential equation involving gas radiation or (iii) both (i) and (ii). From the foregoing example, we can come up with a broad framework for depicting the solution to a parameter estimation problem.

From the schematic shown in Fig. 9.3, one can see that the inverse problem often involves a repeated solution to the forward model. The minimization of $\sum R^2$ converts an inverse problem to a minimization problem in optimization.

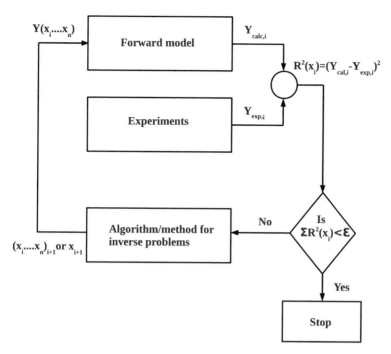

Fig. 9.3 Flow chart for solving an inverse problem

9.2 The Bayesian Approach to Inverse Problems

In the previous section, we learned the basic ideas involved in an inverse problem in thermal sciences and saw an example of single parameter estimation. It is now intuitively apparent that in a multiparameter problem, there could be several combinations of parameters, all of which may lead to the same 'effect', namely, the temperature distribution or a transient response, depending on the problem. Hence, the inverse problem is essentially ill-posed and suffers from a lack of uniqueness. Mathematically, several techniques have been developed, which try to address this problem and these are explained in several texts and journals. However, as engineers, invariably we have some more information about the parameter than what is seen from the measurements or data. If, then, we are able to systematically inject these prior beliefs of ours about the parameters in the estimation process, then what we are attempting is a Bayesian approach.

9.2.1 Bayesian Inference

This is based on Baye's conditional probability theorem and employs probability to characterize all forms of uncertainty in the problem. Baye's theorem to relate the experimental data Y(in the previous problem this was T(t)) and the parameters(this was ε in the previous example) is

$$P(x/Y) = \frac{P(Y/x)\, P(x)}{P(Y)} = \frac{P(Y/x)P(x)}{\int P(Y/x)P(x)dx} \tag{9.16}$$

where $P(x/Y)$ is called the Posterior Probability Density Function (PPDF), $P(Y/x)$ is the likelihood density function, $P(x)$ is the prior density function and $P(Y)$ is the normalizing constant.

In the above equation, the first term on the RHS represents the probability of getting Y for an assumed value of x. This can be obtained from a solution to the direct problem for an assumed x and we convert the $S = \sum_{i=1}^{N}(Y_{exp,i} - Y_{sim,i})^2$ in to a PDF (Probability density function). Invariably, a Gaussian distribution for the measurement errors is assumed for doing this. The $P(x)$ is our prior knowledge belief about x, even before the measurements are made or calculations are done. One can call this as 'expert knowledge'or 'domain knowledge'. For example, if the goal of an inverse problem is to determine the thermal conductivity of a metal using an inverse methodology, if the material looks like steel, we can construct a Gaussian for P(k) with a mean (μ), say 50 W/mK and a standard deviation (σ), 5 W/mK. This means 99% of the time our prior belief is that the thermal conductivity lies between $50 \pm 15 W/mK$. This is a very reasonable assumption and often reduces the ill-posedness. For instance, there is no need to conduct searches starting from, say, k=0.1 W/mK to k=400 W/mK for solving this problem, if a sensible and rational prior is used. This is the hallmark of the Bayesian approach.

Figure 9.4 presents an overview of the Bayesian approach to inverse problems. Construction of $P(x|Y)$ requires data, model, and a model for the distribution of errors. $P(x)$ is a quantification of our beliefs and here again a Gaussian distribution has been used. $P(x|Y)$, the PPDF is a joint PDF of the likelihood and the prior.

9.2.2 Steps Involved in Solving a Problem Using Bayesian Approach

The Bayesian method to solve an inverse problem involves three steps:

1. Experimental data collection. In Example 9.1, the data collected is in the form of temperatures.
2. Modeling-Likelihood and priori.
3. Estimation of x.

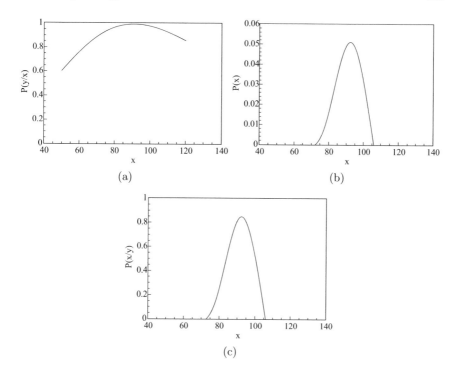

Fig. 9.4 Schematic of **a** likelihood density function $P(Y|x)$ **b** prior density function $P(x)$ and **c** posterior probability density function $P(x|Y)$

The first step is done by conducting experiments. In so far as the likelihood is concerned, we exploit the idea of measurement error in temperature as follows

$$Y_{measured} = Y_{simulated} + \omega \qquad (9.17)$$

ω is a random variable from a normal distribution with mean "0" and standard deviation σ, where σ is the standard deviation of the measuring instrument (thermocouple). Assuming that the uncertainty ω follows a normal or Gaussian distribution, the likelihood can be modeled as

$$P(Y/x) = \frac{1}{(\sqrt{2\pi\sigma^2})^n} \exp\left(-\frac{(Y - F(x))^T (Y - F(x))}{2\sigma^2}\right) \qquad (9.18)$$

where Y is a vector of dimension n, i.e, n measurements are available and F(x) is the solution to the forward model with the parameter vector x (x can consist of several variables)

$$P(Y/x) = \frac{1}{(\sqrt{2\pi\sigma^2})^n} \exp\left(\frac{-\chi^2}{2}\right) \tag{9.19}$$

$$\text{where } \chi^2 = \sum_{i=1}^{n} \frac{(Y_{meas,i} - Y_{sim,i})^2}{\sigma^2} \tag{9.20}$$

In the above equation, $Y_{sim,i}$ are the simulated values of Y for an assumed x (set of parameters).

The posterior PDF (PPDF) then becomes

$$P(x/Y) = \frac{\left[\frac{1}{(\sqrt{2\pi\sigma^2})^n} \exp\left(\frac{-\chi^2}{2}\right)\right][P(x)]}{\int \left[\frac{1}{(\sqrt{2\pi\sigma^2})^n} \exp\left(\frac{-\chi^2}{2}\right)\right][P(x)]dx} \tag{9.21}$$

The prior probability density P(x) typically follows a uniform, normal, or log-normal distribution. In the case of a uniform prior, P(x) is the same for all values of x, i.e., we have absolutely no selective preference. This happens in some cases where we have no knowledge of x. Such a prior is called a **non-informative prior**. Needless to say, it is also an objective prior, as there is no subjective input to the prior.

Let us say P(x) follows a normal distribution with mean μ and standard deviation σ_p. Mathematically, P(x) is given by

$$P(x) = \frac{1}{(\sqrt{2\pi\sigma_p^2})^n} exp \frac{-(x-\mu_p)^2}{2\sigma_p^2} \tag{9.22}$$

Hence, the PPDF becomes

$$P(x/Y) = \frac{\frac{1}{(2\pi)^{\frac{n+1}{2}}(\sigma^n\,\sigma_p)} exp(-)\left[\frac{\chi^2}{2} + \left(\frac{(x-\mu)^2}{2\sigma_p^2}\right)\right]}{\left[\int \frac{1}{(2\pi)^{\frac{n+1}{2}}(\sigma^n\,\sigma_p)} exp(-)\left[\frac{\chi^2}{2} + \left(\frac{(x-\mu)^2}{2\sigma_p^2}\right)\right]\right]dx} \tag{9.23}$$

Therefore, for every assumed value of the data vector $X(x_1, x_2 \ldots x_n)$, P(x/Y) can be worked out. From this posterior distribution, one can use two possible estimators (i) Mean estimate also known as expectation or (ii) Maximum a Posteriori (MAP), that is, the value of x for which P(x/Y) is maximum. Usually, a sampling algorithm is used to generate samples of x consecutively. In a multiparameter problem, the marginal PDF of every parameter must be worked out. It is pertinent to note that P(x/Y) is a joint PDF. The estimators and the concept of sampling will become clear on further consideration of the example problem that was considered in the earlier part of the chapter.

$$P(x/Y) = \frac{\frac{1}{(2\pi)^{\frac{n+1}{2}}(\sigma^n \sigma_p)} exp(-) \left[\frac{x^2}{2} + \left(\frac{(x-\mu)^2}{2\sigma_p^2} \right) \right]}{\left[\int \frac{1}{(2\pi)^{\frac{n+1}{2}}(\sigma^n \sigma_p)} exp(-) \left[\frac{x^2}{2} + \left(\frac{(x-\mu)^2}{2\sigma_p^2} \right) \right] \right] dx} \tag{9.24}$$

Therefore

$$P(x/Y) = \frac{exp(-) \left[\frac{x^2}{2} + \frac{(x-\mu)^2}{2\sigma_p^2} \right]}{\int \left[exp(-) \left[\frac{x^2}{2} + \frac{(x-\mu)^2}{2\sigma_p^2} \right] \right] dx} \tag{9.25}$$

The mean estimate of x then becomes

$$\bar{x} = \frac{\int x \, exp(-) \left[\frac{x^2}{2} + \frac{(x-\mu)^2}{2\sigma_p^2} \right] dx}{\int \left[exp(-) \left[\frac{x^2}{2} + \frac{(x-\mu)^2}{2\sigma_p^2} \right] \right] dx} \tag{9.26}$$

Often, the integral is replaced by a summation when only discrete values of x are used.

$$\bar{x} = \frac{\sum_i x_i \, exp(-) \left[\frac{x^2}{2} + \frac{(x-\mu)^2}{2\sigma_p^2} \right] \Delta x_i}{\sum \left[exp(-) \left[\frac{x^2}{2} + \frac{(x-\mu)^2}{2\sigma_p^2} \right] \right] \Delta x_i} \tag{9.27}$$

If Δx_i are the same, then

$$\bar{x} = \frac{\sum_i x_i \, exp(-) \left[\frac{x^2}{2} + \frac{(x-\mu)^2}{2\sigma_p^2} \right]}{\sum \left[exp(-) \left[\frac{x^2}{2} + \frac{(x-\mu)^2}{2\sigma_p^2} \right] \right]} \tag{9.28}$$

$$\sigma_x^2 = \frac{\sum_i (x_i - \bar{x})^2 \, exp(-) \left[\frac{x^2}{2} + \frac{(x-\mu)^2}{2\sigma_p^2} \right]}{\sum \left[exp(-) \left[\frac{x^2}{2} + \frac{(x-\mu)^2}{2\sigma_p^2} \right] \right]} \tag{9.29}$$

Now, we can use this framework to estimate the emissivity ε in the example problem. In the above equation, σ_x is the standard deviation of the estimated parameter, which is very diagnostic of the potency of the estimation process.

Example 9.2 Consider Example 9.1, wherein an aluminum foil was heated in an evacuated enclosure. Using the same data and the same samples, determine the mean of the estimate using a Bayesian approach. Take $P(\varepsilon)$ to be a Gaussian with $\mu_p = 0.8$ and $\sigma_p = 0.05$. The standard deviation of the uncertainty in the temperature measurement is ± 0.3 K.

Table 9.3 Estimation of ε using the Bayesian method (no priors)

S.No.	ε_i	$S(\varepsilon_i)$	$\varepsilon_i \exp^{-\left(\frac{S(\varepsilon_i)}{2\sigma^2}\right)}$	$\exp^{-\left(\frac{S(\varepsilon_i)}{2\sigma^2}\right)}$	$(\varepsilon_i - \bar{\varepsilon})^2 \exp^{-\left(\frac{S(\varepsilon_i)}{2\sigma^2}\right)}$
1	0.6	358.02	0	0	0
2	0.65	229.76	0	0	0
3	0.7	132.59	8.70×10^{-321}	1.24×10^{-320}	2.85×10^{-322}
4	0.75	65.19	3.87×10^{-158}	5.16×10^{-158}	5.16×10^{-160}
5	0.8	21.57	7.25×10^{-53}	9.06×10^{-53}	2.26×10^{-55}
6	0.85	2.94	6.85×10^{-8}	8.06×10^{-8}	6.81×10^{-21}
7	0.90	5.11	4.22×10^{-13}	4.69×10^{-13}	1.17×10^{-15}
8	0.95	27.04	5.46×10^{-66}	5.75×10^{-66}	5.75×10^{-68}
Σ			6.85×10^{-8}	8.06×10^{-8}	1.17×10^{-15}

Fig. 9.5 PPDF of ε using the Bayesian method (without prior)

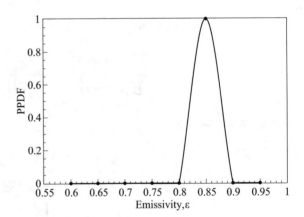

Solution:

Using the Bayesian formulation presented above, for the no prior case, we have the following results presented in Table 9.3.

$$\bar{\varepsilon} = \frac{6.85 \times 10^{-8}}{8.06 \times 10^{-8}} = 0.85, \quad \sigma_\epsilon = \sqrt{\frac{1.17 \times 10^{-15}}{8.06 \times 10^{-8}}} = 1.21 \times 10^{-4} \qquad (9.30)$$

The typical PPDF for the case without prior is shown in Fig. 9.5.

The mean estimate without the prior is given by 0.85 with $\sigma_\varepsilon = 1.21 \times 10^{-4}$. Such an estimate is frequently referred to as the **maximum likelihood estimate**. Now if we include the Gaussian prior, we get results presented in Table 9.4.

Table 9.4 Estimation of ε using the Bayesian method (with a Gaussian prior)

ε_i	$S(\varepsilon)$	$A=\frac{S(\varepsilon_i)}{2\sigma^2}$	$B=\frac{(\varepsilon-\mu_{\varepsilon,prior})^2}{2\sigma_p^2}$	$\varepsilon_i exp^{-(A+B)}$	$exp^{-(A+B)}$	$(\varepsilon_i - \bar{\varepsilon})^2 exp^{-(A+B)}$
0.6	358.02	1989	8	0	0	0
0.65	229.76	1276.44	4.5	0	0	0
0.7	132.59	736.61	2	1.176×10^{-321}	1.68×10^{-321}	4.00×10^{-323}
0.75	65.19	362.17	0.5	2.35×10^{-158}	3.13×10^{-158}	3.13×10^{-160}
0.8	21.57	119.83	0.0	7.25×10^{-53}	9.06×10^{-53}	2.26×10^{-55}
0.85	2.94	16.33	0.5	4.16×10^{-8}	4.89×10^{-8}	2.056×10^{-22}
0.9	5.11	28.39	2.0	5.71×10^{-14}	6.34×10^{-14}	1.59×10^{-16}
0.95	27.04	150.22	4.5	6.06×10^{-68}	6.38×10^{-68}	6.38×10^{-70}
Σ				4.16×10^{-8}	4.89×10^{-8}	1.59×10^{-16}

Fig. 9.6 PPDF of ε using the Bayesian method (with a Gaussian prior)

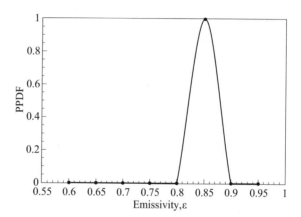

From Table 9.4, we can obtain the mean of ε and also the standard deviation of the estimate as

$$\bar{\varepsilon} = \frac{4.16\times10^{-8}}{4.89\times10^{-8}} = 0.85, \quad \sigma_{\epsilon} = \sqrt{\frac{1.59\times10^{-16}}{4.89\times10^{-8}}} = 5.69 \times 10^{-5} \quad (9.31)$$

The typical PPDF for the case with prior is shown in Fig. 9.5 (Fig. 9.6).

We can see that after the incorporation of the Gaussian prior, the standard deviation of the estimate of ε has gone down. Hence, the subjective informative Gaussian prior has helped us considerably in the estimation process. Again, instead of taking uniform samples of ε from 0.6 in steps of 0.05, one can employ a Markov chain, wherein the next sample of x (ε in this case) depends on only the current value of x. This can be done by drawing the new sample from a Gaussian distribution with its mean being the current value of "x" and "σ" being typically 5% of the current mean. The new sample can be accepted or rejected stochastically, in a manner analogous to what we did in the Simulated Annealing method. The procedure elucidated above is frequently referred to as a "Markov chain Monte Carlo (MCMC)" technique. Interested readers may look up advanced statistics text or journals to learn more about MCMC methods for parameter estimation in thermal sciences.

Problems

9.1 Consider a vertical rectangular mild steel fin of constant area of cross section that is 3 mm in thickness, and 100 mm long. The depth of the fin is 300 mm in the direction perpendicular to the plane of the paper. The fin stands on a horizontal base. The heat transfer coefficient at the fin surface is 8 W/m²K, the base temperature is 370 K and the ambient temperature is 300 K. The fin temperature distribution can be assumed to be one dimensional, varying only along the height. Steady-state temperature excess as $(\theta = T - T_\infty)$ at various locations along the height of the fin is measured and tabulated below. For the case of an adiabatic fin tip, estimate the thermal conductivity of the fin material by using an exhaustive equal interval search, in the range $20 \leq k \leq 60$ W/mK with an interval of 5 W/mK and then switching to a Lagrangian interpolation formula by using a least square approach.

x (mm)	0	10	20	35	50	65	80	100
θ (K)	70	63.9	59.8	52.7	49.1	44.6	43.4	41.3

9.2 Consider the problem of determination of thermal conductivity(k) of a fin with a measured temperature distribution that was discussed in exercise Problem 9.1. With the same data and sampling, determine the mean of the estimate of "k" using a Bayesian approach with (i) Uniform prior and (ii) A Gaussian prior with $\mu = 50$ W/mk and $\sigma_p = 5$ W/mK. The total uncertainty in the temperature measurement (which arises as a consequence of the thermocouple error and the error in accurately determining the position of the thermocouple) is ± 0.5 K.

Summary

Workable Design and Optimum Design

- A workable design is one that satisfies all the performance criteria, has reasonable capital and maintenance costs, and respects the constraints.
- There could be several workable designs for a system. All of them are not equally "desirable". These designs can be graded on the basis of our goal or objective.
- Objective is often stated mathematically.
- The objective, together with a set of constraints and bounds for the variables, constitutes an "optimization problem".

System Simulation

- To determine a workable design, where several components are joined together, first, modeling is required followed by system simulation.
- System simulation mimics the performance of a system, usually on a computer.
- Several techniques like successive substitution, Newton–Raphson method, Gauss–Seidel method, finite difference, finite volume, and finite element methods are available to perform system simulation.
- The choice of a method to accomplish system simulation depends on the nature of the resulting equations of a multicomponent system.
- Nonlinear algebraic equations are handled by Newton–Raphson or successive substitution method.
- Large systems of linear equations are handled by techniques like matrix inversion or Gauss–Seidel method. The latter is preferred if the number of variables exceeds 100.
- Nonlinear partial differential equations encountered in fluid flow and heat transfer (often in thermal systems) are handled by finite difference, finite volume, or finite element techniques.

© The Author(s) 2021
C. Balaji, *Thermal System Design and Optimization*,
https://doi.org/10.1007/978-3-030-59046-8

Curve Fitting

- To perform system simulations, the performances of various components in a system and also the thermodynamic and thermophysical properties of the media should be available in the form of equations. These equations relate quantities like pump efficiency, head loss, or properties like viscosity and conductivity to the operating variables like flow rate, pressure, temperature, mole fractions, time, and spatial coordinates.
- Obtaining equations for the above from a set of points is known as curve fitting. Curve fitting is essentially of two types. (a) Exact fit-curve passes through every point. Useful for a limited number of measurements or a limited number of parameters. Some examples include polynomial interpolation and Lagrange interpolation. (b) Best fit-curve does not pass through every point, invariably used in curve fitting. Usually based on minimization of the sum of squares of the difference between actual data and data generated by the fit. This procedure is known as Least Squares Regression (LSR).
- Many forms like $y = ax^b$, $y = ax_1 + bx_2 + c$, $y = ae^{bt}$ and so on can be regressed using LSR. Forms like $\theta = a[1 - e^{-bt}]$ cannot be linearized. a and b in the above equation have to be regressed using nonlinear least squares.
- Gauss–Newton Algorithm (GNA)- powerful nonlinear regression tool. We start with initial guess values of the parameters and iteratively obtain the parameters (as, for example, a, b of the previous bullet). Marquardt algorithm and Levenberg–Marquardt algorithm are improvements to the basic GNA, where a damping term is introduced.

Optimization

- The process of finding a condition that gives maximum or minimum.
- May not always be feasible or possible, because of the time, labor, or money involved.
- For small projects, the cost, time, and effort may not justify optimization.
- For complex systems, the design is too complex to optimize. One strategy is to subdivide the system subsystems and proceed.

Objective Function

- The important decision is what is to be optimized.
- For e.g., for aircraft and racing cars, it could be the weight.
- For automobiles, it could be the size, cost, or specific fuel consumption.
- For refrigerators, it could be the initial cost when we buy it in the market, while for an air conditioner, the more important parameter is the running cost.

Mathematical Representation of Optimization

Let y be the objective function where

- $y = y(x_1, x_2, \ldots x_n)$.

- $x_1, x_2, \ldots x_n$ are the independent variables.
- constraints can be equality as well as inequality constraints,
- Equality constraints are represented as
  $\phi_1(x_1, x_2, \ldots x_n) = 0, \phi_2(x_1, x_2, \ldots x_n) = 0 \ldots, \phi_m(x_1, x_2, \ldots x_n) = 0.$
- Inequality constraints are represented as
  $\psi_1(x_1, x_2, \ldots x_n) \leq r_1, \psi_2(x_1, x_2, \ldots x_n) \leq r_2, \psi_j(x_1, x_2, \ldots x_n) \leq r_j.$
  . These are more difficult to handle.
- The constraints actually bind the solution.
- The following relations hold.
  $Min(a + y) = a + Min(y);$
  $Max(y) = Min(-y)$

Classification of Optimization Problems

- Single variable,
- Multi-variable, unconstrained, and
- Multi-variable, constrained.

Classification of Optimization Techniques

- Calculus methods,
- Search methods.

Calculus Method

- Most powerful calculus method is the Lagrange multiplier method.
- Uses derivatives to work out the optimum.
- The existence of derivatives is mandatory.

Lagrange Multiplier Method (Most Important Calculus Method)
Optimum is reached when

- $\nabla y - \lambda \nabla \phi = 0$ for $\phi_1, \phi_2 \ldots, \phi_m = 0.$
- $\lambda = \frac{\nabla y}{\nabla \phi}$; gives us the sensitivity of the solution: the change in objective function to the change in constraints. Also the known as shadow price.
- For m constraints and n variables ($m \leq n$), we require $n + m$ equations to solve for $n + m$ variables x_1 to x_n and λ_1 to λ_m and in the Lagrange method (n+m) equations are available.
- λ's are called the Lagrange multipliers.
- If $m = n$, the values of $x''s$ are fixed by the constraints and no optimization is possible.
- For an unconstrained problem, Lagrange multiplier equations reduce to $\frac{\partial y}{\partial x_1} = \frac{\partial y}{\partial x_2} \cdots = \frac{\partial y}{\partial x_n} = 0.$
- The optimum can be a maximum, minimum, or an inflection point. The Hessian of second derivatives has to be positive definite for a minimization problem.

Search Methods

- Based on eliminating a portion of the interval (Elimination method).
- Based on systematically climbing to the top (Hill climbing method).

Key Point in Search Method—Interval of Uncertainty

- The final solution lies between two limits. The precise point of optimization is never known.
- Only the final interval of uncertainty can be specified.
- Reduction ratio $= \frac{\text{original interval of uncertainty}}{\text{new interval of uncertainty}} = \frac{I_0}{I_n}$

Exhaustive Search Technique

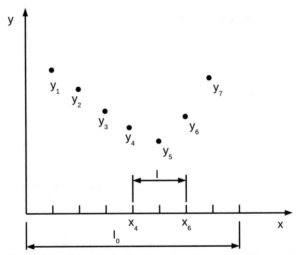

Depiction of interval of uncertainty in an equal interval

The figure is a simple depiction of how to use the two-point test method. The function is monotonic and hence we can only say the optimum lies somewhere between y_4 and y_6.

$I = \frac{2I_0}{(n+1)}$, where n is the number of observations.

Efficient Search Techniques

- For dichotomous search, $I = \frac{I_0}{2^{\frac{n}{2}}}$ (choose two points at $\frac{I_0 + \epsilon}{2}$ from two ends).
- For Fibonacci method, the Fibonacci series is used ($F_n = F_{n-1} + F_{n-2}$ (choose two points at $I_0 \frac{F_n}{F_{n-1}}$ from both ends).
- For the golden section search, the golden mean $= 0.618$ or 1.618 is used (choose two points at $0.618 I_0$ from either end of the interval).

Multi-variable Unconstrained Optimization

- The steepest ascent/descent method.

– At each trial point, the gradient vector is calculated and search proceeds along this direction.

$$\frac{\Delta x_1}{\left(\frac{\partial y}{\partial x_1}\right)} = \frac{\Delta x_2}{\left(\frac{\partial y}{\partial x_2}\right)} = \cdots = \frac{\Delta x_n}{\left(\frac{\partial y}{\partial x_n}\right)} = \alpha$$

Choose Δx_1 and, all the other $\Delta x''s$ can be obtained or we can simultaneously choose all Δxs by defining an α and solving for it.

Multivariable Constrained Optimization

• Penalty function method

– The constrained problem is converted into an unconstrained problem by creating a composite objective function, which takes care of both the objective function and the constraint.
– The penalty parameter penalizes the objective function for violating the constraints.
– The resulting unconstrained problem is solved using known techniques.
– The problem is solved with different values of the penalty parameter.
– If there is no significant change in the optimum with a change in the penalty, one can assume that the final solution has been reached.

Multi-objective Optimization

• Multi-objective optimization problems are a kind of problems in which multiple objectives are present which invariably are in conflict with each other.
• Preference method weights the objectives and solves for an equivalent single-objective optimization problem.

Other Optimization Techniques

• Linear programming: Applicable only if the objective function and the constraints are linear combinations of the independent variables. The graphical method can be used for two-variable problems. LP problems can be algebraically solved by introducing slack variables.
• Simplex method is a very systematic algebraic solver that makes use of an iterative approach to determine the optimum solution for the LP problem.
• Integer programming is a type of LP problem where all the variables must be integers. This type of problems can be solved using cutting plane or searching methods.
• Nontraditional optimization techniques like genetic algorithms and simulated annealing are search techniques that are calculus free, robust, and use stochastic principles. GAs use evolution and SA slow cooling or annealing as the basics for optimization.

Inverse Problems

• If the effect (temperature) is known but the causes, say (k, ε, h) are not known, such a problem is known as an inverse problem.

- An inverse problem is typically ill-posed, as several causes could lead to the same effect.
- For simple problems (for example, those involving one parameter) least square minimization can be used to estimate the parameter. In this method, for guess values of the parameter, the effect (say, temperature distribution) is calculated and the sum of the squares of the difference between calculated and measured qualities (temperature) is minimized.
- Parameter estimation problems (an important category of the inverse problem) are hence, eventually minimization problems. For multiple parameters, ill-posedness can be handled by injecting prior information about the parameters in the estimation process.
- A systematic injection of parameters is afforded by the Bayesian framework.
- In a Bayesian framework, measurements, measurement errors, forward (or mathematical) model, and prior benefits are all synergistically combined to work out the conditional probability of a set of parameters, being the cause of an effect (e.g., temperature distribution).
- These probabilities are worked out and by post-processing them, the mean of the estimate and maximum a posteriori are determined.

Appendix

Problem 4.4 Computer – Based Exercise

(a) An overhead tank of a big apartment complex has a capacity of 12000 l. It is desired to select a pump and piping system to transport water from the sump to the tank. The distance between the two is 270 m and the tank is at a level 22 m above the sump. For operational convenience, the time to fill the tank shall be 60 min. Losses in the expansions, contractions, bends, and elbows have to be calculated appropriately. Design a workable system for the above and sketch the layout.

(b) Using data for PVC pipes from local market sources, assign realistic values for the cost of the pipe (PVC), pump, and running cost including maintenance costs. With the help of any method known to you, obtain the value of ΔP developed by the pump at which the total cost will be minimum. The pump is expected to work every day and the average daily consumption of water is 24000 l. The cost of electricity may be assumed to be Rs[1]0.5.50 per unit and invariant with respect to time. The life of the system may be assumed to be 20 years. Let x the increase in electricity cost per year in percentage. Consider two values of x–6% and 7%

Solution

(a) Design of a workable system and layout
Data given:

Capacity of the tank $= 12000$ liter $= 12$ m^3,
Time to fill the tank $= 60$ min $= 3600$ s,
Distance between sump and tank (L) $= 270$ m, and
Height from sump to tank (h') $= 22$ m.
Properties of the water:
Density, $\rho = 1000$ kg/m^3,
Dynamic viscosity, $\mu = 8 \times 10^{-4}$ N s/m^2.

---

[1](1 USD $\approx$ Rs. 70 (as of April 2019))

© The Author(s) 2021
C. Balaji, *Thermal System Design and Optimization*,
https://doi.org/10.1007/978-3-030-59046-8

Let four bends exist in the pipe

Total head loss in bends $= 4 \times 0.15 \times \dfrac{v^2}{2g}$

Assuming the diameter(d) of the pipe as 1.5 inch $= 38$ mm $= 0.038$ m

Discharge(Q) $= \dfrac{12}{3600} = 3.33 \times 10^{-3}$ m^3/s

Velocity(v) $= \dfrac{4 \times Q}{\pi \times d^2} = 2.94$ m/s

The Reynolds number $(\text{Re}_d) = \dfrac{\rho v d}{\mu} = \dfrac{1000 \times 2.94 \times 0.038}{8 \times 10^{-4}} = 138650$

As $\text{Re}_d > 2300$, the flow is turbulent.

For a turbulent flow, the friction factor in a pipe, f is given by

$$f = 0.182 \times \text{Re}_d^{-0.2} = 0.182 \times (138650)^{-0.2} = 0.017 \qquad \text{(A.1)}$$

Heat loss due to friction

$$(h_l) = \frac{f \times L \times v^2}{d \times 2 \times g} = \frac{0.017 \times 270 \times (2.94)^2}{0.038 \times 2 \times 9.81} = 53.21 \text{ m} \qquad \text{(A.2)}$$

Heat loss due to bends

$$(h_b) = 4 \times 0.15 \times \frac{v^2}{2 \times g} = 0.2643 \text{ m} \qquad \text{(A.3)}$$

Total heat loss

$$H = h' + h_l + h_b = 22 + 53.21 + 0.2643 = 75.47 \text{ m} \qquad \text{(A.4)}$$

Pressure loss

$$\Delta P = \rho g H = 1000 \times 9.81 \times 75.47 = 7.40 \times 10^5 \text{ Pa} \qquad \text{(A.5)}$$

Assuming a pump of efficiency, $\eta = 85\%$

$$Power = \frac{Q \times \Delta P}{\eta \times 746} = \frac{3.33 \times 10^{-3} \times 7.40 \times 10^6}{0.85 \times 746} = 3.88 \text{ hp} \qquad \text{(A.6)}$$

This is a working system, but not an optimum design.

A sketch of the layout of the pump and piping arrangement is given in Fig.A.1.

(b) Optimization

Costs of different pumps are collected from market sources and are tabulated in Table A.1

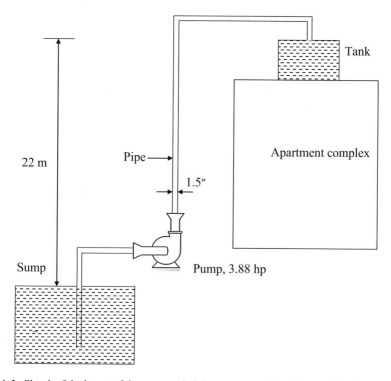

Fig. A.1 Sketch of the layout of the pump and piping arrangement (a design solution)

Table A.1 The costs of different pumps collected from market sources

S.No.	Power (hp)	ΔP (Pa)	Cost (Rs.)
1	0.25	47605.5	2816
2	0.5	95210.2	3088
3	1	190420.4	3551
4	1.5	285630.6	6708
5	2	380840.8	15825
6	3	571261.2	18535
7	5	952102.1	19102
8	7	1332942.9	25000
9	8	1523363.3	26000
10	10	1904204.2	38082
11	12.5	2380255.2	44212
12	15	2856306.3	46622
13	20	3808408.4	65047
14	25	4760510.5	85204

Table A.2 The cost of PVC pipes of different diameters available in market

S.No.	Diameter (mm)	Cost/m (Rs.)
1	20	28.07
2	25	42.32
3	32	69.11
4	40	107.63
5	50	166.68

Sample Calculation

Consider a 0.5 hp (373 W) pump.
Let the efficiency of the pumps, $\eta = 85\%$
Capacity of the tank $= 24000 \ l = 24 \ m^3$
Time taken to fill the tank is 120 min $= 7200$ s
Discharge, $Q = \dfrac{24}{(120 \times 60)} = 3.33 \times (10^{-3}) \ m^3/s$

$$\text{Power} = \frac{\Delta P \times Q}{\eta}$$

$$\Delta P = \frac{(Power(in \ hp) \times 746 \times \eta)}{Q} \ (Pa)$$

From column 2 in Table A.1, we have

$$\Delta P_1 = \frac{0.25 \times 746 \times 0.85}{3.33 \times 10^{-3}} = 47605.11 \ Pa$$

$$\Delta P_2 = \frac{0.5 \times 746 \times 0.85}{3.33 \times 10^{-3}} = 95210.21 \ Pa$$

$$\Delta P_3 = \frac{1 \times 746 \times 0.85}{3.33 \times 10^{-3}} = 190420.42 \ Pa$$

$$\Delta P_4 = \frac{1.5 \times 746 \times 0.85}{3.33 \times 10^{-3}} = 285630.63 \ Pa$$

The data in Table A.1 is fit linearly to obtain the pump cost in terms of ΔP as follows.

From Fig. A.2, the cost of the pump in terms of ΔP is given by the following equation:

Pump cost $=$ Rs. $0.016 \times \Delta P + 3526$

Pipe cost:

In a similar fashion, the pipe costs are also collected from market sources and are tabulated in Table A.2.

Fig. A.2 Linear fit for cost of the pump

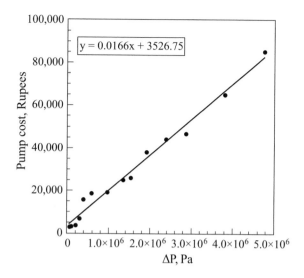

Fig. A.3 Linear fit for cost of the pipe

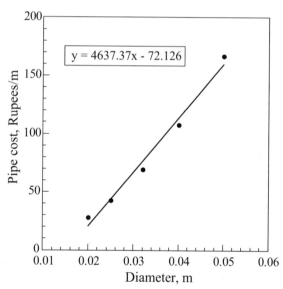

Again using a linear fit for the cost of the pipe, we obtain the pipe cost in terms of diameter d as follows:

pipe cost per meter = $4637.37 \times d - 72.126$ (It is shown in Fig. A.3)

The pipe cost now needs to be obtained as $f(\Delta P)$

$$\Delta P' = \Delta P - \text{static pressure} \tag{A.7}$$

$$\Delta P' = \frac{f \times l \times u^2 \times \rho \times g}{2 \times d \times g} \qquad (A.8)$$

$$f = 0.182 \times (Re)^{-0.2} = \frac{0.0327}{d^{-0.21}} \qquad (A.9)$$

$$\Delta P' = 0.0794 \times d^{-4.8} \qquad (A.10)$$

$$d = 0.5899 \times (\Delta P')^{-0.21} \qquad (A.11)$$

$$\text{Static pressure} = \rho g h = 215820 \text{ Pa} \qquad (A.12)$$

$$d = 0.5899 \times (\Delta P' - 215820)^{-0.21} \qquad (A.13)$$

Cost of the pipe per meter

$$= \text{Rs. } 2735.36(\Delta P - 215820)^{-0.21} - 72.12 \qquad (A.14)$$

Cost of the pipe

$$= \text{Rs. } 738547.2(\Delta P - 215820)^{-0.21} - 19472.4 \qquad (A.15)$$

Running Cost

Case 1: Running cost for 20 years with 6% increase in unit cost per year. Based on the data given, the running costs are calculated and tabulated in Table A.3.

Sample Calculation for Generation of Table A.3

Based on the data given in the problem, the life of the system is assumed to be 20 years. The unit cost of electricity is Rs. 5.50 per unit and this increases by 6% every year and the pump has to work two hours daily.

Sum of the electricity costs per unit over the-20 year period

$$= [5.5 + 5.5 \times 1.06 + (5.5 \times 1.06) \times 1.06 + \cdots \text{ up to 20 years}] = \text{Rs. } 202.3196$$

Running cost = Power × Cost of electricity per unit over 20 years

$$\text{Running cost} = \frac{(\Delta P \times Q)}{(\eta \times 1000)} \times 202.3196 \times 2 \times 365 \qquad (A.16)$$

Table A.3 Running cost for different pumps over 20 years with 6% increase in unit cost per year

S.No	Power (hp)	ΔP (Pa)	Running cost (Rs.)
1	0.25	47605.5	27544.8
2	0.5	95210.2	55088.6
3	1	190420.4	110177.3
4	1.5	285630.6	165265.8
5	2	380840.8	220354.5
6	3	571261.2	330531.7
7	5	952102.1	550886.3
8	7	1332942.9	771240.8
9	8	1523363.3	881418.1
10	10	1904204.2	1101772.5
11	12.5	2380255.2	1377215.7
12	15	2856306.3	1652658.8
13	20	3808408.4	2203545.1
14	25	4760510.5	2754431.4

$$\text{Running cost (RC)} = \Delta P \times 0.5786 \tag{A.17}$$

For a 0.25 hp pump power, the running cost is

$$RC_1 = 47605.1 \times 0.5786 = \text{Rs. } 27544.8 \tag{A.18}$$

For a 0.5 hp pump power, this will

$$RC_2 = 95210.2 \times 0.5786 = \text{Rs. } 55088.6 \tag{A.19}$$

Using a linear fit for life time running cost, we obtain the following relation (Fig. A.4).
So,

$$\text{Running cost} = \text{Rs. } 0.578(\Delta P) + 0.071 \tag{A.20}$$

Optimization

Total life time cost (C) of the pump and piping arrangement

$$= \text{Pump cost} + \text{Running cost} + \text{Pipe cost} \tag{A.21}$$

$$= 0.594(\Delta P) + 738547.2(\Delta P - 215820)^{-0.21} - 15946.33 \tag{A.22}$$

Fig. A.4 Linear fit for the running cost of the pump

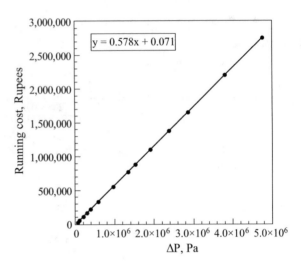

For optimum value of total life time cost (C),

$$\frac{dC}{d\Delta P} = 0 \tag{A.23}$$

$$\Delta P = 249288\,Pa \tag{A.24}$$

So, the optimum point is $(\Delta P, C) = (249288, 214964)$

The optimum diameter of the pipe then turns out to be

$$d = 0.5899 \times (\Delta P - 215820)^{-0.21} = 66.16 \text{ mm}$$

The optimum power of the pump $= \dfrac{Q \times \Delta P}{\eta \times 746} = 1.31$ hp

The nearest rating of the pump available in the market is 1.50 hp.

Therefore, the optimum pipe diameter and pump power for a life of 20 years with 6% increase in unit cost per year are 66.16 mm and 1.5 hp, respectively.

Case 2: Running cost for 20 years with 7% increase in unit cost per year. Using the approach detailed below, the running costs are calculated and presented in Table A.4.

Sample Calculation for Generation of Data in Table A.4

The life of the system is 20 years. The unit cost of electricity is Rs. 5.50 per unit and this increases by 7% every year and the pump has to work two hours daily.

Table A.4 Running costs for different pumps over 20 years with 7% increase in unit cost per year

S.No.	Power (hp)	ΔP (Pa)	Running cost (Rs.)
1	0.25	47605.5	30695.7
2	0.5	95210.2	61391.5
3	1	190420.4	122783.0
4	1.5	285630.6	184174.6
5	2	380840.8	245566.2
6	3	571261.2	368349.3
7	5	952102.1	613915.4
8	7	1332942.9	859481.6
9	8	1523363.3	982264.7
10	10	1904204.2	1227830.8
11	12.5	2380255.2	1534788.6
12	15	2856306.3	1841746.3
13	20	3808408.4	2455661.7
14	25	4760510.5	3069577.2

Total electricity cost over 20 years

$$= [5.5 + 5.5 \times 1.07 + (5.5 \times 1.07) \times 1.07 + \cdots \text{up to 20 year}] = \text{Rs. } 225.47$$

Running cost = Power $\times$ (Cost of electricity per unit over 20 years)

$$= \frac{(\Delta P \times Q)}{(\eta \times 1000)} \times 225.47 \times 2 \times 365 \tag{A.25}$$

$$\text{Running cost (RC)} = \Delta P \times 0.6448 \tag{A.26}$$

For a 0.25 hp pump power, the running cost is

$$RC_1 = 47605.1 \times 0.6448 = \text{Rs. } 30695.77 \tag{A.27}$$

For a 0.5 hp pump power, this will

$$RC_2 = 95210.2 \times 0.6448 = \text{Rs. } 61391.5 \tag{A.28}$$

Using a linear fit for lifetime running cost, the running cost is given by
Running cost of pump for 20 years with 7% increase per unit cost (Fig. A.5).

$$= \text{Rs. } 0.644(\Delta P) + 0.0155 \tag{A.29}$$

Fig. A.5 Linear fit for the running cost of the pump

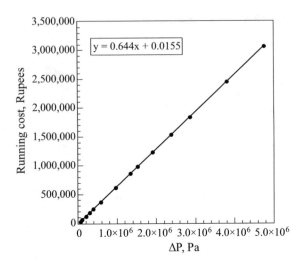

$$y = 0.644x + 0.0155$$

Running cost, Rupees vs ΔP, Pa

Optimization

Total life time cost (C) of the pump and piping arrangement

$$= \text{Pump cost} + \text{Running cost} + \text{Pipe cost} \tag{A.30}$$

$$= 0.66(\Delta P) + 738547.2(\Delta P - 215820)^{-0.21} - 15946.38 \tag{A.31}$$

The optimum value of total life time cost (C) is obtained as follows:

$$\frac{dC}{d\Delta P} = 0 \tag{A.32}$$

$$\Delta P = 243292 \, Pa \tag{A.33}$$

The optimum point is (ΔP, C) = (243292, 230966)
The optimum diameter of pipe d

$$= 0.5899 \times (\Delta P - 215820)^{-0.21} = 68.96 \text{ mm} \tag{A.34}$$

The optimum power of the pump

$$= \frac{Q \times \Delta P}{\eta \times 746} = 1.28 \text{ hp} \tag{A.35}$$

The nearest rating of the pump available is 1.50 hp.

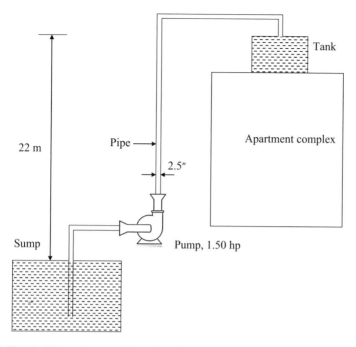

Fig. A.6 Sketch of the optimum configuration of the pump and piping arrangement (case 2)

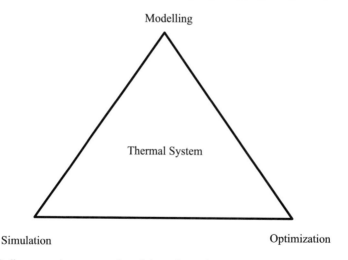

Fig. A.7 A diagrammatic representation of the tools required to optimize a thermal system

Therefore, the optimum pipe diameter and pump power for a life of 20 years with 7% increase in unit cost per year are 68.96 mm and 1.5 hp respectively.

A sketch of the final configuration of the pump and piping arrangement for case 2 is given in Fig. A.6 .

In summary, this example epitomizes the three key aspects involved in an optimizing thermal system and this can be visualized as shown in Fig. A.7.

Random Number Table

0.001213	0.898980	0.578800	0.676216	0.050106
0.499629	0.282693	0.730594	0.701195	0.182840
0.108501	0.386183	0.769105	0.683348	0.551702
0.557434	0.799824	0.456790	0.216310	0.876167
0.092645	0.589628	0.332164	0.031858	0.611683
0.762627	0.696237	0.170288	0.054759	0.915126
0.032722	0.299315	0.308614	0.833586	0.517813
0.352862	0.574100	0.265936	0.859031	0.433081
0.941875	0.240002	0.655595	0.385079	0.908297
0.199044	0.936553	0.888098	0.817720	0.369820
0.339548	0.543258	0.624006	0.091330	0.416789
0.155062	0.582447	0.858532	0.887525	0.337294
0.751033	0.239493	0.535597	0.333813	0.493837
0.634536	0.199621	0.650020	0.745795	0.791130
0.227241	0.191479	0.406443	0.081288	0.734352
0.721023	0.222878	0.072814	0.641837	0.442675
0.789616	0.052303	0.106994	0.558774	0.141519
0.760869	0.120791	0.277380	0.657266	0.792691
0.805480	0.826543	0.294530	0.208524	0.429894
0.585186	0.986111	0.344882	0.343580	0.115375

© The Author(s) 2021
C. Balaji, *Thermal System Design and Optimization*,
https://doi.org/10.1007/978-3-030-59046-8

Bibliography

Arora, J. S. (1989). *Introduction to Optimum Design*. Singapore: Mc Graw Hill.

Assari, A., Mahesh, T., & Assari, E. (2012b). Role of public participation in sustainability of historical city: usage of TOPSIS method. *Indian Journal of Science and Technology*, *5*(3), 2289–2294.

Burmeister Louis, C. (1998). *Elements of Thermal-Fluid System Design*. New Jersey, USA: Prentice Hall.

Chong, E. K. P., & Zak, S. H. (2004). *An Introduction to Optimization*. New York, USA: Wiley.

Deb, K. (1995). *Optimization for Engineering Design-Algorithms and Examples*. New Delhi, India: Prentice Hall India.

Jaluria, Y. (1998). *Design and Optimization of Thermal Systems*. Singapore: Mc Graw Hill.

Madadi, R. R., & Balaji, C. (2008). Optimization of the location of multiple discrete heat sources in a ventilated cavity using artificial neural networks and micro genetic algorithm. *International Journal of Heat and Mass Transfer*, *51*(9–10), 2299–2312.

The Mathworks, Inc. (2015). Natick, Massachusetts, MATLAB version 8.5.0.197613 (R2015a).

© The Author(s) 2021
C. Balaji, *Thermal System Design and Optimization*,
https://doi.org/10.1007/978-3-030-59046-8

Index

© The Author(s) 2021
C. Balaji, *Thermal System Design and Optimization*,
https://doi.org/10.1007/978-3-030-59046-8

Printed in the United States
by Baker & Taylor Publisher Services